THE ARRL
ANTENNA VOL5
COMPENDIUM

EDITOR
R. Dean Straw, N6BV

ASSISTANT EDITOR
Rich Roznoy, KA1OF

PRODUCTION
Shelly Bloom, WB1ENT
Dan Wolfgang
Jodi Morin, KA1JPA
Steffie Nelson, KA1IFB
Joe Shea

TECHNICAL ILLUSTRATIONS
Dave Pingree, N1NAS
Joe Costa

COVER DESIGN
Sue Fagan

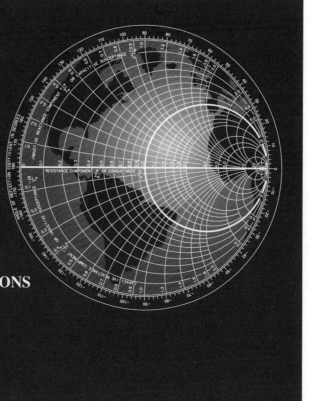

Foreword

Let's face it: as transceivers have become more and more complex, individual hams have a tougher and tougher time coming up with truly "state-of-the-art" electronic hardware. That's not too surprising, considering that large multidisciplinary teams are required to make today's sophisticated products. (Anyone remember how it used to be possible to tune, or even soup-up, an automobile by yourself?)

However, two places where hams still can, and do, contribute significantly to advancing the frontiers of knowledge are in software and antenna design. **Yes, hams love their antennas!** In this, the fifth volume of the immensely popular *ARRL Antenna Compendium* series, the combination of powerful software and powerful antenna knowledge will provide you with entertainment and lots of practical ideas.

This book contains 41 previously unpublished articles, covering a wide range of antenna-related topics—from a heavy-duty discussion by Al Christman, KB8I, on the nitty-gritty of elevated radials for low-band antennas to a somewhat whimsical treatise on practical portable 6-meter antennas by Markus Hansen, VE7CA.

Are you fascinated with how a signal is propagated through the ionosphere? So is Carl Luetzelschwab. Check out his revealing article in this volume. Are you heavily into computer modeling of antennas? So are a lot of antenna specialists, like Brian Beezley, Al Christman, and John Stanley, to name a few prominent practitioners. They all have contributed to this volume.

Or are you a low-frequency diehard? Eight articles are devoted to 80 and 160-meter antennas, some of them pretty impressive. Take Glenn Rattmann's 80-meter two-element Yagi, for example, or editor Dean Straw's two-element 80-meter quad. Perhaps you may decide to duplicate one of them and become a "big gun" on 80/75 meters!

Just like the fourth volume in this popular series, we've bundled a disk containing the source data used to model many of the antennas. You really can gain a lot of insight into the modeling process by looking over source data created by experts. Once again, the latest version of the nifty *PLOT* program by K6STI is included for interactively viewing the many plot files on the disk.

The data files included are meant to be used with several different commercial modeling programs (for example, *AO* or *NEC/Wires* by Brian Beezley, *EZNEC* by Roy Lewallen, or *NEC2*, available for free on numerous bulletin boards, including the ARRL Hiram BBS.) Please remember that we do not provide the modeling programs themselves, only the data that work in them.

In short, there should be something for any antenna enthusiast in this volume! Perhaps you may be inspired to write an antenna article of your own. We'd love to see it, as we prepare next for Volume 6.

David Sumner, K1ZZ
Executive Vice President

Newington, Connecticut
October 1996

Instructions for Accompanying Diskette

Instructions for Accompanying Diskette

The diskette bundled in the back of this volume includes numerous data and plot files created by the authors of *The ARRL Antenna Compendium, Vol 5* to analyze their antennas, using commercially available antenna modeling software such as *NEC2*, *NEC/Wires*, *EZNEC* and *AO*.

The ARRL does not include the modeling software itself on the diskette, only data for these programs. Note also that *NEC-4.1* is currently not publicly available because of security restrictions by the US government, although certain institutions have access to this program.

Installing the Software

To install the software, first make a working copy of the disk and store the original in a safe place. Insert the working disk into your floppy drive A and type: A:INSTALL [Enter] from the DOS prompt. You can also do this from the Program Manager in Windows, using File|Run menu selection. Then follow INSTALL's on-screen prompts.

The *PLOT* Program

Brian Beezley, K6STI, has again allowed the ARRL to include his latest *PLOT* program on the diskette. This program is used to view the many plotting files generated by the authors in the book. (Plot files are identified by their ".PF" filename extensions.) *PLOT* will only work on computers with a VGA display system and having a '386DX (with numeric coprocessor), '486DX or a Pentium processor. All other computers will fail to run *PLOT*.

Be sure to read the PLOT.DOC file for complete documentation on the *PLOT* program. You may use the *V* (View) program or a word processor to read the ASCII text. For example, you can at the DOS prompt type: V PLOT.DOC [Enter] to use *V*. You can also print the documentation by typing at the DOS prompt: PRINT PLOT.DOC [Enter].

If you can't see anything on the screen using *PLOT* or if the plot looks extremely distorted on your computer, use the command SET VBE=NO to disable the automatic use of VESA by *PLOT*.

A comment about the antenna modeling data files: Even if you are an experienced antenna modeler, you will gain valuable insight into how the experts work by examining their data files. Some very interesting techniques are displayed in a number of the data files, and it certainly beats typing in the data manually when you wish to see if you can possibly improve or "tweak" a design any further!

The *GENPF* Program

GENPF generates plot files from measured data you supply. You can make accurate measurements of antenna patterns using your receiver's S-meter and a calibrated attenuator, create a plot file with *GENPF*, and then view the patterns with *PLOT* and compare them with others. *GENPF* will prompt you for all information; just follow the prompts.

The *V* (View) Program

V is a program for viewing ASCII files. It runs under DOS or in a full-screen window under Windows, not in a partial window. To exit from *V*, hit the [Esc] key.

Other Programs

Several authors also wrote special analysis programs for their articles. Executable versions of these are also on the diskette, together with source code, when applicable. Where a program can be customized by the reader, the source code is supplied in BASIC. All programs are written for the IBM PC, or fully compatible computers.

Organization of the Disk

The root directory of the disk contains three programs (*PLOT*, *V* and *GENPF*), a documentation file (PLOT.DOC), some sample PF plot files and this README.DOC file.

The other data on the diskette is organized into 19 separate subdirectories, each named using the author's amateur call sign. For example, the data corresponding to the article by Al Christman, KB8I, is found in the \KB8I subdirectory, while the article by Peter Dodd, G3LDO, refers to disk files found in the \G3LDO subdirectory. Each data file has a distinct filename extension corresponding to the antenna-analysis program in which it is used. The filename extensions on the disk are:

*.PF   plot file used with *PLOT* program
*.NEC   used for the *NEC2* or *NEC4.1* program
*.ANT   used for the *NEC/Wires* or *AO* programs by K6STI
*.EZ   used for the *EZNEC* program by W7EL
*.BAS   BASIC source code
*.EXE   executable file

When you wish to examine or use a particular data file, change to the appropriate subdirectory from the DOS prompt using the "CD" (Change Directory) command, or using Windows FileManager. For example, to get into the \K6STI subdirectory, you would type CD \K6STI [Enter]. Individual antenna data files with an *.ANT or *.NEC filename extension may be examined using the *V* program or a word-processing program, since each such file contains ASCII data. The *.EZ files are binary files and can only be examined inside the EZNEC program itself.

Assistance

For assistance with the *PLOT* program, please **do not** contact K6STI directly. Contact the TIS (Technical Information Service) group at ARRL HQ: by voice at 860-594-0214; by fax at 860-594-0259; at the ARRL BBS 860-594-0306; e-mail at tis@arrl.org; CompuServe at 70007,3373; America On Line (HQARRL1).

Notes:

The *NEC2* program is available on the ARRL BBS without instruction manual, or from the Applied Computational Electromagnetics Society, c/o Dr. Richard W. Adler, Code 62AB, Naval Postgraduate School, Monterey, CA 93943.

NEC/Wires and *AO* are available from Brian Beezley, K6STI, 3532 Linda Vista Dr, San Marcos, CA 92069, 619-599-4962.

EZNEC is available from Roy Lewallen, W7EL, PO Box 6658, Beaverton, OR 97007.

Contents

80 and 160-Meter Antennas

VHF/UHF Antennas

Antenna Modeling

Multiband Antennas

Propagation and Ground Effects

Measurements and Computations

Special Antennas

Antenna Tuners, Baluns and Transmission Lines

About the American Radio Relay League

The seed for Amateur Radio was planted in the 1890s, when Guglielmo Marconi began his experiments in wireless telegraphy. Soon he was joined by dozens, then hundreds, of others who were enthusiastic about sending and receiving messages through the air—some with a commercial interest, but others solely out of a love for this new communications medium. The United States government began licensing Amateur Radio operators in 1912.

By 1914, there were thousands of Amateur Radio operators—hams—in the United States. Hiram Percy Maxim, a leading Hartford, Connecticut, inventor and industrialist saw the need for an organization to band together this fledgling group of radio experimenters. In May 1914 he founded the American Radio Relay League (ARRL) to meet that need.

Today ARRL, with more than 170,000 members, is the largest organization of radio amateurs in the United States. The League is a not-for-profit organization that:
- promotes interest in Amateur Radio communications and experimentation
- represents US radio amateurs in legislative matters, and
- maintains fraternalism and a high standard of conduct among Amateur Radio operators.

At League headquarters in the Hartford suburb of Newington, the staff helps serve the needs of members. ARRL is also International Secretariat for the International Amateur Radio Union, which is made up of similar societies in more than 100 countries around the world.

ARRL publishes the monthly journal QST, as well as newsletters and many publications covering all aspects of Amateur Radio. Its headquarters station, W1AW, transmits bulletins of interest to radio amateurs and Morse code practice sessions. The League also coordinates an extensive field organization, which includes volunteers who provide technical information for radio amateurs and public-service activities. ARRL also represents US amateurs with the Federal Communications Commission and other government agencies in the US and abroad.

Membership in ARRL means much more than receiving QST each month. In addition to the services already described, ARRL offers membership services on a personal level, such as the ARRL Volunteer Examiner Coordinator Program and a QSL bureau.

Full ARRL membership (available only to licensed radio amateurs) gives you a voice in how the affairs of the organization are governed. League policy is set by a Board of Directors (one from each of 15 Divisions). Each year, half of the ARRL Board of Directors stands for election by the full members they represent. The day-to-day operation of ARRL HQ is managed by an Executive Vice President and a Chief Financial Officer.

No matter what aspect of Amateur Radio attracts you, ARRL membership is relevant and important. There would be no Amateur Radio as we know it today were it not for the ARRL. We would be happy to welcome you as a member! (An Amateur Radio license is not required for Associate Membership.) For more information about ARRL and answers to any questions you may have about Amateur Radio, write or call:

ARRL Educational Activities Dept
225 Main Street
Newington CT 06111-1494
(860) 594-0200
Prospective new amateurs call:
800-32-NEW HAM (800-326-3942)
You can also contact us via e-mail: **ead@arrl.org**
or check out our World Wide Web site: **http://www.arrl.org/**

80 and 160-Meter Antennas

Elevated Vertical Antennas for the Low Bands: Varying the Height and Number of Radials

By Al Christman, KB8I
Grove City College
100 Campus Drive
Grove City, PA 16127-2104

In response to my August 1988 *QST* article on elevated verticals,[1] I received many inquiries about the effects of changing the height of the antenna above ground, or increasing the number of radials. This paper will describe the results of some analyses I performed for 80 and 160 meters, utilizing W7EL's new *EZNEC*[2] software package and *NEC-4.1*, the latest, security-restricted *NEC* modeling program.[3] As expected, the use of more radials and/or a higher height above ground will produce an increase in low-angle power gain. However, there are some trade-offs for us to consider.

KB8I revisits the subject of low-frequency verticals using elevated radials, including detailed comparisons between elevated and buried radial systems.

Background

A single vertical monopole, with elevated horizontal radials, was selected as the test antenna. I used aluminum wire (#12 AWG) for all conductors, since *EZNEC* does not allow more than one type of metal in any particular model. All of the radials, and the monopole itself, had a fixed length of 0.25 λ. Operating frequencies were set at 3.75 MHz (0.25 λ = 65.576 feet) on 80 meters and 1.84 MHz (0.25 λ = 133.647 feet) on top

band. For the sake of simplicity, I utilized the same set of ground constants at both frequencies (conductivity = 0.004 Siemens/meter and dielectric constant = 13), although this assumption may be somewhat unrealistic. Research indicates that the apparent soil conductivity can *increase* from 15% to 45% as the frequency is raised from 2 to 4 MHz, depending upon the type of ground. Similarly, the dielectric constant may *decrease*

from 5% to 25% over this same frequency range.[4]

I assumed "isolated feed" for each antenna, since my earlier modeling work had shown that it produced slightly more gain than "direct feed." To achieve isolation, the inner ends of the radials should be bonded only to the outer (shield) conductor of the coaxial feed line, with no electrical connection to ground. A choke balun or external

Table 1

Summary of Performance for Elevated Vertical Monopole Antenna, H = 5 Feet, on 3.75 MHz

	Number of Elevated Radials							
	4	*8*	*12*	*16*	*20*	*24*	*30*	*36*
Max Gain (dBi)	−0.145	−0.04	−0.04	−0.03	−0.03	−0.02	−0.02	−0.01
Take-off Angle (°)	24	24	24	24	24	24	24	24
Half-Power Beamwidth (°)	40	40	40	40	40	40	40	40
Half-Power Points (°)	9, 49	9, 49	9, 49	9, 49	9, 49	9, 49	9, 49	9, 49
R_{in} (Ω)	38.61	37.66	37.52	37.43	37.33	37.24	37.12	37.00
X_{in} (Ω)	8.61	9.34	10.52	11.28	11.79	12.16	12.54	12.82

Table 2

Summary of Performance for Elevated Vertical Monopole Antenna, H = 10 Feet, on 3.75 MHz

| | Number of Elevated Radials | | | | | | | |
	4	8	12	16	20	24	30	36
Max Gain (dBi)	−0.055	0.00	0.01	0.02	0.03	0.03	0.04	0.04
Take-off Angle (°)	23	23	23	23	23	23	23	23
Half-Power Beamwidth (°)	39	39	39	38	38	38	38	38
Half-Power Points (°)	8, 47	8, 47	8, 47	8, 46	8, 46	8, 46	8, 46	8, 46
R_{in} (Ω)	35.66	35.16	35.02	34.91	34.81	34.72	34.60	34.51
X_{in} (Ω)	3.65	6.84	8.54	9.49	10.09	10.51	10.96	11.28

Table 3

Summary of Performance for Elevated Vertical Monopole Antenna, H = 15 Feet, on 3.75 MHz

| | Number of Elevated Radials | | | | | | | |
	4	8	12	16	20	24	30	36
Max Gain (dBi)	−0.005	0.03	0.05	0.05	0.06	0.07	0.08	0.08
Take-Off Angle (°)	22	22	22	22	22	22	22	22
Half-Power Beamwidth (°)	36	36	36	36	36	36	36	36
Half-Power Points (°)	8, 44	8, 44	8, 44	8, 44	8, 44	8, 44	8, 44	8, 44
R_{in} (Ω)	33.28	32.93	32.78	32.67	32.58	32.49	32.39	32.31
X_{in} (Ω)	1.51	5.52	7.34	8.33	8.97	9.42	9.91	10.26

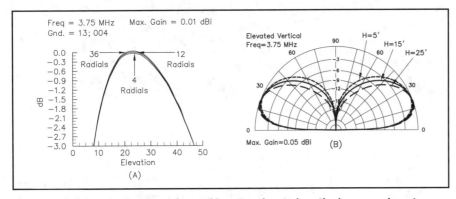

Fig 1—At A, elevation patterns for an 80-meter elevated vertical monopole antenna with 4, 12 and 36 radials, at height of 10 feet above the ground. At B, elevation-plane radiation patterns with four elevated horizontal radials, at heights of 5, 15, and 25 feet.

ferrite beads can also help to reduce the flow of unwanted current on the outer surface of the coax shield. Lightning protection is always advisable, and may be accomplished with a suitable arrestor mounted some distance from the antenna itself, nearer the shack end of the line.

In practice, it is likely that a grounded tower section or heavy pipe mast will be installed to support the base-insulated vertical radiator. This is fine as long as the structure is not connected to the coax shield or the inner ends of the radials. Small posts used to hold up the far ends of the elevated radials may be composed of metal, plastic, or wood, although nonconductive posts

seem to produce a *small* performance increase. It appears that the optimal configuration is a grounded metallic central mast, isolated from the inner ends of the radials, in combination with nonconductive supports for the ends of the radials.

80-Meter Tests

I ran a total of 48 different models for the 80-meter portion of the analysis. The height of the elevated horizontal radials (and thus, the base height of the vertical monopole) was adjusted in 5-foot increments, from 5 to 30 feet. The number of radials was varied over the range from 4 to 36. In all cases, the radials were uniformly spaced around the

feedpoint of the monopole.

Table 1 shows a summary of the results for the 3.75-MHz antenna when the base is just five feet above ground. [Fully detailed tables listing the power gain as a function of take-off angle in 5° steps are located on the diskette accompanying this book in the file \KB8I\TABLES.TXT—*Ed.*] Eight different radial configurations were investigated, where the number of radials was varied from a minimum of 4 to a maximum of 36. When only four radials are used, the azimuthal-plane patterns at all heights are *slightly* non-circular. They have a bit more gain off the ends of the radials than midway between them. Otherwise, the patterns are completely nondirectional for any particular take-off angle.

NEC modeling reveals that increasing the number of radials always yields more gain, although the improvement may be just a few thousandths of a dB—too small to show up in the tabulated data. When the base height (H) of the elevated vertical is only five feet above ground, the gain increases at all take-off angles as more radials are added. In other words, the efficiency of the antenna is actually improving as we install more radials. It is likely that excessive coupling to lossy earth is present at this low elevation height.

In Table 1, I have indicated both the maximum gain and the take-off angle at which it occurs. At H = 5 feet, increasing the number of radials by a factor of nine has no impact whatsoever on the elevation angle, while the

Table 4

Summary of Performance for Elevated Vertical Monopole Antenna, H = 20 Feet, on 3.75 MHz

	Number of Elevated Radials							
	4	8	12	16	20	24	30	36
Max Gain (dBi)	0.02	0.06	0.07	0.08	0.09	0.09	0.10	0.11
Take-Off Angle (°)	21	21	21	21	21	21	21	21
Half-Power Beamwidth (°)	33	33	33	33	33	33	33	33
Half-Power Points (°)	8, 41	8, 41	8, 41	8, 41	8, 41	8, 41	8, 41	8, 41
R_{in} (Ω)	31.24	30.94	30.80	30.69	30.60	30.53	30.43	30.36
X_{in} (Ω)	0.40	4.73	6.60	7.62	8.28	8.76	9.28	9.66

Table 5

Summary of Performance for Elevated Vertical Monopole Antenna, H = 25 Feet, on 3.75 MHz

	Number of Elevated Radials							
	4	8	12	16	20	24	30	36
Max Gain (dBi)	0.04	0.07	0.09	0.10	0.10	0.11	0.12	0.12
Take-Off Angle (°)	20	20	20	20	20	20	20	20
Half-Power Beamwidth (°)	31	31	31	31	31	31	31	31
Half-Power Points (°)	8, 39	8, 39	8, 39	8, 39	8, 39	8, 39	8, 39	8, 39
R_{in} (Ω)	29.45	29.18	29.05	28.94	28.86	28.79	28.70	28.64
X_{in} (Ω)	−0.16	4.31	6.20	7.25	7.93	8.43	8.98	9.38

Table 6

Summary of Performance for Elevated Vertical Monopole Antenna, H = 30 Feet, on 3.75 MHz

	Number of Elevated Radials							
	4	8	12	16	20	24	30	36
Max Gain (dBi)	0.05	0.08	0.10	0.11	0.11	0.12	0.13	0.13
Take-Off Angle (°)	19	19	19	19	19	19	19	19
Half-Power Beamwidth (°)	30	30	30	30	30	29	29	29
Half-Power Points (°)	7, 37	7, 37	7, 37	7, 37	7, 37	7, 36	7, 36	7, 36
R_{in} (Ω)	27.89	27.64	27.50	27.40	27.33	27.27	27.19	27.13
X_{in} (Ω)	−0.35	4.18	6.09	7.15	7.86	8.37	8.94	9.36

gain improves by less than 0.14 dB. The half-power beamwidth is the angular width of the main lobe (in the elevation plane) within which the gain is no more than 3 dB below its peak value. The half-power points are the two angles marking the upper and lower limits of the main lobe. We can see that adding more radials does not alter the overall shape of the radiation pattern, since both of the half-power points (and thus the half-power beamwidth) remain constant at a height of only H = 5 feet.

R_{in} and X_{in} are the real and imaginary parts, respectively, of the input impedance. Since I utilized #12 wire for the vertical monopole, these impedance values will probably not mean much in a practical sense. However, it is evident that the feed-point resistance goes down and the reac-tance goes up as extra radials are installed at H = 5 feet.

Table 2 reveals what happens when the antenna is raised to a height of 10 feet. Now, the addition of more radials subtly changes the distribution of radiated power as a function of take-off angle. **Fig 1A** compares elevation patterns for 4, 12 and 36 elevated radials at a height of 10 feet. Notice that the gain at a take-off angle of 45° is essentially independent of the number of radials. There is a bit of vertical focusing, as evidenced by a 1° narrowing in the elevation-plane beam-width shown in Table 2.

Tables 3 through **6** show a continual im-provement in performance as the antenna system is raised farther and farther above ground. The peak gain increases slightly, while the corresponding take-off angle de-creases. In fact, the elevation angle at which maximum gain occurs seems to be deter-mined entirely by the height above ground, rather than by the number of radials. Notice that each five-foot increase in the base-height of the antenna leads to a 1° drop in the take-off angle at 3.75 MHz.

Fig 1B illustrates the changes which occur in the elevation-plane radiation pat-tern as an antenna with four elevated hori-zontal radials (NR = 4) is raised succes-sively from H = 5 to 15 to 25 feet. **Fig 2** shows the results for a vertical at heights of 10, 20, and 30 feet, when NR is held con-stant at 12. The half-power beamwidth (in the elevation plane) continually shrinks as more and more RF energy is compressed into the lower portion of the main lobe, at the expense of higher-angle signal strength.

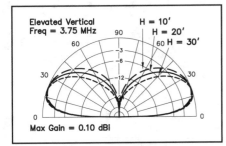

Fig 2—Elevation-plane radiation patterns for an 80-meter elevated vertical monopole antenna with 12 elevated horizontal radials, at heights of 10, 20, and 30 feet above the ground.

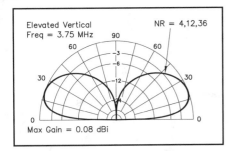

Fig 4—Elevation-plane radiation patterns for an 80-meter elevated vertical monopole antenna at a height of 15 feet above the ground, when the number of elevated horizontal radials (NR) is 4, 12, or 36.

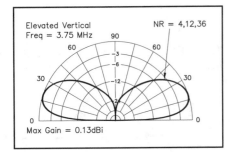

Fig 5—Elevation-plane radiation patterns for an 80-meter elevated vertical monopole antenna at a height of 30 feet above the ground, when the number of elevated horizontal radials (NR) is 4, 12, or 36.

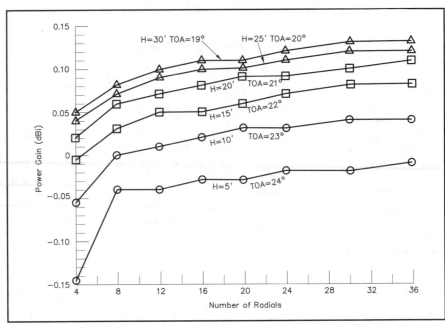

Fig 3—Maximum gain versus number of radials and height above ground, for 80-meter elevated vertical-monopole antenna systems. The take-off angle at which peak gain occurs is listed adjacent to each base-height.

Fig 3 summarizes the information supplied in the first six tables, by comparing the maximum gain to the number of radials (NR) and their height above ground (H). For convenience, the take-off angles at which peak gain occurs are also listed on the graph. The trade-off between "more radials" and "more height" is clearly shown. It appears that a combination of eight radials at a height of 10 feet provides the most "bang for the buck" on 80 meters. Increasing the number of elevated radials from four to eight provides the largest incremental improvement, with diminishing returns after that point. Similarly, raising the base height of the antenna from 5 to 10 feet yields the greatest marginal increase in gain (roughly 0.1 dB). Of course, those operators seeking to squeeze out the last small fraction of a decibel will probably want to install more radials at a greater height above ground.

To put this subject of "maximum gain"

into perspective, **Fig 4** illustrates the effect of increasing the number of radials from 4 to 12 to 36, when the base height of the elevated vertical is fixed at 15 feet. Notice that the three individual plots are virtually indistinguishable from one another, because the differences in gain are so small. **Fig 5** gives the results when H = 30 feet, as NR is varied over the range from 4 to 36.

160-Meter Tests

As before, 48 different models were run to complete the analysis of the elevated antenna on top-band. This time, the base height of the vertical radiator (and the height of the elevated horizontal radials) was varied in 10-foot increments from 10 to 60 feet, while the number of radials again spanned the range from 4 to 36.

Table 7 shows the results for the 1.84-MHz antenna when the base height is 10 feet. As was true on 80 meters, the azimuthal-plane ra-

---

Table 7

Summary of Performance for Elevated Vertical Monopole Antenna, H = 10 Feet, on 1.84 MHz

	Number of Elevated Radials							
	4	8	12	16	20	24	30	36
Max Gain (dBi)	0.485	0.59	0.60	0.60	0.60	0.61	0.61	0.61
Take-Off Angle (°)	22	22	22	22	22	22	22	22
Half-Power Beamwidth (°)	40	40	40	40	40	40	40	39
Half-Power Points (°)	8, 48	8, 48	8, 48	8, 48	8, 48	8, 48	8, 48	8, 47
R_{in} (Ω)	40.46	39.48	39.34	39.26	39.19	39.13	39.03	38.95
X_{in} (Ω)	10.17	10.98	12.13	12.88	13.38	13.73	14.09	14.34

Table 8

**Summary of Performance for Elevated Vertical Monopole Antenna,
H = 20 Feet, on 1.84 MHz**

	Number of Elevated Radials							
	4	8	12	16	20	24	30	36
Max Gain (dBi)	0.565	0.62	0.63	0.64	0.65	0.65	0.66	0.66
Take-Off Angle (°)	21	21	21	21	21	21	21	21
Half-Power Beamwidth (°)	38	38	38	38	38	38	38	38
Half-Power Points (°)	7, 45	7, 45	7, 45	7, 45	7, 45	7, 45	7, 45	7, 45
R_{in} (Ω)	37.50	36.97	36.83	36.73	36.64	36.57	36.47	36.39
X_{in} (Ω)	4.91	8.06	9.74	10.66	11.24	11.64	12.06	12.35

Table 9

**Summary of Performance for Elevated Vertical Monopole Antenna,
H = 30 Feet, on 1.84 MHz**

	Number of Elevated Radials							
	4	8	12	16	20	24	30	36
Max Gain (dBi)	0.605	0.65	0.66	0.67	0.68	0.68	0.69	0.70
Take-off Angle (°)	20	20	20	20	20	20	20	20
Half-Power Beamwidth (°)	35	35	35	35	35	35	35	35
Half-Power Points (°)	7, 42	7, 42	7, 42	7, 42	7, 42	7, 42	7, 42	7, 42
R_{in} (Ω)	35.08	34.69	34.54	34.43	34.34	34.27	34.17	34.09
X_{in} (Ω)	2.46	6.40	8.20	9.16	9.78	10.20	10.66	10.98

Table 10

**Summary of Performance for Elevated Vertical Monopole Antenna,
H = 40 Feet, on 1.84 MHz**

	Number of Elevated Radials							
	4	8	12	16	20	24	30	36
Max Gain (dBi)	0.63	0.67	0.68	0.69	0.70	0.71	0.71	0.72
Take-off Angle (°)	19	19	19	19	19	19	19	19
Half-Power Beamwidth (°)	33	33	33	33	33	33	33	33
Half-Power Points (°)	7, 40	7, 40	7, 40	7, 40	7, 40	7, 40	7, 40	7, 40
R_{in} (Ω)	32.96	32.62	32.46	32.35	32.26	32.19	32.10	32.02
X_{in} (Ω)	1.09	5.35	7.18	8.17	8.80	9.26	9.74	10.10

Table 11

**Summary of Performance for Elevated Vertical Monopole Antenna,
H = 50 Feet, on 1.84 MHz**

	Number of Elevated Radials							
	4	8	12	16	20	24	30	36
Max Gain (dBi)	0.64	0.68	0.70	0.71	0.71	0.72	0.73	0.73
Take-off Angle (°)	18	18	18	18	18	18	18	18
Half-Power Beamwidth (°)	30	30	30	30	30	30	30	30
Half-Power Points (°)	7, 37	7, 37	7, 37	7, 37	7, 37	7, 37	7, 37	7, 37
R_{in} (Ω)	31.06	30.74	30.59	30.48	30.39	30.32	30.23	30.16
X_{in} (Ω)	0.32	4.71	6.56	7.58	8.23	8.70	9.22	9.60

diation pattern is slightly noncircular at all heights if only four radials are installed. Once more, the addition of extra radials produces an increase in gain at all take-off angles, revealing an improvement in antenna efficiency, and suggesting that the antenna should be raised higher above ground.

When the top-band ground plane is raised to a height of 20 feet, **Table 8** shows the outcome. There is a slight change in the distribution of the radiated power—more RF energy is concentrated into the lower take-off angles at the expense of high-angle signal strength. **Tables 9** through **12** display the computer-generated results for base heights

Table 12

Summary of Performance for Elevated Vertical Monopole Antenna, H = 60 Feet, on 1.84 MHz

	Number of Elevated Radials							
	4	8	12	16	20	24	30	36
Max Gain (dBi)	0.64	0.68	0.70	0.71	0.72	0.72	0.73	0.73
Take-off Angle (°)	17.5	17	17	17	17	17	17	17
Half-Power Beamwidth (°)	28	28	28	28	28	28	28	28
Half-Power Points (°)	7, 35	7, 35	7, 35	7, 35	7, 35	7, 35	7, 35	7, 35
R_{in} (Ω)	29.37	29.06	28.91	28.80	28.72	28.65	28.57	28.50
X_{in} (Ω)	−0.04	4.41	6.28	7.31	7.98	8.47	9.01	9.41

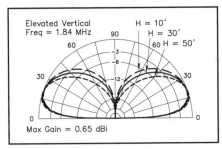

Fig 6—Elevation-plane radiation patterns for a 160-meter elevated vertical monopole antenna with four elevated horizontal radials, at heights of 10, 30, and 50 feet above the ground.

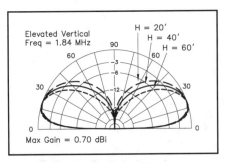

Fig 7—Elevation-plane radiation patterns for a 160-meter elevated vertical monopole antenna with 12 elevated horizontal radials, at heights of 20, 40, and 60 feet above the ground.

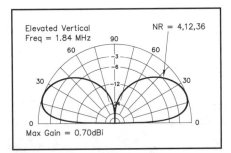

Fig 9—Elevation-plane radiation patterns for a 160-meter elevated vertical monopole antenna at a height of 30 feet above the ground, when the number of elevated horizontal radials (NR) is 4, 12, or 36.

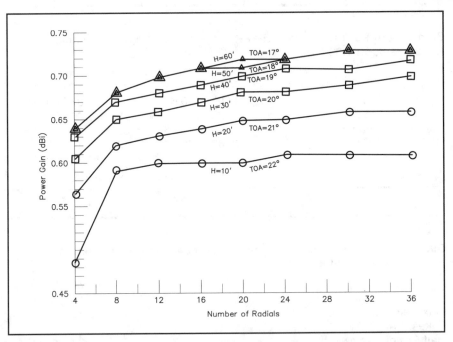

Fig 8—Maximum gain versus number of radials and height above ground, for 160-meter elevated vertical-monopole antenna systems. The take-off angle at which peak gain occurs is listed adjacent to each base-height.

ranging from 30 to 60 feet. On the 160-meter band, the take-off angle for maximum gain decreases by 1° for each 10-foot increase in the elevation of the antenna.

Fig 6 illustrates the variations in radiation pattern that occur when a four-radial vertical is raised from 10 to 30 to 50 feet above ground. **Fig 7** shows the outcome for a 12-radial antenna at H = 20, 40, and 60 feet. We see once again that the elevation-plane beamwidth continually diminishes as the base height of the system is raised. The performance at low take-off angles gets better, while high-angle signal strength is attenuated.

Fig 8 summarizes all of the 160-meter data, and the results are quite interesting. Notice that, according to *EZNEC*, raising the base height of the system from 50 to 60 feet produces almost no benefit in terms of peak gain, although the take-off angle at H = 60 feet is 1° lower. As on 80 meters,

"moving up" from four to eight radials produces the largest incremental improvement in gain. The optimal number of radials seems to be about eight, although 12 may be better if H = 50 or 60 feet. Selecting the best height, however, is more difficult than it was earlier, since the spacing between the "lines of constant height" on this graph is different from that of Fig 3. Now, we pick up an extra 0.03 dB of gain if the base-height is increased from 10 to 20 feet (assuming NR = 8), but we can add yet another 0.03 dB by moving up still farther, from H = 20 to H = 30 feet.

Fig 9 shows how the radiation pattern changes when NR is varied from 4 to 12 to 36, for a top-band antenna whose base height is 30 feet. **Fig 10** displays the same information when the height above ground is held constant at 60 feet. As before, the three separate plots appear almost as one,

Table 13

Summary of Performance for Ground-Mounted Vertical Monopole Antenna on 3.75 MHz

	Number of Buried Radials					
	4	8	16	32	64	120
Max Gain (dBi)	−2.20	−1.31	−0.62	−0.10	+0.18	+0.29
Take-off Angle (°)	25	25	25	25	25	25
Half-Power Beamwidth (°)	43	43	43	43	43	43
Half-Power Points (°)	9, 52	9, 52	9, 52	9, 52	9, 52	9, 52
R_{in} (Ω)	70.91	58.71	50.62	44.68	40.97	39.24
X_{in} (Ω)	35.34	31.15	28.60	26.13	23.69	22.08

Note: Radials are buried 2 inches in the ground.

Table 14

Summary of Performance for Ground-Mounted Vertical Monopole Antenna on 1.84 MHz

	Number of Buried Radials					
	4	8	16	32	64	120
Max Gain (dBi)	−1.06	−0.41	+0.07	+0.46	+0.72	+0.83
Take-off Angle (°)	23	23	23	23	23	23
Half-Power Beamwidth (°)	44	44	44	44	44	44
Half-Power Points (°)	7, 51	7, 51	7, 51	7, 51	7, 51	7, 51
R_{in} (Ω)	64.77	56.08	50.36	45.95	42.77	41.01
X_{in} (Ω)	36.61	31.86	29.05	26.81	24.73	23.22

Note: Radials are buried 2 inches in the ground.

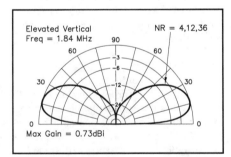

Fig 10—Elevation-plane radiation patterns for a 160-meter elevated vertical monopole antenna at a height of 60 feet above the ground, when the number of elevated horizontal radials (NR) is 4, 12, or 36.

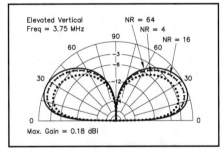

Fig 11—Elevation-plane radiation patterns for a conventional 80-meter ground-mounted quarter-wave vertical-monopole antenna, when the number of buried quarter-wave radials (NR) is 4, 16, or 64.

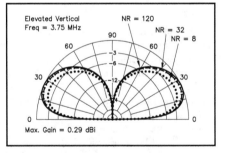

Fig 12—Elevation-plane radiation patterns for a conventional 80-meter ground-mounted quarter-wave vertical-monopole antenna, when the number of buried quarter-wave radials (NR) is 8, 32, or 120.

because the gains are so similar.

Comparisons

The "standard reference" vertical antenna is typically a base-insulated λ/4 monopole mounted at ground level over 120 buried λ/4 radials. This design is often used by commercial medium-wave AM-broadcast stations in the USA, but many hams are either unwilling or unable to install such an extensive radial system. **Tables 13** and **14** illustrate the maximum gain and take-off angle for conventional ground-mounted antennas utilizing from 4 to 120 buried radials, on 80 and 160 meters respectively. (In the *NEC4.1* models, these #12 AWG aluminum radials were placed at a depth of 2 inches beneath the surface of the earth.) As expected, the gain increases as more buried radials are added, but repeatedly doubling the number of radials quickly leads to a point of diminishing returns.

Fig 11 displays the radiation patterns for a conventional ground-mounted buried-radial antenna on 80 meters, when the number of radials is changed from 4 to 16 to 64, while **Fig 12** shows the results for the same antenna when NR = 8, 32, or 120. Clearly, the addition of more radials produces higher gain at all take-off angles, as the efficiency of the system steadily improves. The outcome is exactly the same on 160 meters, with the results illustrated in **Figs 13** and **14**.

Interestingly, none of the elevated-radial antennas described in this article is able to develop as much peak gain as the 120-buried-radial configuration, on either 80 or 160 meters. However, an el-

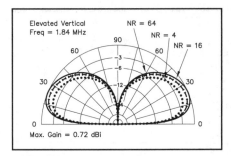

Fig 13—Elevation-plane radiation patterns for a conventional 160-meter ground-mounted quarter-wave vertical-monopole antenna, when the number of buried quarter-wave radials (NR) is 4, 16, or 64.

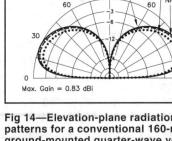

Fig 14—Elevation-plane radiation patterns for a conventional 160-meter ground-mounted quarter-wave vertical-monopole antenna, when the number of buried quarter-wave radials (NR) is 8, 32, or 120.

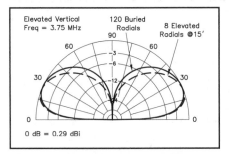

Fig 15—Elevation-plane radiation patterns for a conventional 80-meter ground-mounted antenna with 120 buried radials and an elevated vertical-monopole antenna system with eight radials at a height of 15 feet.

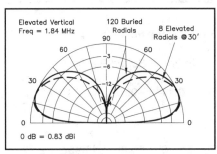

Fig 16—Elevation-plane radiation patterns for a conventional 160-meter ground-mounted antenna with 120 buried radials and an elevated vertical-monopole antenna system with eight radials at a height of 30 feet.

evated vertical at a modest height concentrates more radiation at low take-off angles, which can be very desirable for low-band DXing. A careful examination of the various tables reveals the trade-off between height above ground, number of radials, and gain versus elevation angle. For example, compare data from Table 3 with Table 13—an elevated 80-meter antenna with eight radials at a height of 15 feet can outperform a conventional ground-mounted vertical with 120 buried radials at all take-off angles below about 17° or 18°. This information is shown graphically in **Fig 15**. The same behavior is true for an elevated 160-meter antenna with eight radials at a height of 30 feet (using Table 9 and Table 14), as indicated by **Fig 16**.

All of the buried-radial antennas described here were modeled with *NEC-4.1*, since *EZNEC* (*NEC-2*) was not designed to handle metallic structures which contact or penetrate the earth. Two of the elevated verticals were modeled with both *EZNEC* and *NEC-4.1* to see if there were any discrepancies between the results predicted by the two computer codes. Values for feed-point impedance were within 0.2 Ω of one another, which is quite good. The difference in reported gains was never more than 0.02 dBi, except at take-off angles of 0° and 90°. The disparity at these two angles occurs because the mathematical formulations utilized to model "real ground" are not the same in *NEC-2* as in *NEC-4.1*. Since both antennas have essentially zero radiated signal at these take-off angles, the lack of agreement is neither meaningful nor important!

Variations

As mentioned earlier, the elevated-ra-

dial systems that I modeled for this article were constructed entirely of aluminum, because of software restrictions. If copper radials are used in conjunction with an aluminum monopole, the results will be slightly better. By the same token, substituting steel tower sections for the vertical radiator will yield a proportional degradation in performance. In regions with higher soil conductivity, these antennas will produce more gain at a given height than is shown here, while the opposite is true for geographical areas with more resistive (lossy) soil.

Conclusions

Modeling with *NEC*-derived software has shown that the gain of an elevated vertical-monopole antenna system with elevated horizontal radials is dependent upon both the height above ground and the number of radials. As more radials are added, the low-angle gain of the antenna generally increases, but the change is relatively small and is definitely non-linear. In a similar fashion, antenna performance improves as the system is raised higher above ground, but once again the incremental gain is not constant. When modeling elevated vertical antennas, *EZNEC* (and presumably, other *NEC-2*-based software packages) appears to give results which are nearly identical to those derived from the more recently developed *NEC-4.1*, which is much harder to obtain because of security restrictions.

Acknowledgments

The author would like to thank Roy Lewallen, W7EL, for his work on *EZNEC*, which is more user-friendly and accessible to the average ham radio operator than native *NEC-2*. Appreciation is also expressed to the ARRL for assistance with the *NEC-4.1* analyses.

References
[1]Al Christman, KB8I, "Elevated Vertical Antenna Systems," *QST*, Aug 1988, pp 35-42.
[2]*EZNEC* is available from Roy Lewallen, W7EL, PO Box 6658, Beaverton, OR 97009.
[3]NEC-4.1 is presently controlled by Lawrence LIvermore National Laboratory under national security restrictions.
[4]George Hagn, SRI International, Arlington, VA, private communication.

A 75/80 Meter Full-Size $^1/_4$-λ Vertical

By Guy Hamblen, AA7QZ/2
16 Dongan Lane
Newfoundland, NJ 07435
e-mail: tel1gah@is.ups.com

AA7QZ/2 gives details on the design and installation of his killer vertical for 75 and 80 meters.

Verticals have always been fascinating to me. For a ground-mounted antenna, their low angle of radiation and omnidirectional radiation patterns can result in an effective DX antenna. This is particularly true for the 80 and 160-meter bands, since an 80-meter horizontal dipole would have to be higher than a wavelength to perform as well as a ground-mounted vertical. I don't know of anyone with two 250 foot towers!

In low-sunspot years with little activity on the higher frequencies, the 40, 80 and 160-meter bands provide unique opportunities for hams to explore fascinating aspects of propagation. A whole new group of low-band enthusiasts is emerging. Prior to my relocating to the East Coast, city and neighborhood restrictions constrained my antenna efforts to low-profile installations. I tried the GAP Challenger vertical and was impressed with its multiband effectiveness. However, I wondered what a full-size quarter-wave 80-meter vertical might offer in comparison. Boy, was I surprised!

Some History

The *Low Band DXing* series of books by John Develdore, ON4UN,[1] have become the standard text for 80 and 160-meter antenna enthusiasts. Chapter 9 of John's latest book provides a wealth of design data for verticals, but it can be intimidating to the newcomer. Orr and Cowan's *All About Vertical Antennas*[2] is short on details on the design and construction of simple verticals for the newcomer.

If you have tall trees available, wire verticals are easy to erect. Inverted-L antennas enjoy a devoted following. Choosing a configuration for the top part of a vertical antenna is the easy part. The more challenging part is at the bottom: making a simple, effective matching network and putting in the wire radial system required for efficiency. Elevated-radial systems have been popularized by Christman, KB8I,[3] and Russell, N4KG,[4] in recent years. However, I still recommend use of a traditional radial system for your first low-band vertical. [After all, not every 80-meter enthusiast has the space, or inclination, to install four 65-foot long radials elevated 10 feet off the ground and in plain sight of the neighbors and your own family!—*Ed.*]

Verticals require a good RF grounding system. Concrete or asphalt are not good conductors of RF currents and will definitely diminish the overall effectiveness of a vertical. Sea water is best. If you live near the ocean, verticals are ideal antennas. Seasonally wet, farm-like and tillable soil is above average for RF conductivity and will give good results with a minimal wire radial system. An excellent discussion of ground losses, radiation resistance and antenna efficiencies may be found on pages 9-1 through 9-5 of *Low Band DXing*. Paul Lee, N6PL, in Chapter 9 of his *Vertical Antenna Handbook*[5] has an excellent discussion on the effects of the earth on the efficiency of vertical antennas.

Design Goals

I wanted to construct a full-size, high-efficiency quarter-wave vertical for 80/75 meters, using aluminum tubing and non-conducting guys made of $^1/_8$-inch Philystran. I also wanted to design a simple, but easy-to-tune matching network, without bulky variable capacitors. It had to be capable of handling a kW. I planned to use a relay to change resonance from the phone to the CW portion of the band and I intended to use a low-power SWR tester from MFJ or Autek to tune the system in my backyard.

I also wanted to design the vertical so it could easily be taken down during the late spring and summer months, when static levels and daytime absorption make the 80-meter band nearly unusable. Part of this was for aesthetic reasons, but I also needed an easy-to-install portable antenna for Field Day or portable contest operations.

I planned to use insulated stranded wire radials because of their flexibility, ease of installation and because they could be rolled up in the springtime. Cost of the materials was a close secondary requirement to a design that is simple, modular, and easy to install and disassemble.

Construction Details

I followed Chapter 9 of ON4UN's book closely. Even though most of his construction details are oriented to using a 60-plus foot tower on 80 or 160 meters, I found with some interpolation that his design formulas are surprisingly accurate for smaller diameter aluminum tubing. Orr and Cowan covered matching networks in some detail in Chapter 3 of their book.

I used six telescoping lengths of 12-foot long aluminum tubing[6] with 0.058-inch wall thickness. My base section was 1.500 inch OD aluminum tubing, with sub-

Table 1

Telescoping Aluminum Tubing, Lengths for AA7QZ Vertical

Tubing OD (Inches)	Cut Length (Feet)	Inst. Length (Feet)	Cumulative Height (Feet)	Hole for Guy Plate (Inches)
0.875	5.5	4	58	
1.000	12	10.5	54	0.875
1.125	12	10.5	43.5	
1.250	12	10.5	33	
1.375	12	10.5	22.5	1.250
1.500	12	12	12	1.500

Note: overlap of telescoping joints is 18 inches.

sequent sections with ODs of 1.375, 1.250, 1.125, 1.000 and finally 0.875 inches. See **Table 1** for details.

The 0.058-inch wall thickness for each section allows for a tight fit between sections. The top section had 48 inches sticking out above the joint, to give an overall height of 58 feet. Although smaller-diameter sections could be used at the expense of some rigidity in the overall vertical, during installation you'll have lower anxiety levels with larger tubing. It always amazes me how elastic a 58-foot length of aluminum tubing is!

Most aluminum tubing has some very small burrs on the inside. When inserting smaller sections, these burrs can catch and gall and generally stymie the "ease of installation" requirement. I borrowed an 18-inch electric drill extension from a neighbor and used a small fiber grinding attachment. In five minutes I had removed all burrs on the inside of the tubing. Using the same grinding tool, I rounded all tubing edges to minimize any roughness. A 400-grit wet-grade sandpaper on the 18-inch extention finished off the mechanical portion of the vertical. I inserted each section and drilled a single 0.125-inch hole one inch down from the top of each lower section for a pinning bolt. I used stainless steel bolts, nuts and star lockwashers for each section in order to ensure easy disassembly in the springtime.

The overall length of the monopole can be figured from the formula:

$$L(\text{feet}) = \frac{245.9}{f(\text{MHz})} \times P \qquad \text{(Eq 1)}$$

where P can be interpolated from Figure 9-43 on page 9-34 of *Low Band DXing*. I used a P of 91%, giving a length of 58 feet at a resonant frequency of 3.850 MHz. The feedpoint impedance will be capacitive if the vertical element is shorter than the target operating frequency. The series capacitive reactance of the antenna and a shunt inductance to ground form an L-network that transforms the nominal 36-Ω feedpoint impedance to 50 Ω.

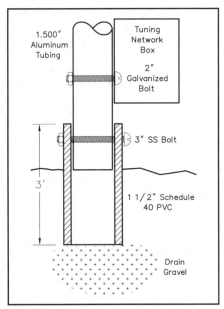

Fig 2—Details of base mount for AA7QZ/2 vertical. The 1.5-inch schedule 40 PVC pipe is placed in the hole on a bed of gravel for drainage. The 1.500-inch tubing is bolted to the PVC pipe with a 3-inch long stainless steel bolt. The tuning network is mounted to and electrically connected to the 1.500-inch tubing with a 2-inch long galvanized bolt to minimize electrolytic corrosion.

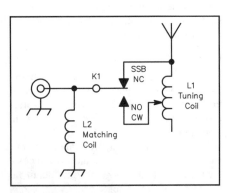

Fig 1—Schematic of tuning network, showing relay switching required to change from phone to CW bands. One setting of L2 provides a match for both phone and CW.

My target operating frequencies were 3.800 MHz for phone and 3.550 MHz for CW. Adding a coil in series with the vertical element allows me to tune the vertical system to the CW band. This is much easier and less expensive than using a vertical that is longer than the target operating frequencies and then tuning it with a series variable capacitor. The mode change can be done with a simple RF-rated relay.[7] See **Fig 1** for the schematic of my tuning/matching network.

I then assembled and bolted all the sections together. I also marked each section at the joint for proper bolt alignment during installation. I placed a small PVC endcap on the top section to seal it off. This was installed by carefully heating the tubing with a small propane torch and "sweating" it onto the PVC piece. This will keep rain out and will prevent wind from turning your vertical into a large, very annoying whistle.

I guyed the vertical at the 12, 33 and 54 foot levels, corresponding to tubing section changes. Three 4-inch round, $^1/_{16}$-inch thick, aluminum plates were drilled with center holes of 1.500, 1.125 and 0.875-inch diameter. Smaller holes were drilled on the periphery of the plates to attach guylines.

After disassembly, I painted all the sections the same bark-brown as the tree trunks in my backyard. This turned out so well that the vertical is nearly indistinguishable from other trees in the backyard and certainly adds to the neighborhood aesthetic appeal. Be sure to use an enamel or lacquer-based paint.

For a base, I used a 3-foot piece of 1.5-inch schedule 40 PVC pipe placed 2 feet into the ground. See **Fig 2** for construction details of the base. The pipe has a nominal ID of 1.625 inches. This insulates the vertical element from ground and provides sufficient stability. An aluminum radial mounting plate drilled with a clearance hole was mounted over the PVC base. I drilled equally spaced holes at its periphery to attach the radial wires. Stainless steel self-tapping metal screws facilitated the installation of radials.

The matching network relay and coils were housed in a heavy-duty Rubbermaid container. The price, assortment of various sizes, and the waterproof nature of these containers make them an ideal, cost-effective choice. The container was bolted to the base of the vertical, close to the radial plate. See the photo in **Fig 3**.

I eventually ended up with 40 radials, 64-feet long and made of #14 stranded and insulated wire because of easy availability at any electrical supply store. Smaller wire may be used also. Radial wires operate in the near field of the vertical antenna system and provide a path for return RF currents. They contribute to the overall efficiency of the antenna. The number required is depen-

Fig 3—Photo of inside of tuning network box, along with aluminum plate used to mount the ground radials. Note the ferrite-core choke balun on the feed coax to help keep RF off the coax shield back to the shack.

dent on your ground conductivity. Lengths and required number of radials have been controversial topics of discussion over the years; see ON4UN's Chapter 9 for an excellent synopsis on this topic.

Installation and Tuning Details

In the next steps I describe the actual installation of my antenna. I cannot emphasize strongly enough that you must make *absolutely sure* you have adequate clearance from powerlines in case the vertical should fall for any reason—either during installation or later!

I set the radial mounting plate over the PVC base insulator, drove a ground rod (for lightning protection) and connected it to the radial plate. Install the bottom 1.500-inch diameter aluminum tubing first and guy it in place with the lower set of guys. I premeasured and cut the guy-line lengths. A 12-foot stepladder was placed standing next to the bottom section.

I assembled the top three sections (the 1.125, 1.000 and 0.875-inch tubing), with the top set of guys attached. This made the top assembly 25 feet long. Prior to inserting each section, I coated each with a conductive grease to facilitate disassembly. The top assembly was positioned so that its bottom was next to the PVC base.

We then raised the top section to the vertical position, with one person at each of the three guy tie-down points. All guy lines were held at just enough tension to keep the mast from bending. I then pushed the top

section upwards off ground and placed the bottom tubing temporarily on one of the upper stepladder steps. Standing on the stepladder, I placed the middle guy ring and lines at the bottom of the 1.125-inch tubing, temporarily holding it from sliding downwards with tape. Then, standing on the stepladder, I pushed the top assembly up until my assistant could guide it into the 1.250-inch tubing from below. I bolted the two sections together.

The middle set of guys was walked out to the three guy handlers, and the 12-foot long 1.375-inch diameter tubing was brought to the stepladder. Again, I pushed up the whole assembly until it could be slipped into and then bolted to the 1.375-inch tubing. The whole assembly was pushed upwards until it could be fit into the top of the bottom 1.500-inch bottom section and was secured with the pinning bolt.

All throughout, tension was maintained on the guy lines for stability of the vertical, but not so much that I couldn't push the sections upwards! Of course, this is easier to say than do—good communication and planning with your guy-line team makes this whole job a 30 minute effort.

All the radials were then laid in place. For safety reasons, I used a sidewalk edger garden tool to make a slit in the ground to bury each radial. Otherwise, my wife had warned me that our lawnmower and 2000 feet of radial wires above the grass would make a memorable story! This is a time-consuming effort, so plan accordingly.

From formulas in Orr and Cowan's book, I knew I needed 2.0 to 2.5 µH of shunt loading coil and a net 1700 to 1800 pF of series capacitance to form an L-network to bring the feedpoint impedance up to 50 Ω. As explained earlier, the shorter-than-resonant vertical element itself provided the series capacitance and the shunt inductance was made up of a 2-inch diameter (B&W 3026, 8 TPI, #14 wire) air-wound coil. A series inductance would be needed for the CW part of the band, as well as providing fine-tuning of the vertical for phone.

To tune the system, I used an Autek RF Analyst, connected with a short length of coax to the SO-239 connector. On the ground side of the shunt matching coil, I attached a jumper wire to an alligator clip and connected it about 3 turns from the top. I did the same with a jumper wire and an alligator clip for the series resonating coil. I connected the RF Analyst and experimentally determined the series coil tap to resonate the vertical at 3.8 MHz. The tap should be at or near the top of the series coil. Then I experimentally adjusted the matching shunt-coil's tap for lowest SWR, readjusting the resonating series-coil tap as necessary. These became my final tap points for the phone band. Then I determined the se-

ries-coil tap to resonate the vertical for 3.550 MHz to cover the CW portion of the band, using the same shunt-coil tap as was used on phone. I switch between the series-coil taps with a 12-V relay to cover most of the 75 and 80-meter bands with one vertical.

On-the-Air Results

I used a remote coax switch to give me "antenna A versus antenna B" testing ability. On-the-air testing revealed the full-length vertical was 2 to 3 S-units stronger on receive and transmit than the GAP Challenger. It was remarkable how much difference was observed in received signal strengths—stations not readable over the noise became discernible. This doesn't mean the GAP is not doing the job it was designed for. The GAP is a linear-loaded $1/8$-λ vertical on 75 meters; it can't possibly be better than a full-size $1/4$-λ vertical with an extensive ground radial system. In fairness, the $1/4$-λ vertical doesn't work at all on any other band, whereas the GAP is an efficient multiband antenna.

Although my sample size is small and I continue to collect field reports, I am quite encouraged by the effectiveness of this relatively simple vertical system. It is a pleasure to consistently work European DX stations and bust pileups with a minimum of calls.

Next...

I am so enthusiastic about the effectiveness of this system, I am now planning to assemble a second vertical and build a variable phase angle network to "steer" the resultant radiation pattern. Modeling shows a 4 to 5 dB gain over a single vertical and a significant 18+ dB front-to-back ratio. On 80 meters, this should be a FUN antenna system on the East Coast!

Notes and References

[1] John Devoldere, ON4UN, *Antennas and Techniques for Low-Band DXing* (Newington: American Radio Relay League, Inc, 1994)

[2] William I. Orr, W6SAI and Stuart D. Cowan, W2LX, *All About Vertical Antennas* (Lakewood, NJ: Radio Amateur Callbook, 1993)

[3] A. Christman, "Elevated Vertical Antenna Systems," *QST*, Aug 1988, pp 35-42; Feedback, *QST*, Oct 1988, p 44.

[4] T. Russell, "Simple, Effective, Elevated Ground-Plane Antennas," *QST*, Jun 1994, pp 45-46.

[5] Paul H. Lee, N6PL, *The Amateur Radio Vertical Antenna Handbook* (Hicksville, NY: CQ Publishing, 1984)

[6] Aluminum tubing is available from national suppliers who advertise in *QST*. Try to organize or pool an order with others to reduce shipping charges.

[7] I used B&W coil stock (part number 3026, 2-inch OD, 8 TPI, #14 wire) and RF-rated 12-V relays available from Nebraska Surplus Sales.

The K6NA 80-Meter Wire Beam

By Glenn Rattmann, K6NA
14250 Calle de Vista
Valley Center, CA 92082

Ever get crunched in a DX pileup on 80/75 meters by K6NA? Lots of folks have! K6NA reveals the secrets behind his two-element "killer" Yagi.

This article describes a practical design for a two-element, horizontal wire Yagi for the 75/80-meter band using just two support towers. Designed with DXing and contesting in mind, this radiator-and-reflector combination features moderate gain and good front-to-back ratio, combined with instant beam reversal and band-segment (mode) switching. It uses tuned feeders and a balanced-output antenna tuner at the operating position for matching at the transmitter. Remote relays at ground level handle reversal and parasitic-tuning requirements, taking advantage of the convenient properties of half-wave transmission lines.

I designed this antenna empirically, before PC-modeling programs were readily available. I've confirmed that it really works, through years of successful 80-meter operation. Later, I verified this with computer modeling. **Fig 1** shows the general layout of my system.

Background

At my former location I had two towers, 70 and 100-feet high, fairly close together on a half-acre of property. There was an assortment of quads and Yagis for 10-40 meters,[1] but as is often the case with all-band DXers, the 80-meter antenna was an afterthought. Through the years I experimented with many types, looking for the "silver bullet" antenna for 80 that would finally make this challenging band a pleasure to operate.

Each season numerous magazine articles and endless discussions by experts extolled the virtues of this or that magic 80-meter DX antenna. I even tried some of them: a diamond quad loop, a delta loop (bottom-fed or corner-fed), a pair of parasitic deltas, phased quarter-wave slopers (tilted ground planes), a $\lambda/4$ vertical with lots of radials. I always came back to a basic truth: with the tower heights available, and faced with the reality of rough, rocky, lossy earth, my 95-foot high inverted-V dipole was usually as good as—or better than—any of the "trick" antennas I had spent hundreds of hours building. As a bonus, usually the dipole was quieter on receive.

Even Jim Lawson, W2PV, in an early *QST* article about broadbanding an 80-meter antenna, mentioned almost in passing that his high (110-foot) inverted-V was superior to his four-element phased vertical array, at least on transmit.[2] What I observed about 80-meter DXing in the 1970s was that most attempts using vertically polarized systems to achieve low-angle radiation were not satisfactory. This was due to excessive ground losses (both near-field return-current losses and far-field reflective losses); in addition, verticals were generally noisy on receive.

Neither vertically polarized antennas nor relatively low horizontal antennas were really getting the job done. I resolved to construct my next station thinking about the 80-meter band from the start. "High and horizontal" would be the goal.

Design Criteria

My new station would have two main towers, each 140 feet high. Because I had all-band contesting in mind, there would be numerous other antennas on the towers. I wanted to put a three-element 40-meter beam on top of one tower and a large 20-meter Yagi on the top of the other. Though it likely would be an excellent performer, a rotating beam of some type for 80 meters was ruled out because I did not want to dedicate the top of one tower to it. Also, there was the near-certainty of difficult maintenance problems with such a gigantic antenna. Instead, I settled on a design for a fixed-wire beam. These would be the main criteria for reliable, day-to-day operation:

- Oriented properly for best results in both DX and domestic contests
- Moderate gain, with a wide lobe for good azimuth coverage
- Instant beam reversal with a decent front-to-back ratio
- Near-instant band-segment (phone or CW mode) switching, especially for casual, daily DXing
- Easy, low-loss matching anywhere in the band
- Horizontally polarized to minimize both ground-reflection losses and noise on receive
- High enough to produce reasonably low-angle radiation
- Reliable and easy to maintain without disrupting other antennas on the towers.

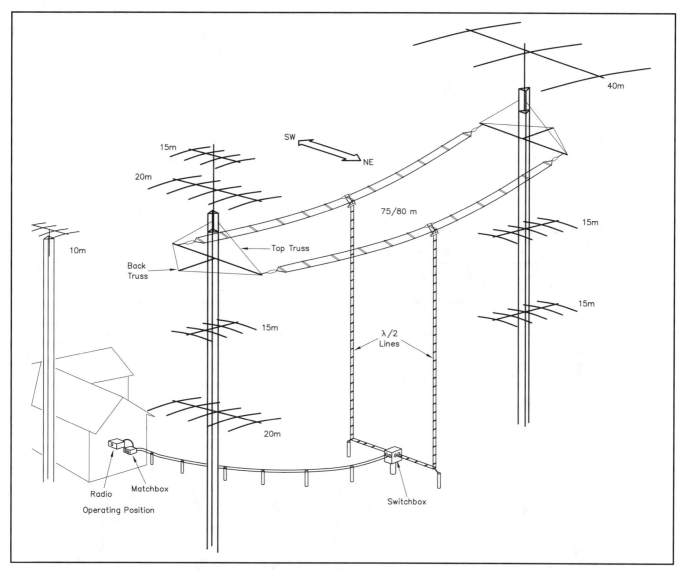

Fig 1—The antenna farm at K6NA, showing the 80-meter wire beam installed on the pair of 140-foot towers, using 35-foot long crossarms made of aluminum tubing and fiberglass. The rearward truss and upper truss wires make each crossarm rigid. For clarity, the tower guy wires are not shown.

Design Discussion

Some discussion of these interrelated design criteria will aid in understanding the trade-offs and choices I made while planning this antenna. For example, from San Diego the typical short path to Europe on 80 meters *is not* the expected true Great-Circle path of about 25°. Rather, the short path toward Europe is almost always a "bent" or skewed path, where signals pass to the south or southeast of the highly absorptive auroral oval, propagating by means of a scatter mode that uses ionized patches over the central Atlantic Ocean. Consequently, our pseudo short-path heading toward Europe is commonly 50° to 90° in azimuth. Occasionally, European signals on 80 meters arrive in California from the southeasterly direction. Rarely, around the time of the Equinox, they may even arrive by scatter path from straight south, as reported by some rotary beam users in California.

I decided to orient the two support towers so that the reversible wire beam, slung between the towers on a pair of horizontal crossarms, would point at 60°/240°. This would provide good coverage across the USA and into Europe (both short and long path), Africa, the Caribbean and South Pacific areas, which together make up about 85% of my annual QSOs on 80 meters.

I could build a simple two-element parasitic design, using λ/2 elements on a reasonable boomlength of about 35 feet, or λ/8.[3] **Fig 2B** shows that such a design has a half-power azimuth beamwidth of about 74° in the azimuth plane. I was willing to accept the slightly lower gain and front-to-back ratio compared to a three-element design (which would be more difficult mechanically), or a two-wire phased array using double-extended Zepp elements. The Zepp would require the towers to be placed about 350 feet apart. Either of these

antennas conceivably would provide more gain, but at the price of reduced azimuthal coverage. A reversible, all-driven array has the additional disadvantage that its phasing and matching networks become overly complicated to make it work properly at both ends of the band.

To achieve the twin requirements of easy mode-switching and beam reversal, I decided to construct a symmetrical, two-element Yagi beam. The elements are identical dipoles, self-resonant near 3.675 MHz. This frequency is about 3% to 4% lower than the phone operating area around 3.800 MHz, meaning that each element could be used as a reflector in the phone band with no additional loading. Each dipole is fed with λ/2 of open-wire line. These lines come together in a relay box at ground level. Power is applied to one or the other driven element, and a short is applied across the base of the re-

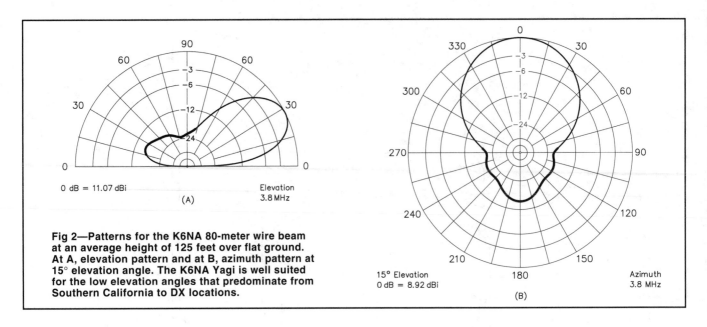

Fig 2—Patterns for the K6NA 80-meter wire beam at an average height of 125 feet over flat ground. At A, elevation pattern and at B, azimuth pattern at 15° elevation angle. The K6NA Yagi is well suited for the low elevation angles that predominate from Southern California to DX locations.

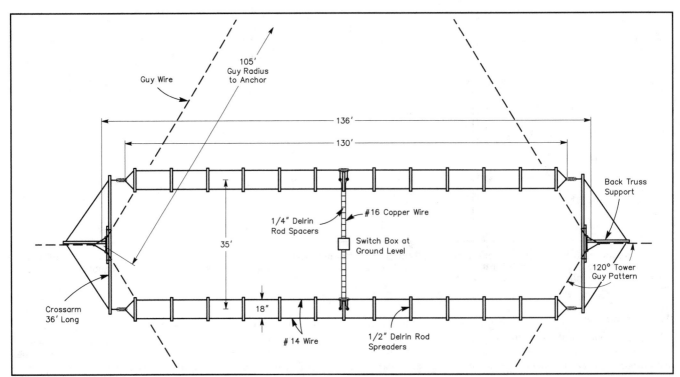

Fig 3—Top view of the 80-meter wire beam mounted on the towers, with details of the crossarms and rearward truss system that prevents the tensioned antenna wires from bending the crossarms inward. The tower-guy placement geometry provides several advantages for antenna-farm management! See text.

maining feeder to force the remaining element to act as a parasitic reflector.

To operate near 3.5 MHz, the reflector simply is retuned to about 3.4 MHz, using an inductor in place of the direct short in the relay box. Supplying RF to the relay box is accomplished using low-loss, open-wire feeders. A balanced-output tuner in the shack is simply touched up to move between phone and CW subbands, and to accommodate the change in SWR. There is no requirement for an exact match at the line input to the driven element because of the low-loss transmission lines.

With the crossarm supports mounted at 130 ft, the dipoles are about λ/2 high and thus produce fairly low-angle radiation. A horizontally polarized antenna is less responsive to man-made noise sources, which have a large vertical component. The height allows λ/2 lines to reach down to the switchbox near ground level. All switching relays are in this weatherproof box, so the system is reliable and easy to maintain. No inaccessible relays or matching networks are up at the elements,

where failures might occur.

Construction

No doubt, the 80-meter wire beam can be built and integrated into many existing multi-tower systems. I had the luxury of planning the tower and guy arrangement well in advance of building the antenna. This helped make erection and maintenance simpler. Such planning contributes to convenient erection of other antennas too!

Fig 3 shows the recommended two-tower

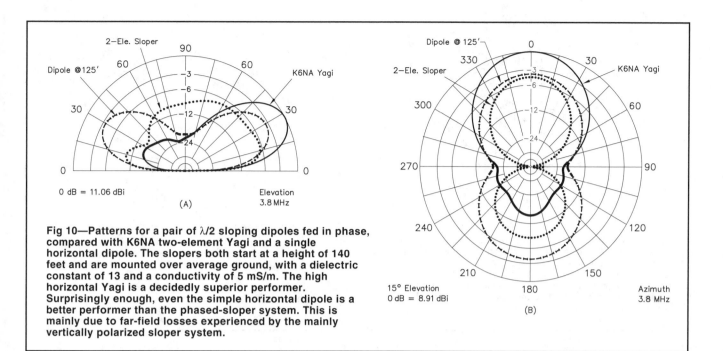

Fig 10—Patterns for a pair of λ/2 sloping dipoles fed in phase, compared with K6NA two-element Yagi and a single horizontal dipole. The slopers both start at a height of 140 feet and are mounted over average ground, with a dielectric constant of 13 and a conductivity of 5 mS/m. The high horizontal Yagi is a decidedly superior performer. Surprisingly enough, even the simple horizontal dipole is a better performer than the phased-sloper system. This is mainly due to far-field losses experienced by the mainly vertically polarized sloper system.

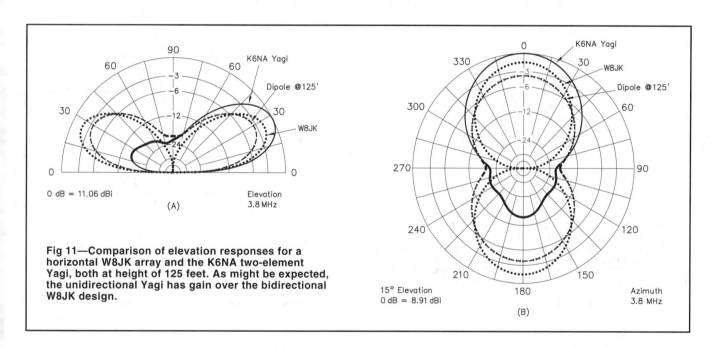

Fig 11—Comparison of elevation responses for a horizontal W8JK array and the K6NA two-element Yagi, both at height of 125 feet. As might be expected, the unidirectional Yagi has gain over the bidirectional W8JK design.

tower. In addition, a high dipole has better rejection of local (high-angle) signals, so it will be a superior DX receiving antenna. Inspection of Fig 8A indicates that the high wire beam probably will outperform a 70-foot high dipole by nearly 10 dB at low elevation angles.

Some people might say: "With 140-foot towers, you should be using quad loops!" Not so fast. **Fig 9A** shows that with the top of a diamond-like, full-wave loop at 135 feet, the λ/2 dipole at 125 feet is superior by at least 1 dB at all angles lower than 40°, in spite of the small stacking gain expected from the loop. The expected gain from the

bigger loop cannot be realized fully because the upper and lower wires are at drastically different heights (in terms of fractional wavelengths) relative to each other and the earth. This prevents proper phase addition. Further, the average height of the loop is lower, which results in a higher peak angle of radiation. The dipole has better high-angle rejection, too, and should be better on receive. If both antennas were up a couple of wavelengths on 80 meters, the quad loop would beat the dipole...but who has a 550-foot tower?

How about a pair of half-wave slopers (that is, tilted vertical dipoles) tied from the

tops of the 140-foot towers one λ/2 apart, and fed in phase for broadside gain? See **Fig 10A**. All the antenna books show the phased-sloper array to be a 4 dB gain, low-angle monster. The author had this exact antenna up for a few months in 1980. It performed fairly well, but was very responsive to powerline noise. For comparison, a single horizontal dipole was hung between the towers a few feet above the top ends of the slopers. With this single dipole, my local noise level dropped by 8 to 10 dB, and stations in New Zealand reported a consistent transmit advantage over the sloper array. This was a real eye-opener. As seen in

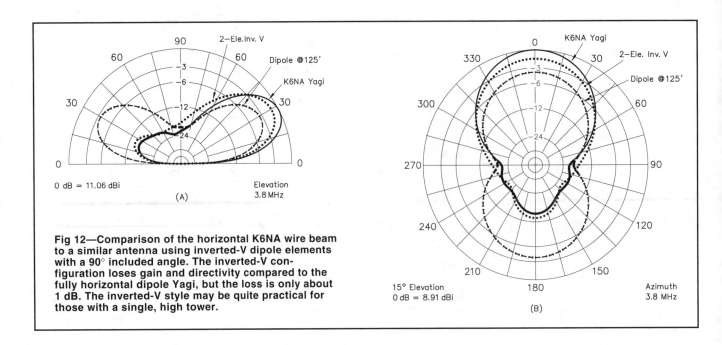

Fig 12—Comparison of the horizontal K6NA wire beam to a similar antenna using inverted-V dipole elements with a 90° included angle. The inverted-V configuration loses gain and directivity compared to the fully horizontal dipole Yagi, but the loss is only about 1 dB. The inverted-V style may be quite practical for those with a single, high tower.

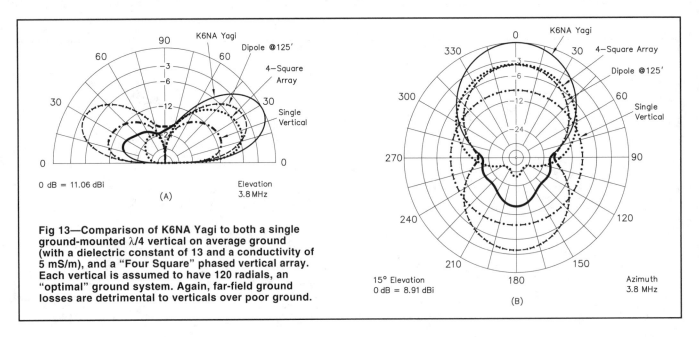

Fig 13—Comparison of K6NA Yagi to both a single ground-mounted λ/4 vertical on average ground (with a dielectric constant of 13 and a conductivity of 5 mS/m), and a "Four Square" phased vertical array. Each vertical is assumed to have 120 radials, an "optimal" ground system. Again, far-field ground losses are detrimental to verticals over poor ground.

Fig 10A, the high wire beam *buries* the phased slopers by 7 dB at the peak angle of 28°, and by 5 dB at 15° elevation!

The famous W8JK is a bi-directional, all-driven wire beam made from two dipoles fed 180° out of phase. At 125 feet in the air, this antenna has excellent rejection of high-angle signals, and is only slightly down from the peak gain of the Yagi configuration. See **Fig 11A**. It requires no relay box. For DXing, it is inferior to the Yagi because there is no way to reject signals (noise) off the back. Years ago, the author fed the two new folded dipoles as a W8JK for a few months prior to building the switchbox. It performed much like the patterns show, but once the switchbox was available to config-

ure the array as a Yagi with good F/B ratio, the W8JK configuration was abandoned. However, a W8JK antenna 100-feet high in Kansas City would be an outstanding Sweepstakes antenna.

Fig 12 compares the horizontal wire beam to a similar antenna with inverted-V elements with a 90° included angle. The V-style gain is down a bit and the pattern is not as clean. But if you have a single tall tower and install this antenna (only one crossarm is required), you will be in the 95th percentile of effective 80-meter DXing antennas.

In **Fig 13** the K6NA Yagi is compared to a quarter-wave vertical mounted over average ground. Since there are less losses due to ground reflection characteristics, hori-

zontal polarization allows the Yagi to outperform substantially the vertical at all angles of interest, even down to 10°. In most locations, the vertical will be noisy on receive. *An extensive local ground screen (or elevated radials) will not make any significant change in the pattern relationship depicted here.* Only by installing the vertical in, say, a saltwater marsh extending perhaps 100 λ, would its low-angle performance be improved substantially.[6]

The "Four-Square" has become popular in recent years. Fig 13 also shows the beautiful pattern of this array, along with that of the two-element beam and a single vertical. In a quiet location this vertical system with a switching matrix makes a terrific receiving antenna, allowing full azimuth coverage in four directions. However, the horizontal wire Yagi still has superior gain at all elevation angles of interest—note the 4 dB advantage at 15°—and the Yagi likely will be quieter on receive in all but the most remote locations. The remarks in the paragraph above about attempts to improve the low-angle performance of the $\lambda/4$ vertical also apply to the Four-Square array.

Conclusions

Someone once said: "All antennas work... some more so." The sometimes-large differences seen during model comparisons can be misleading, as we know from our day-to-day operating. Propagation, pileup dynamics, and operator skills (one skill is picking a quiet location...) are important factors that determine 80-meter DXing success. Virtually every antenna discussed in the comparison section of this report is a "good" antenna. The author's first 180 countries on 80 meters were worked with an inverted-V dipole at 70 feet.

If *tall* supports are available, a horizontally polarized, gain antenna will certainly provide a statistical increase in performance over the others. In lieu of really tall towers, a well-installed vertical system can be an excellent antenna, especially in a quiet receiving location. As the computer models show, height above ground is the most significant variable when designing a low-band station using horizontally polarized antennas. [Hills help too!—*Ed.*]

It is hoped that some of the construction techniques for the crossarms, wire antennas, and feed lines described in this report will stimulate others to build even better horizontal arrays. You will enjoy special satisfaction using homemade open-wire lines, and a 40-year old Johnson Matchbox tuner makes a nice addition to your modern shack.

The author would like to thank H. Shep-

herd, W6US, for his valuable advice regarding wire antennas and open-wire feeder construction; R. Craig, N6ND, for mechanical suggestions; E. Andress, W6KUT, for manuscript review; and N6ND and J. McCook, W6YA, for modeling assistance.

Notes and References

[1]Wayne Overbeck, N6NB, "Quads vs. Yagis Revisited," *Ham Radio*, May 1979, p 12.
[2]James L. Lawson, W2PV, "160/80/75-Meter Broad-Band Inverted-V Antenna," *QST*, Nov 1970, p 17.
[3]James L. Lawson, W2PV, *Yagi Antenna Design*, W. Myers, K1GQ, editor (Newington: ARRL, 1986), p 2-2.
[4]Doug DeMaw, W1FB, editor, *ARRL Electronics Data Book* (Newington: ARRL, 1976), p 82.
[5]Frank Witt, AI1H, "Optimizing the 80-Meter Dipole," *The ARRL Antenna Compendium, Vol 4* (Newington: ARRL, 1995), p 39.
[6]Charles J. Michaels, W7XC, "Some Reflections on Vertical Antennas," *QST*, Jul 1987, p 18.

EWE "Four" Me

By James B. Smith, VK9NS
PO Box 90
Norfolk Island 2899
Australia

VK9NS describes his recipe for good ears on top-band—a switched directional 160-meter receiving array.

Recently, the article "Is this EWE for You," by Floyd Koontz, WA2WVL, in Feb 1995 *QST* caught my interest. It's funny how things turn out sometimes. I had just spent two months looking at the various possibilities for a low-noise 160-meter receiving system and had reached a decision. The local saw mill had delivered, two days before I read Floyd's article, several substantial gum-wood posts to support a dual-direction Beverage antenna.

My first reaction to the article was amazement, coupled with a bit of disbelief. What incredible things could be done from the comfort of the shack, using a computer and a good antenna-modeling program. That approach certainly beats hours of cut-and-try just to get something simple like a sloper working!

My question was: "Would the EWE really work?" After spending some time going back and forth through the article, I suddenly realized that I had a method to try out WA2WVL's system. I could easily prove (or disprove) what Lloyd had written about for myself.

The key was in my existing four-element vertical 40-meter array. I immediately visualized four EWE antennas—easily switched to receive from four directions. I could check the front-to-back ratio (if it existed) and also the front-to-side pattern. Such a test could provide other good insights into the EWE's operation, such as the input impedance, SWR variation and optimum termination values.

But I am jumping the gun a bit. My initial desire was to check front-to-back ratio. For that, two antennas would be fine. If successful, then the two other antennas would rapidly follow.

Fig 1 shows the layout of my four-element 40-meter vertical array with its elevated radials. To make things a bit easier, all spacings are shown as 10 meters (32.8 feet) and the eight posts carrying the radials are labeled from A to H.

A moment with a calculator shows the diagonal spacing is about 22 meters (72 feet)

between any two alternate posts: for example, between Post A and Post C or between post C and E. In other words, I had ready-made supports for up to four EWE antennas, with ideal distances between them. Fig 1 also shows how my 40-meter array is orientated—firing 45°, 135°, 225° and 325°. The directions are approximate. The four elements are labeled 1, 2, 3 and 4.

Now look at **Fig 2**. The use of posts A, C, E and G result in an approximate slew of about 27° counter-clockwise from the array headings. The resultant receive direction for each EWE is shown in the figure. In **Fig 3**, you can see that the use of posts B, D, F and H result in an approximate 27° slew in the other direction from the normal array headings.

You should remember (see Figure 2 in the original *QST* article) that a single EWE is quite wide in the azimuth plane. Exactly which set of posts to use is therefore somewhat academic. In any case, no matter which posts are used, they are a bit short to meet the 10-feet design height of the EWE antennas. I had to do further "ground" work. I decided to replace posts A, C, G and E with four of the longer gum-wood Beverage posts mentioned earlier. The hardest part of the job was getting the original posts out of the ground.

I thought it made sense to make arrangements for some cut-and-try work, just in case. Since 160 meters was the band of most interest to me, I decided to use a height of 10 feet and to feed all my EWE antennas at the bottom. Table 2 in the original *QST* article

was used as a starting point, with a length of 21 meters (68.9 feet) to use all of the available distance between posts. Shorter lengths give better front-to-back ratios, but also more signal loss.

My first task was to make four Input Matching Units (IMUs). I used four small plastic junction boxes used by electricians for conduit work. Each has a waterproof cover. I used the conduit inlet to house the SO-239 coax connector, held in place with a good epoxy-resin adhesive. (Ground and inner wires were first soldered in place on the connector before gluing.) The box now had a

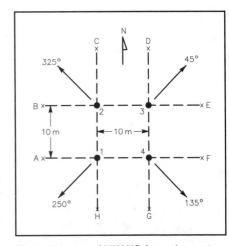

Fig 1—Layout of VK9NS four-element vertical array for 40 meters. Posts labeled A through H support the elevated radials for this array. The 40-meter elements are labeled 1 through 4.

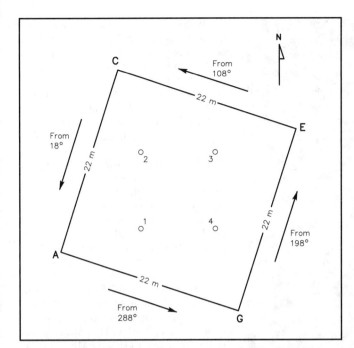

Fig 2—Directional characteristics of four EWE antennas using posts A, C, E and G supporting existing VK9NS 40-meter array. This arrangement provides broad forward lobes centered on 18°, 105°, 198° and 288°.

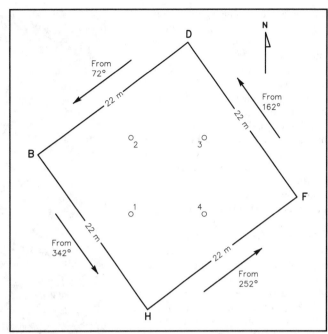

Fig 3—Directional characteristics of four EWE antennas using posts B, D, F and H supporting existing VK9NS 40-meter array. This arrangement provides broad forward lobes centered on 72°, 162°, 252° and 342°.

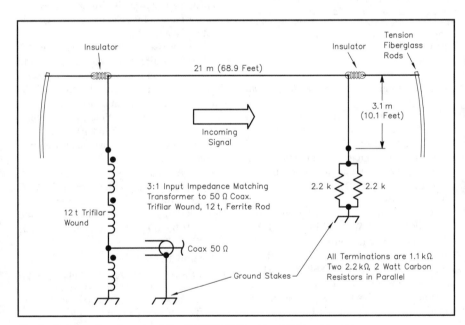

Fig 4—Schematic of each of the VK9NS 160-meter EWE antennas.

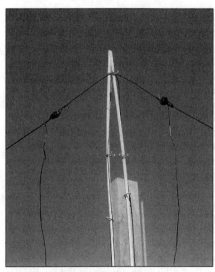

Fig 5—Photo of top of one of the support posts, showing fiberglass fishing rods that help tension the wires.

coax socket and connection "tails." I used a piece of braid for the ground connection.

I then made four 12-turn trifilar-wound coils using short lengths of ferrite rod. I did not have four toroids to make each transformer identical. (I expected to eventually epoxy a transformer in place inside each box.) See **Fig 4**.

I drilled two holes in each box—one opposite the coaxial connector input—this was to be the connection to the EWE antenna. A

hole drilled at right angles to the plane of the antenna and the coax connector was used for the ground connection. I used stainless steel hex-headed bolts, tightening the nuts really tight. About 2.5 cm (one inch) then protruded from the box. I could then make ground and antenna connections very easily with suitable terminal lugs, washers and nuts.

I quickly constructed the first two antennas. One was strung between posts A and C,

and this would be terminated to favor reception from an azimuth of 18°. (See Fig 2 again.) The other was strung between posts E and G. I terminated this one to favor reception from 198°. Switching between these two antennas would establish front-to-back ratio, if it existed.

To tension the antenna wires, I used fiberglass fishing rod extensions. These may be seen in the photograph of **Fig 5**. The horizontal length of the antenna is shorter than the actual distance between posts (by about a meter [3 feet]), so this allows some tension to be applied. The length of each antenna is 21 meters (68.8 feet).

A relay box was built to house the four 12-V relays, which I control from the shack. Each relay switched one antenna through to the common output coaxial connector. The signal then makes its way to the shack through coaxial cable. The box also had five coax connectors (four coax inputs and one output) and one four-pin connector to carry the four control lines. Common return is through the coaxial outer shield. With no relay energized (switch box off), no antenna is connected to the feed coax.

The shack antenna-selector box contains the actual control switches, plus a 12-V dc power supply. The switches are wired so that the relays switch from North, East, South and West—actually, 18°, 108°, 198° and 289° in azimuth. Any one of the four EWE antennas may be selected instantly.

Each post holds the ends of two antennas, and so each has a ground stake associated with it. This allows me to ground the input end of one EWE antenna and the output end of the other. I initially terminated the two antennas using 1.5 kΩ, 2 W carbon resistors.

The initial test results were a revelation! The antennas were certainly quiet compared to my vertical. They also exhibited marked directional properties. I noted that the EWE signals dropped quite dramatically; in fact, by about 3 or 4 S units compared to the vertical. However, the antenna was much quieter for static crashes. The net result was that signals were much easier to read. I consistently found about a 3 S-unit drop in signal "off the back." It also became quickly apparent that I had lost a few "birdies" from local TV sets. In was then that Lloyd had a new believer! These excellent results prompted me to do a few more things.

I first completed the other two EWE antennas. I now had four EWEs, and could rapidly switch the pattern in four directions. In the antenna switchbox in the shack I installed a homebrew 160-meter preamplifier that I salvaged from an old loop project

Fig 6—Photo of junction box used to house matching transformer at the input of each EWE antenna. Coaxial connector is epoxied into the inside of the box's input flange.

many years ago. (See *The ARRL Antenna Anthology.*[2])

The output from the preamplifier is controlled by the gain switch to provide 0, 10, 20, 30 and 40 dB of gain. The highest level is more than enough to overcome the signal loss of the EWE antenna, which is typically about 20 dB. I usually run the preamp at the 10-dB gain level.

I wanted to determine the optimum termination for my EWEs. Using a MFJ-249 SWR Analyzer, I measured the SWR at the input of each antenna. The lowest SWR was about 1.5:1, right around 1.800 MHz, which was about the lowest I could go with my instrument. With Kirsti, VK9NL, at one end of the antenna using the MFJ-249 and my-

self at the terminated end, I made some checks using a variable resistor with a couple of clip leads. Kirsti set the MFJ-249 to 1.830 and at the other end I varied the termination for minimum SWR. An excellent SWR was readily achieved; the variable resistor measured 1.1 kΩ. I then substituted two 2.2 kΩ, 2 W resistors in parallel. Each antenna was then rechecked and the SWR was good on each one.

Back in the shack, on-the-air checks continued on an almost daily basis. I find it very interesting to be able to prove that a signal from Europe is coming long path, or that the US station I just worked was coming short path. I can also frequently say with certainly where the static is coming from—the antennas are certainly directional enough to show this.

I believe that the EWE antenna is sure to go places. It is so simple in concept! Even the limited experiments I have done on termination are interesting. The antenna is frequency conscious—using the MFJ-249, it is easy to sweep a wide range of frequencies and look at SWR.

In some ways my "EWE Four Me" arrangement might not be optimum—there could be a parasitic effect from the 40-meter array elements or radials, for example. However, this is not noticeable, if it exists at all.

I am impressed with the results from my EWE array. A directional receiving antenna opens up another dimension on 160 meters. I intend to try an arrangement with two EWE antennas side-by-side in phase. (See Figure 8 in the original *QST* article.) I also intend to build another dual-band EWE, located away from everything, just to make a couple of other checks.

Finally, I have a secret to divulge—computer modeling in the form of *ELNEC* will be with me soon. I am indebted to Lloyd Koontz for his EWE design. See you on 160 meters. I hope I can hear you!

Notes and References

[1]Floyd Koontz, WA2WVL, "Is This EWE for You?" Feb 1995 *QST*, pp 31 to 33.
[2]*The ARRL Antenna Anthology* (Newington: ARRL, 1978), p 78 [out of print].

Using the Half-Square Antenna For Low-Band DXing

By Rudy Severns, N6LF
32857 Fox Lane
PO Box 589
Cottage Grove, OR 97424

A ntennas widely used by amateurs have a few basic characteristics in common. They provide modest performance and good efficiency, are simple in design, inexpensive to fabricate and very flexible with regard to height, shape and construction materials. There is a very wide range of differences between QTHs, resources and personal circumstances. It is vital that the basic performance of an antenna be preserved even for significant variations in dimensions and materials if it is to be widely useful.

The dipole antenna fits these requirements admirably and is probably the most widely used antenna of all. Unfortunately, on the low frequency bands (80 and 160 meters) it is increasingly difficult to get good DX performance from a dipole due to the problem of getting the antenna high enough (in terms of wavelength). The landmark work by N6BV on HF propagation clearly illustrates this.[1, 2] **Fig 1** shows one of his graphs to illustrate the range of radiation angles most likely to be usable on an 80-meter path from New England to Europe. Over 90% of the time the angles are between 17° and 24°. Other longer paths (and those from different locations) show similar patterns, except that the longer paths have lower peak angles, in the range of 10° to 18°. For DX work on 80 meters, the desirable radiation angles are generally between 10° and 20°.

Also shown in Fig 1 are the radiation patterns for dipoles at 100 feet and 200 feet. At 200 feet the pattern is great, but lowering the antenna to 100 feet reduces radiation at the desired angles significantly. For most hams 100-foot dipoles are not possible and 200-foot dipoles not even a fantasy.

Can't put up a really high horizontal antenna for 80-meter DXing? Maybe the vertically polarized "Half-Square" might be the antenna for you.

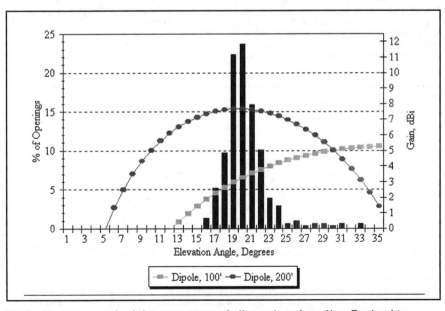

Fig 1—80-meter graph of the percentage of all openings from New England to Europe versus elevation angles, together with overlay of elevation patterns over flat ground for dipoles at two different heights. The 200-foot high dipole clearly covers the necessary elevation angles better than does the 100-foot high dipole. *(From The ARRL Antenna Book, 17th edition, Fig 30.)*

Heights in the range of 40 to 80 feet are much more typical, with the emphasis more towards 40 than 80 feet. This further degrades performance.

Another problem with low dipoles, from a DXing point of view, is that they have great response at high angles. This brings in local and US stations S9+ while you are trying to copy an S3 DX station.

Is there a way to improve on the dipole's DX performance while retaining most of its practical advantages? The answer is "Yes." The half-square antenna can provide 3 to 10 dB of improvement at angles between 10° and 20°, depending on the available height and soil conductivity in the ground reflection zone. In addition, the high-angle radiation can be suppressed. The shape, dimensions and feed point options are also more flexible than previous descriptions have indicated.

The half-square and its close cousin, the "bobtail curtain," have been known to amateurs for nearly 50 years.[3] For the most part, articles describing the half-square have been relatively brief and have not attempted

to examine many of the finer points.[4, 5, 6] This very simple antenna has many subtle details and more than a few surprises. You can get very good results without great effort, but it is also possible to obtain very poor performance if moderate care is not taken!

The purpose of this article is to take a careful look at this antenna including:
• Comparison to a dipole at comparable heights, over different grounds.
• The effect of changing shape and dimensions on performance.
• Useful bandwidth, including both impedance and pattern effects.
• Different feed and matching schemes.
• Multiband operation

Modeling Notes

Much of the work presented here was done using computer modeling. Because these antennas are close to ground (in terms of wavelengths) and different parts of the antenna are at different heights, *NEC2* rather than *MININEC* modeling programs were used.[7,8,9] To maximize the accuracy, I included the wire losses and all wires connected at a corner used segment tapering. I assumed real ground, us-

ing the high accuracy (Norton-Sommerfeld) ground model. I carefully observed the proscription against grounding wires directly to a real ground. The accuracy of the modeling should be very good.

The Half-Square Antenna

A simple modification to a dipole would be to add two λ/4 vertical wires, one at each end, as shown in **Fig 2**. This is a *half-square antenna*. The antenna can be fed at one corner (low impedance, current fed) or at the lower end of one of the vertical wires (high impedance, voltage fed). Other feed arrangements are also possible.

The "classical" dimensions for this antenna are λ/2 (131 feet at 3.75 MHz) for the top wire and λ/4 (65.5 feet) for the vertical wires. However, there is nothing sacred about these dimensions! You can vary them over a wide range and still obtain nearly the same performance.

This antenna is two λ/4 verticals, spaced λ/2, fed in-phase by the top wire. The current maximums are at the top corners. The theoretical gain over a single vertical, for two in-phase verticals, is 3.8 dBi.[10] An important advantage of this antenna is that it does not require the extensive ground system and feed arrangements that a conventional pair of phased λ/4 verticals would.

Comparison To A Dipole

In the past, one of the things that has turned off potential users of the half-square on 80 and 160 meters is the perceived need for λ/4 verticals. This forces the height to be > 65 feet on 80 meters and > 130 feet on 160 meters. That's not really a problem. If you don't have the height there are several things you can do. For example, just fold the ends in, as shown in **Fig 3**. This compromises the performance surprisingly little.

Let's look at the examples given in Figs 2 and 3, and compare them to dipoles at the same height. For this comparison I have selected two heights, 40 and 80 feet, and average, very good and sea-water grounds. I have also assumed that the lower end of the vertical wires had to be a minimum of 5 feet above ground.

At 40 feet the half-square is really mangled, with only 35 foot high (≈ λ/8) vertical sections. The comparison between this antenna and a dipole of the same height is shown in **Fig 4**. Over average ground the half-square is superior below 32° and at 15° is almost 5 dB better. That is a worthwhile improvement. If you have very good soil conductivity, like parts of the lower Midwest and South, then the half-square will be superior below 38° and at 15° will be nearly 8 dB better. For those fortunate few with saltwater-front property the advantage at 15° is 11 dB! Notice also that above 35°, the response drops off rapidly. This is great for

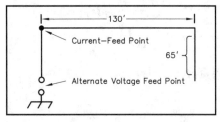

Fig 2—Typical 80-meter half-square, with λ/4-high vertical legs and a λ/2-long horizontal leg. The antenna may be fed at the bottom or at a corner. When fed at a corner, the feed point is a low-impedance, current-feed. When fed at the bottom of one of the wires against a small ground counterpoise, the feed point is a high-impedance, voltage-feed.

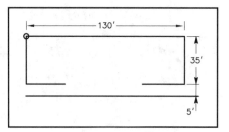

Fig 3—An 80-meter half-square configured for 40-foot high supports. The ends have been bent inward to reresonate the antenna. The performance is compromised surprisingly little.

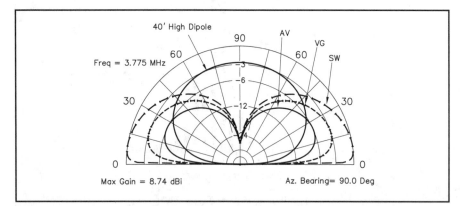

Fig 4—Comparison of 80-meter elevation response of 40-foot high, horizontally polarized dipole over average ground and a 40-foot high, vertically polarized half-square, over three types of ground: average (conductivity σ = 5 mS/m, dielectric constant ε = 13), good (σ = 30 mS/m, ε = 20) and saltwater (σ = 5000 mS/m, ε = 80). The quality of the ground clearly has a profound effect on the low-angle performance of the half-square. However, even over average ground, the half-square outperforms the low dipole below about 32°.

DX but is not good for local work.

If we push both antennas up to 80 feet (**Fig 5**) the differences become smaller and the advantage over average ground is 3 dB at 15°. The message here is that *the lower your dipole and the better your ground, the more you have to gain by switching from a dipole to a half-square*. The half-square antenna looks like a good bet for DXing. However, there are a few other things to consider before replacing your dipole.

Changing the Shape

Just how flexible is the shape? We'll look now at several distortions of practical importance. Some have very little effect but a few are fatal to the gain. Suppose you have either more height and less width than called for in the standard version or more width and less height, as shown in **Fig 6A**.

The effect on gain from this type of dimensional variation is given in **Table 1**. For a top length (L_T) varying between 110 and 150 feet, where the vertical wire lengths (L_v) readjusted to resonate the antenna, the gain changes only by 0.6 dB. For a 1 dB change the range of L_T is 100 to 155 feet, a pretty wide range.

Another variation results if we vary the length of the horizontal top wire and readjust the vertical wires for resonance, while

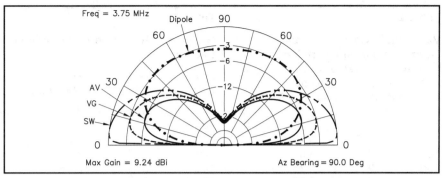

Fig 5—Comparison of 80-meter elevation response of 80-foot high, horizontally polarized dipole over average ground and an 80-foot high, vertically polarized half-square, over same three types of ground as in Fig 4: average, good and saltwater. The greater height of the dipole narrows the gap in performance at low elevation angles, but the half-square is still a superior DX antenna, especially when the ground nearby is saltwater! For local, high-angle contacts, the dipole is definitely the winner, by almost 20 dB when the angle is near 90°.

Table 1

Variation in Gain with Change in Horizontal Length, with Vertical Height Readjusted for Resonance. See Fig 6A.

L_T (feet)	L_V (feet)	Gain (dBi)
100	85.4	2.65
110	79.5	3.15
120	73.7	3.55
130	67.8	3.75
140	61.8	3.65
150	56	3.05
155	53	2.65

Fig 6—Varying the horizontal and vertical lengths of a half-square. At A, both the horizontal and vertical legs are varied, while keeping the antenna resonant. At B, the height of the horizontal wire is kept constant, while its length and that of the vertical legs is varied to keep the antenna resonant. At C, the length of the horizontal wire is varied and the legs are bent inwards in the shape of "vees." At D, the ends are sloped outwards and the length of the flattop portion is varied. All these symmetrical forms of distortion of the basic half-square shape result in small performance losses.

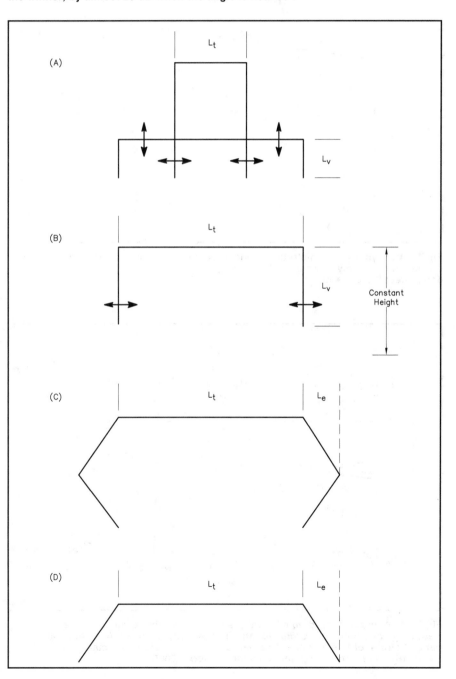

Table 2

Variation in Gain with Change in Horizontal Length, with Vertical Length Readjusted for Resonance, but Horizontal Wire Kept at Constant Height. See Fig 6B.

L_T (feet)	L_V (feet)	Gain (dBi)
110	78.7	3.15
120	73.9	3.55
130	68	3.75
140	63	3.35
145	60.7	3.05

Table 3

Gain for Half-Square Antenna, Where Ends Are Bent Into V-Shape. See Fig 6C.

Height ⇒	H=40'	H=40'	H=60'	H=60'
L_T (feet)	L_e (feet)	Gain (dBi)	L_e (feet)	Gain (dBi)
40	57.6	3.25	52.0	2.75
60	51.4	3.75	45.4	3.35
80	45.2	3.95	76.4	3.65
100	38.6	3.75	61.4	3.85
120	31.7	3.05	44.4	3.65
140	-	-	23	3.05

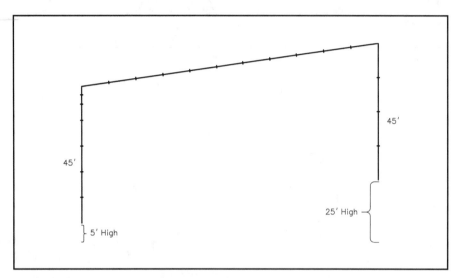

Fig 7—An asymmetrical distortion of the half-square antenna, where the bottom of one leg is purposely made 20 feet higher than the other. This type of distortion does affect the pattern!

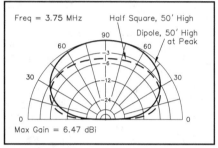

Fig 8—Elevation pattern for the asymmetrical half-square shown in Fig 7, compared with pattern for a 50-foot high dipole. This is over average ground, with a conductivity of 5 mS/m and a dielectric constant of 13. Note that the zenith-angle null has filled in and the peak gain is lower compared to conventional half-square shown in Fig 5 over the same kind of ground.

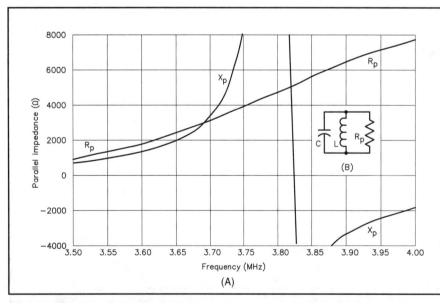

Fig 9—At A, graph of feed point shunt resistance and shunt reactance versus frequency for a half-square with voltage-feed at bottom corner. At B, equivalent parallel circuit of this antenna. This particular half-square is resonant at about 3.820 MHz, where its feed point resistance is about 5000 Ω.

keeping the top at a constant height. See Fig 6B. **Table 2** shows the effect of this variation on the peak gain. For a range of L_T = 110 to 145 feet, the gain changes only 0.65 dB.

The effect of bending the ends into a V shape, as shown in Fig 6C, is given in **Table 3**. The bottom of the antenna is kept at a height of 5 feet and the top height (H) is either 40 or 60 feet. Even this gross deformation has only a relatively small effect on the gain! Sloping the ends outward as shown in Fig 6D and varying the top length also has only a small effect on the gain. While this is good news because it allows you to dimension the antenna to fit different QTHs, not all distortions are so benign.

Suppose the two ends are not of the same height, as illustrated in **Fig 7**, where one end of the half-square is 20 feet higher than the other. The radiation pattern for this antenna is shown in **Fig 8** compared to a dipole at 50 feet. This type of distortion does affect the pattern. The gain drops somewhat and the zenith null goes away. The nulls off the end of the antenna also go away, so that there is some end-fire radiation. In this example the difference in height is fairly extreme at 20

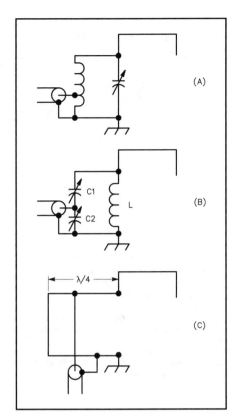

Fig 10—Typical matching networks used for voltage-feeding a half-square antenna.

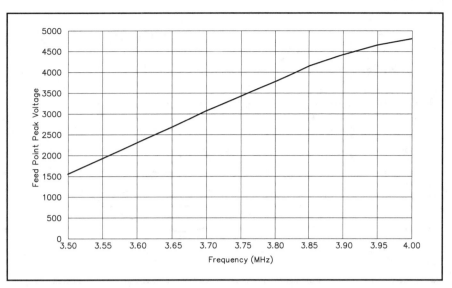

Fig 11—Graph of peak RF voltage at feed point of voltage-fed half-square antenna with 1500 W power.

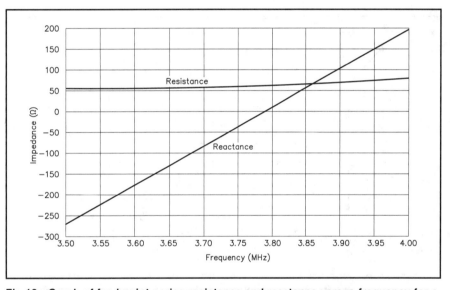

Fig 12—Graph of feed point series resistance and reactance versus frequency for a half-square with current-feed at one corner. Note that the resistive component changes slowly with frequency. This particular antenna is resonant at just under 3.8 MHz.

feet. Small differences of 1 to 5 feet do not affect the pattern seriously.

If the top height is the same at both ends but the length of the vertical wires is not the same, then a similar pattern distortion can occur. The antenna is very tolerant of *symmetrical* distortions but it is much less accepting of *asymmetrical* distortion.

What if the length of the wires is such that the antenna is not resonant? Depending on the feed arrangement that may or may not matter. We will look at that issue later on, in the section on patterns versus frequency. The half-square antenna, like the dipole, is very flexible in its proportions.

Feed-Point Impedance

There are many different ways to feed the half-square. Traditionally the antenna has been fed either at the end of one of the vertical sections, against ground, or at one of the upper corners as shown in Fig 2.

A typical example of the impedance variation for voltage feed is shown in **Fig 9A**. The impedance generated from the modeling program represents the parallel-equivalent impedance (Fig 9B) when driven at one end. This form is most informative when using a parallel L-C matching network, such as the one shown in **Fig 10**.

In addition to the variation in reactance (X_p), the resistance (R_p) varies from 1200 to 5700 Ω. This very high impedance means

that the voltage at the feed point will be quite high. A graph of peak voltage for 1.5 kW drive power is given in **Fig 11**. The feed point voltage will be over 4 kV! This must be kept in mind when designing matching networks. Because of the large range of impedances, simple matching schemes yield relatively narrow SWR bandwidths.

For current feed, the impedance is much lower, as shown in **Fig 12**. The resistive component doesn't change very much but the reactive component does. This is a relatively high-Q antenna (Q ≈ 17). **Fig 13** shows the SWR variation with frequency for this feed arrangement. Again, the bandwidth is quite narrow. An 80-meter dipole is not par-

ticularly wideband either, typically exhibiting an SWR range of about 6:1 over the whole band. A dipole will have less extreme variation in SWR than the half-square.

Patterns Versus Frequency

Impedance is not the only issue when defining the bandwidth of an antenna. The effect on the radiation pattern of changing frequency is also a concern. For an end-fed half-square, the current distribution changes with frequency. For an antenna resonant near 3.75 MHz, the current distribution is nearly symmetrical. However, above and below resonance the current distribution increasingly becomes asymmetrical. In effect,

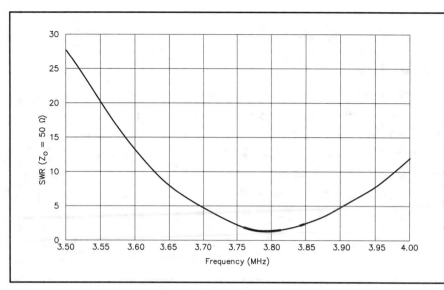

Fig 13—Variation of SWR with frequency for current-fed half-square antenna. The SWR bandwidth is quite narrow.

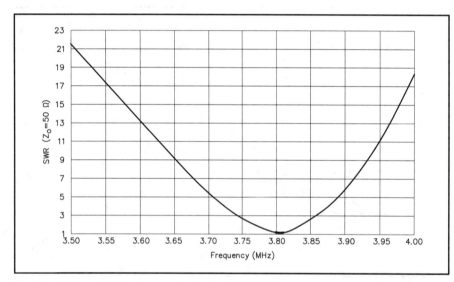

Fig 15—SWR versus frequency for voltage-fed half-square antenna, using matching network shown in Fig 10B, with L = 15 µH, C_1 = 125 pF and C_2 = 855 pF. The SWR bandwidth is less than 100 kHz at the 2:1 SWR points.

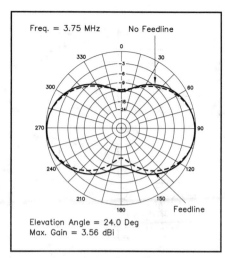

Freq. = 3.75 MHz   No Feedline

Elevation Angle = 24.0 Deg
Max. Gain = 3.56 dBi

Fig 14—Effect of feed line on azimuth radiation pattern for current-fed half-square antenna. The feed line introduces only small distortions in symmetrical radiation pattern. The coaxial feed line was modeled as being brought out straight for 30 feet from the corner, then brought down close to ground level and led away for 50 feet more, where it was grounded.

the open end of the antenna is constrained to be a voltage maximum but the feed point can behave less as a voltage point and more like a current maxima. This allows the current distribution to become asymmetrical.

The effect is to reduce the gain by −0.4 dB at 3.5 MHz and by −0.6 dB at 4 MHz. The depth of the zenith null is reduced from −20 dB to −10 dB. The side nulls are also reduced. Note that this is exactly what happened when the antenna was made physically asymmetrical. Whether the asymmetry is due to current distribution or mechanical arrangements, the antenna pattern will suffer. In my model, I used four ground wires, 10 feet long. These represent an adequate ground for the antenna when

operated not too far from resonance. Even shorter wires could be used.

When corner-feed is used, the asymmetry introduced by off-resonance operation is much less, since both ends of the antenna are open circuits and constrained to be voltage maximums. The resulting gain reduction is only −0.1 dB. It is interesting that the sensitivity of the pattern to changing frequency depends on the feed scheme used!

Of more concern for corner feed is the effect of the transmission line. The usual instruction is to simply feed the antenna using coax, with the shield connected to vertical wire and the center conductor to the top wire. Since the shield of the coax is a conductor, more or less parallel with the

radiator, and is in the immediate field of the antenna, you might expect the pattern to be seriously distorted by this practice. This arrangement seems to have very little effect on the pattern!

A number of different feed-line arrangements were modeled. An example of the patterns for one of them is shown in **Fig 14**. The wire, representing the outside of the coax feeding the antenna at the corner, was brought out straight for 30 feet, then brought down close to ground and led away for 50 feet more and grounded. The effect at resonance was barely detectable, as shown in Fig 14. At 3.5 MHz the gain was down by −0.5 dB and at 4 MHz was actually up by +0.1 dB. Other lengths and feed-line arrangements were tried with similar lack of effect. The greatest effect came when the feed-line length was near λ/2. Such lengths should be avoided.

Frankly, this result came as a considerable surprise. There are at least two possible explanations. First, the feed line is connected to a low-voltage point. Second, the feed line is located off the end of the antenna, where the field is canceled to some extent by the phasing of the radiators. Whatever the reason, this is very good news. It means that the antenna can be kept just as simple as a dipole.

Of course, you may use a balun at the feed point if you desire. This might reduce the coupling to the feed line even further but it doesn't appear to be worth the trouble. In fact, if you use an antenna tuner in the shack to operate away from resonance with a very

40

high SWR on the transmission line, a balun at the feed point would take a beating.

Voltage-Feed at One End of Antenna: Matching Schemes

Several straightforward means are available for narrow-band matching. However, broadband matching over the full 80-meter band is much more challenging. Voltage feed with a parallel-resonant circuit and a modest local ground, as shown in Fig 10, is the traditional matching scheme for this antenna. Matching is achieved by resonating the circuit at the desired frequency and tapping down on the inductor in Fig 10A or using a capacitive divider (Fig 10B). It is also possible to use a 1/4λ transmission-line matching scheme, as shown in Fig 10C.

If the matching network shown in Fig 10B is used with L = 15 μH, C_1 = 125 pF and C_2 = 855 pF, you will obtain the SWR characteristic shown in **Fig 15**. At any single point the SWR can be made very close to 1:1 but the bandwidth for SWR < 2:1 will be very narrow at <100 kHz. Altering the L-C ratio doesn't make very much difference. This antenna has a well-earned reputation for being narrowband. If you only want to DX on phone or CW then that may be acceptable, but most users want to do both.

It is possible to change the capacitors or tune the inductor, either with switches, manual adjustment or a motor drive. However, that level of complexity is unacceptable, especially since we are trying to replace a dipole with something equally simple. It is also possible to design wideband matching networks with multiple elements, but again that approach is relatively complex.

Current-Feed: Matching Schemes

The antenna can be current-fed at points other than the upper corners. Some possibilities are shown in **Fig 16**. As the feed point is moved away from the current maxima, the voltage increases and it becomes necessary to use a balun to decouple the transmission line. For narrowband use or if there is a matching network at the feed point this may be acceptable and may result in a more convenient feed point. As shown in Fig 16A, the feed point can be moved down the vertical wire to a higher impedance point and a 4:1 or 9:1 balun used. If the ends of the antenna are bent back toward the center, then a convenient feed point would be the lower corner, as shown in Fig 16B. By making the ends symmetrical as shown in Fig 16C even better decoupling could be obtained and the symmetry of the antenna is maintained.

Another possibility that has been used in the past is to invert the antenna, as shown in **Fig 17** and feed it at a lower corner. The problem with this approach is that the losses are higher because the current maxima are

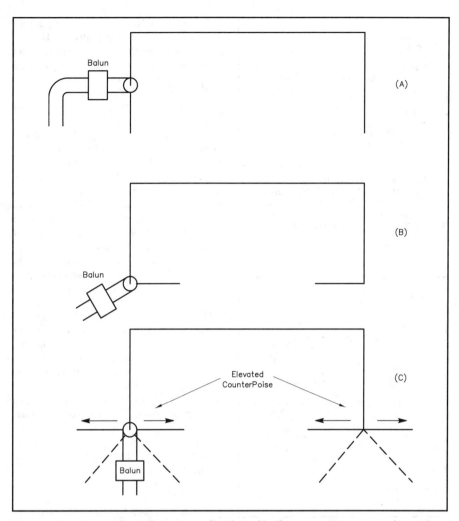

Fig 16—Possible methods for current-feeding of half-square antenna at points other than the upper corners. At A, a balun is used to decouple the feed line from the feed point at the center of one of the vertical legs of the antenna. At B, the ends of the vertical legs are both bent back horizontally to provide a feed point. At C, an elevated counterpoise is used to provide a feed point at the bottom of a vertical leg.

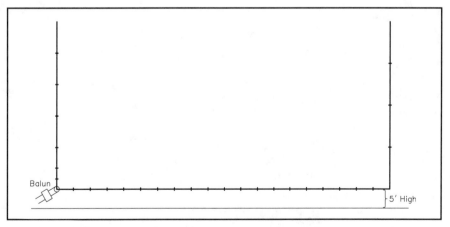

Fig 17—An "inverted half-square" antenna, current-fed at a lower corner. The losses in this configuration are excessive unless the ground under the antenna is exceptionally good, RF-wise.

close to ground. A comparison between a normal half-square and an inverted one, 5 feet over average ground, is made in **Fig 18**.

The difference is over 2 dB. For greater height or better ground, the loss would be lower. The killer antenna built by Tom

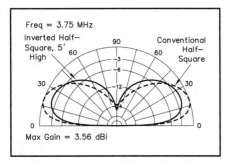

Fig 18—Elevation pattern for a conventional half-square, compared with an "inverted half-square" whose horizontal wire is located 5 feet over average ground. The difference is more than 2 dB.

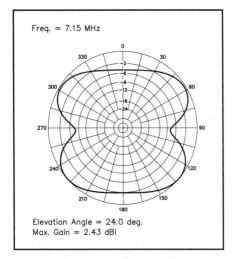

Fig 19—An attempt to load an 80-meter half-square antenna on 7 MHz. The pattern is badly distorted. The half-square is a monoband antenna!

Erdmann, W7DND, used this configuration but it was installed over a saltwater beach.[11] As a consequence the losses were very low and the feed point very conveniently located.

Multiband Operation

An 80-meter half-square can be used on other bands but the pattern and the drive-point impedance will change. A current-fed, 80-meter half-square will have a radiation pattern like that shown in **Fig 19** when driven at 7.15 MHz. On 40 meters the pattern has four lobes and the feed-point impedance is approximately $3300 + j 1500 \ \Omega$. If end-feed is used, the impedance will be in the region of $450 + j 110 \ \Omega$. With end-feed, the pattern will be somewhat asymmetric.

If the antenna is used on 20 meters the pattern will have eight lobes and the impedance at 14.2 MHz will $\approx 1100 + j 900 \ \Omega$. If a tuner is available this antenna can be used at higher frequencies but it will have a multi-lobed pattern typical of a harmonic antenna.

On the higher bands (40 meters and up), the height in wavelengths is greater for a given physical antenna height. Over average ground, the advantage of the half-square over a typical dipole thus becomes smaller and the half-square may even become inferior to the dipole. When the antenna is installed over very good ground or seawater, then the half-square may still be a contender on the higher bands.

Conclusion

The half-square antenna has some definite advantages. It is a simple and effective alternative to a typical dipole on the 80 and 160-meter bands, where the half-square radiates a stronger signal at the low angles

most appropriate for DX work. The height and shape of the antenna are quite flexible and can be tailored to fit the needs of a given QTH. As a DX receiving antenna, it has the advantage of discriminating against strong high-angle signals arriving from stations within 1500 miles.

One disadvantage of the half-square is that it is more narrowband than a dipole—for DX work this may not be a serious disadvantage, since the ranges of frequencies for the DX "windows" are quite small. The antenna is also vertically polarized, which means more noise pickup when receiving.

Notes and References

[1]*The ARRL Antenna Book*, 17th edition (ARRL, Newington: 1994), Chapter 23.
[2]Dean Straw, N6BV, *All The Right Angles*,1993, Published by LTA, PO Box 77, New Bedford, PA.
[3]Woodrow Smith, W6BCX, "Bet My Money On A Bobtail Beam," *CQ*, Mar 1948, pp 21-23, 92.
[4]Ben Vester, K3BC, "The Half-Square Antenna," *QST*, Mar 1974, pp 11-14.
[5]Paul Carr, N4PC, "A Two Band Half-Square Antenna With Coaxial Feed," *CQ*, Sep 1992, pp 40-45.
[6]Paul Carr, N4PC, "A DX Antenna For 40 Meters," *CQ*, Sep 1994, pp 40-43.
[7]*EZNEC* is available from Roy Lewallen, W7EL, PO Box 6658, Beaverton, OR 97007.
[8]*NEC-Wires* is available from Brian Beezley, K6STI, 3532 Linda Vista Drive, San Marcos, CA 92069, tel 619-599-4962.
[9]*NEC-WIN Basic* is available from Paragon Technology, 200 Innovation Blvd, Suite 240, State College, PA 16803, tel 814-234-3335.
[10]*The ARRL Antenna Book*, 17th edition (ARRL: Newington), Chapter 8, p 8-6.
[11]Jerrold Swank, W8HXR, "The S-Meter Bender," *73*, Jun 1978, pp 170-173.

Broadbanding the Half-Square Antenna for 80-Meter DXing

By Rudy Severns, N6LF
PO Box 589
Cottage Grove, OR 97424

T he half-square antenna is by nature relatively narrow band.[1] On 80 meters, for example, an SWR below 2:1 can be achieved anywhere in the band, but only over a relatively small range (60 to 100 kHz). The primary reason for using a half-square instead of a dipole is for improved performance on DX contacts.

There are two DX "windows" on 80 meters, 3.500-3.520 MHz and 3.750-3.800 MHz—most CW activity is close to 3.500 and SSB around 3.790 MHz. It is very easy to adjust a normal half-square antenna to have low SWR at either one of these frequencies, but not at both. Practically speaking, any serious DXer will want to be able to use both CW and SSB, so this is a real disadvantage.

It is possible of course to build a matching network of some kind or to use a tuner to load the antenna at both frequencies. However, that may not be as simple as it sounds, because if the SWR is low in one window, it will be very high at the other. It could be 20:1 or more!

The attraction of the half-square is its simplicity. It would be nice to allow operation in both windows while keeping the simplicity. This article shows a way to do that by adding two wires to the classical half-square.

Broadbanding the Half-Square

On 80 meters even a dipole is not a broad-

N6LF discusses a simple way to broadband the classic half-square antenna to operate in both the CW and SSB "DX windows" on 80/75 meters.

band antenna. One trick frequently used to broadband or multiband a dipole is to add additional wires to the dipole to form a fan, as shown in **Fig 1**. The two wires on each side of the feed point have different lengths and are adjusted to produce two resonance points. A variation of this idea works for the half-square. It can provide the desired double resonance and can also provide 3-4 dB of front-to-back ratio if that is desired.

The bi-directional (0 dB front-to-back) version of the half-square is shown in **Fig 2**. The single vertical wires at each end of the antenna have been replaced with two wires, of different lengths (L1 and L2), with the lower ends well separated. Note that the vertical wires are in the plane of the horizontal top wire (L_T). In a bit we will see what happens if the wires are not in this plane. The pattern from this antenna is shown in

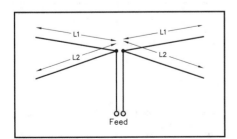

Fig 1—Broadbanding an 80-meter dipole using a fan-shaped pair of unequal-length radiators.

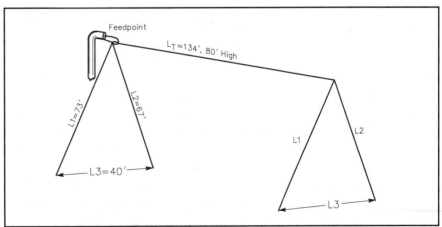

Fig 2—Typical N6LF broadband symmetrical half-square for 80 meters. All wires are in the plane of the horizontal top wire. The vertical wires are spread out 40 feet at the bottom in this case.

$L_T = 134'$, 80' High
$L1 = 73'$
$L2 = 67'$
$L3 = 40'$
Feedpoint

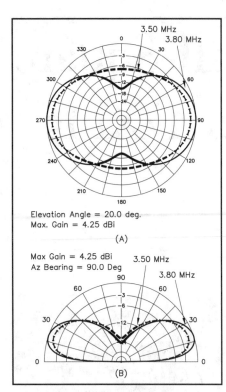

Elevation Angle = 20.0 deg.
Max. Gain = 4.25 dBi

(A)

Max Gain = 4.25 dBi
Az Bearing = 90.0 Deg

(B)

Fig 3—At A, azimuth response of symmetrical broadband 80-meter half-square at 3.8 and 3.5 MHz. At B, elevation response of symmetrical broadband 80-meter half-square at 3.8 and 3.5 MHz.

Fig 3. There is some sacrifice in gain at the lower resonance, but only about 1 dB.

If the vertical wires do not lie in the plane of the top wire, as shown in **Fig 4**, it will still be possible to obtain the double resonance, but the pattern will be affected. As shown in **Fig 5**, the pattern is no longer strictly bi-directional. There can be several dB of front-to-back ratio. The front-to-back ratio improves the gain in one direction; this may be helpful in some situations. More often, however, it is desirable to work long path as well as short path and the bi-directional pattern will be preferred.

Experimental Results

An antenna with the dimensions in Fig 2 was built and the measured SWR is shown in Fig **6**. As expected, there are two resonances, giving acceptable SWR in both the CW and SSB DX windows.

The exact lengths for each wire will depend on the particular installation—the width and height available. If an antenna modeling program such as *EZNEC*[2], *NEC/Wires*[3] or *NEC-WIN*[4] is available, then the antenna can be designed very closely for a particular site, including the ground effects. If the modeling is not available, then it will be necessary to adjust the wire lengths experimentally. Fortunately, all of the adjustments can be made at ground level.

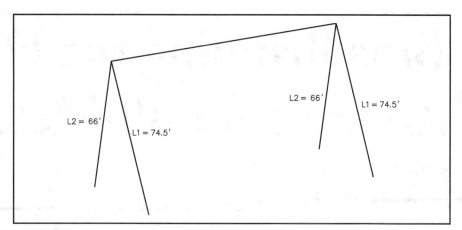

Fig 4—An asymmetrical variation of broadband half-square. Here, the equal-length vertical wires are placed on the same side of a vertical plane cutting through the length of the horizontal top wire.

The length of the top wire (L_T) is set during initial construction and can vary from 120 to 150 feet, depending on the space available. The longer lengths will mean that the vertical wires can be made shorter. This allows for lower heights. More detail of this trade-off can be found in Reference 1. There are three other variables: L_1, L_2 and L_3.

The adjustment begins by setting the spacing between the ends of the vertical wires (L_3), then L_3 is adjusted for resonance at 3.790 MHz. Finally, L_1 is adjusted to resonate at 3.510 MHz. L_2 and L_1 are then adjusted one more time. Usually this will be sufficient to place the resonances in the desired locations. If the SWR is not as low as desired, then L_3 can be changed and L_1 and L_2 readjusted. This process should converge rapidly.

Because L_1 and L_2 may need to be either shortened or lengthened, I usually start with extra wire and fold the excess length back on the wire, rather than cutting it off. That way, extra is available to lengthen the wire, if needed.

Conclusion

The narrow bandwidth of the classical half-square antenna can be overcome by adding another set of vertical wires. With a little adjustment, two resonances, with SWR < 2:1 can be achieved. This will allow operation in both the CW and SSB DX windows on 80 meters.

The principle shown here will, of course, also work on other bands. On 160 meters, for example, it would allow a substantial part of the band to be covered without retuning.

Notes and References

[1]Severns, Rudy, N6LF, "Using the Half-Square Antenna for Low-Band DXing," elsewhere in this book.
[2]*EZNEC* is available from Roy Lewallen, W7EL, PO Box 6658, Beaverton, OR, 97007.
[3]*NEC/Wires* is available from Brian Beezley,
K6STI, 3532 Linda Vista Drive, San Marcos, CA 92069, 619-599-4962.
[4]*NEC-WIN Basic* is available from Paragon Technology, 200 Innovation Blvd, Suite 240, State College, PA 16803, 814-234-3335.

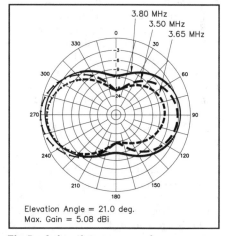

Elevation Angle = 21.0 deg.
Max. Gain = 5.08 dBi

Fig 5—Azimuth response of asymmetrical broadband 80-meter half-square at 3.8, 3.65 and 3.5 MHz, showing how front-to-back ratio changes with frequency.

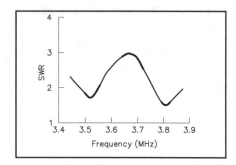

Fig 6—SWR curve versus frequency for symmetrical broadband 80-meter half-square showing characteristic double-resonance.

The N6BV 75/80-Meter Quad

By R. Dean Straw, N6BV
32 Beacon Hill Road
Windham, NH 03087

The low bands get hot when the sunspots are naught! Ever dream about having a big signal on 75/80 meters? N6BV describes his rather unusual quad.

I n previous articles I've made it known that I'm a dyed-in-the-wool contest operator. So, I write this article with a tiny bit of reluctance—after all, a contester always wants to have some sort of "secret weapon" in his arsenal. Hopefully, the benefits to the amateur fraternity at large will outweigh my, admittedly, selfish concerns.

Over more than 25 years of contesting, I've only felt *really strong* on 75/80 meters from two locations. In the mid 1980s, I operated from one of the premier contest stations on the West Coast, N6RO. In the last two years, I've felt loud from my own station in New Hampshire. At both QTHs I used two-element quads on 75/80 meters.

The N6RO 75/80-Meter Quads

I helped design and install the quads at Ken Keeler, N6RO's, QTH in Oakley, CA. Yes, I did say "quads"—Ken had two horizontally polarized quads for 75/80 meters. The loops were diamond-shaped, supported from rope catenaries strung between four widely separated 130-foot tall towers. Ropes pulled the corners of each loop out to make a "diamond." The bottom of each wire loop was 10 feet off the ground, supported by a wooden fence post cemented into the ground. A plastic refrigerator box, with relays and reflector coil, was placed at the top of each post. The operator could switch each quad instantly in either of two directions (Europe or South Pacific for one quad; Japan or South America for the other). Additional relays and loading coils tuned each loop for either CW or SSB. See **Fig 1**.

You might ask: "Why did you choose a quad instead of a wire Yagi?" After all, a Yagi is basically a two-dimensional antenna, while a quad is a much more complicated three-dimensional affair. The problem is that remotely tuning a Yagi for both 75 and 80 meters can be a real challenge. One solution would be to place separate relays and tuning coils at each end of a Yagi boom

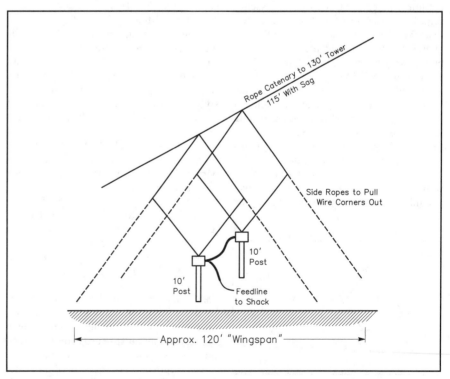

Fig 1—Layout of N6RO 75/80-meter quad, using rope catenaries strung between 130-foot tall towers. Ropes from the corners pull out the wires to form a diamond shape.

Labels in figure: Rope Catenary to 130' Tower 115' With Sag; Side Ropes to Pull Wire Corners Out; 10' Post; 10' Post; 10' Post; Feedline to Shack; Approx. 120' "Wingspan"

to switch between phone and CW operation. However, I considered this to be too complicated mechanically. It would also require running very long dc control cables along the catenaries.

Another possibility would be to drop each rope catenary so that I could reach the ends of the elements to lengthen or shorten them, depending on whether I wanted to operate CW or SSB. I wanted to be able to switch modes and directions from the comfort of the shack! I went with a quad design.

On the night after the installation was completed, I remember tuning around the CW band, using the quad pointed at Europe. Only a few signals were coming through—not surprising, considering that it was July. I had a tough time pulling out the call sign of one weak signal. When I switched to the South-Pacific direction, I was astounded to hear the weak station come up to a solid S7—it was a ZL calling CQ. Trust me, you haven't really lived until you've experienced an instant 20 dB front-to-back ratio on 80 meters!

Evolution of the N6BV 75/80-Meter Quad

I moved back to the East Coast in the late 1980s, forgoing the pleasures of the ARRL Sweepstakes Contest from California for the wonders of DX contests from Cape Cod. I put up a number of different wire antennas there, trying to achieve a strong 75/80-meter signal using a single 70-foot high tower. Although working Europe on 80 meters was definitely much more fun from Massachusetts than from California, I never really felt *loud*. The best antenna was a simple horizontal dipole strung at the 70-foot level between my tower and that of my neighbor Jan Carman, K5MA.

In 1990 I moved farther north, to what must surely be Mecca for contesters. Almost every hilltop in southern New Hampshire sports a set of amateur antennas, some of them very large. One reason is the amateur-friendly zoning regulations in many New Hampshire towns. In my town of Windham, the major restriction is simple: if your tower should fall down, it must fall entirely on

your own property. I received a permit to put up a 120-foot tower, and festooned it with four stacked triband antennas, covering 40 through 10 meters. I described this system in an article in February 1994 *QST*, co-authored with Fred Hopengarten, K1VR.[1]

I am reasonably competitive on the higher bands, especially into Europe. However, on 75/80 meters, the 115-foot high inverted-V dipole I used for two years never really felt "first-tier." Don't get me wrong: I did reasonably well on 75/80 in contests. (My contesting colleagues to the South and West are quick to remind me that this is due, in no small part, to my proximity to Europe!) Still, I was rarely first or second in really big pileups. When I made comparisons with some of the outstanding local stations, such as KC1XX with his 180-foot high two-element 75/80-meter delta quad, I was regularly down 10 to 15 dB. (Modeling by computer showed that this much difference was unreasonable, but time and again actual comparison tests said otherwise. More on this later.)

One option, using vertically polarized antennas, I quickly ruled out. Verticals are far from optimal when surrounded by the rocky soil of New England—and New Hampshire isn't known as the "Granite State" for nothing! While verticals do work very well near the seashore, the closest salt-water is about 25 miles to the East.

After being thoroughly trounced in many such tests using my inverted V, I would fondly reminisce about how strong I used to be using the N6RO quads. But I kept telling myself my situation was different. I had only one tower, and this precluded installing horizontal rope catenaries to hold up wire quads. However, I carefully read and reread the *Ham Radio* article by Bill Myers, K1GQ, describing details for the W2PV-style 75/80-meter quad he had installed at his New Hampshire QTH.[2] Bill's quad used two equilateral "delta" loops, suspended from a 115-foot high, 40-foot long boom. Two ropes out to the side pulled the loop wires away from the boom, just as I had done at N6RO. See **Fig 2**.

Computer modeling using *NEC/Wires* by Brian Beezley, K6STI,[3] showed that the K1GQ quad had appreciable gain over an inverted-V dipole, but that the rearward pattern was not really what I wanted. No matter what loop configuration I modeled, diamond, square or delta, I concluded that a 75/80-meter quad must be really high in the air to be effective at the elevation angles needed for reliable signals into Europe. From New England to Europe on 80 meters, the "average" elevation angle is about 20°.[4]

The original W2PV quad used a 40-foot boom that was 156 feet high. The 115-foot high K1GQ version had the horizontal bottom wire electrically closer to ground, and

this detuned the quad. In fact, no amount of retuning the delta-type reflector loop could achieve my goal of at least 20 dB front-to-back ratio over a selected portion of either the 80-meter CW band or the DX-window portion of the 75-meter SSB band.

Like K1GQ's, my tower is 120 feet high—high, but not really very high on 80 meters. I had enough land to give a "wingspread" of about 120 feet each side of the tower for the ropes used to pull out the corners of the quad to form a diamond shape. I also had access to a heavy-duty 35-foot long boom. So I narrowed the design down to one using a two-element wire quad on a 35-foot boom, 115 feet high. If I aimed the forward lobe toward Europe (azimuth of 45°), I wanted to be able to reverse it (to 225°) for a good signal into the US. The rearward direction would also give an adequate (although not optimal) signal towards South America and the Caribbean.

With a single two-element quad fixed at 45° or 225°, working Japan on 80 meters would be a problem, since the antenna has a null in that direction. Then again, we East Coasters don't work many JAs on the lower bands! So JA and long-path VK (over South America in the early evening) would be sacrificed in favor of a strong signal into either Europe or Texas.

I also needed to keep the wires well away from the many trees on my property. Like Charlie Brown and his kite-eating tree, I knew from past experience that trees have a natural appetite for snagging and eating even simple wire antennas. Imagine how they'd tangle up a quad! Then there was an additional challenge in my single-tower installation—getting everything around not only the three TH7DX triband Yagis at 90, 60 and 30 feet, but also past the guy wires.

Loop Configurations

With these requirements in mind, I now searched for a loop configuration to improve on the delta-quad pattern. An article by Floyd Koontz, WA2WVL, in *The ARRL Antenna Compendium, Vol 4* caught my attention.[5] Floyd's monster 75-meter Yagi used seven Yagi-type directors along a 328-foot long rope catenary that acted as a boom. Floyd employed an interesting looking, quad-like loop for his driven element and reflector. See **Fig 3A**. (Did I mention that Floyd's station was one of those that regularly stomped me in pileups?)

One big advantage of Floyd's quasi diamond-loop configuration is that it keeps the bottom of each loop physically higher in the air than would a conventional diamond or delta shape. This keeps each loop as far as possible from ground-detuning and loss effects. Unlike the equilateral delta loop, which has one leg

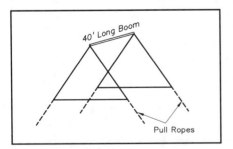

Fig 2—Layout of K1GQ 75/80-meter delta quad, using 40-foot long boom at 115 feet height.

parallel to lossy ground, the WA2WVL loop is more like a diamond pointed down at the ground. This has less mutual coupling between the wires and lossy ground. Further, like the N6RO quads, tuning and direction switching can be done near the ground with Floyd's design, rather than out at the ends of the boom.

I decided to use the WA2WVL quad-like loop design, but with a modification. I didn't want to place tuning boxes at the bottom of both loops for direction reversing and CW/SSB tuning. Instead, I decided to bring the wires from both loops back to the tower at the 50-foot level. See Fig 3B for the layout of the N6BV 75/80-meter quad. This creates what looks like a "tapered transmission line" from the bottom of each loop back to the tower. The loop wires would thus clear the 30-foot high TH7DX, which I had fixed on Europe. Now, the lowest point of each loop would be about 35 feet off the ground.

Computer modeling revealed an additional, serendipitous, benefit to this unusual looking loop design—the feedpoint impedance at resonance is close to 50 Ω, instead of the 100-Ω typical of a conventional two-element quad. This meant that I would not have to design and build matching systems for both phone and CW. I computed the 2:1 SWR bandwidth at about 80 kHz for either phone or CW settings. **Fig 4** shows computed elevation patterns at 3.8 MHz and 3.510 MHz, compared to an inverted-V dipole, 115 feet high with a 120° included angle between the two wires.

I had some concerns about wire stretching, since the side ropes would pull the wire quite taut. The first time around, I built the quad using "silky" multi-strand woven copperweld wire. A word of advice: stick with hard-drawn #14 copper wire—it doesn't rust. I had to replace all the original wire within one year.

Electrical Design Details

The two quad loops are identical. For SSB operation, relays place a 4.5 µH coil in series with one of the loops, making it a reflector. The residual capacitive reactance at the driven element is canceled with 0.4 µH of series inductance. The feed-line coax incorporates a choke balun, using six large Chromerics CNO-SORB 9754 beads placed over the RG-213 feedline. To reverse direction, I switch the reflector coil and the feedline between the loops.

For CW operation, relays place an additional inductance of 11.6 µH in series with the driven element and an additional 10.5 µH in series with the reflector loop. See the schematic of the tuning/switching box in **Fig 5**. I use DPDT open-frame relays, with 15-A contacts rated at 120 V ac.

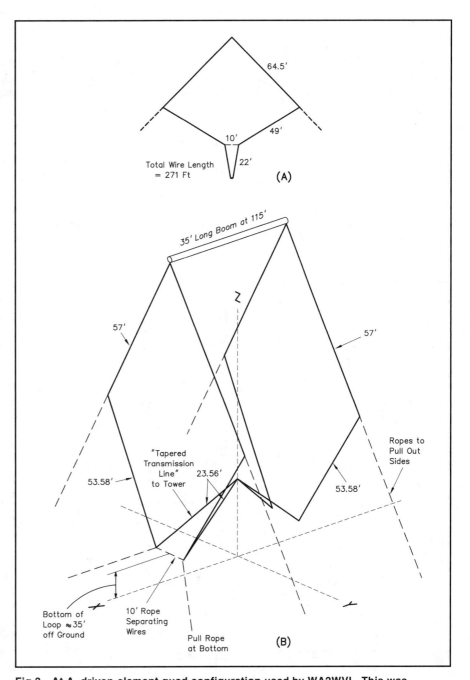

Fig 3—At A, driven-element quad configuration used by WA2WVL. This was adopted to keep the loop as high as possible above lossy ground and to keep the wires more taut. At B, modified-WA2WVL quad-type design by N6BV. The two wires at the bottom of each loop are brought back to the tower at the 50-foot level to a tuning/switching box, forming a "tapered transmission line." The quad can be switched between CW or SSB operation and the main-lobe direction can be reversed, using relays inside the box.

The coils are 24 V dc, and have 0.01-µF bypass capacitors and quenching diodes to help control surge voltages whenever a relay is de-energized. I mounted all components in a plastic refrigerator box for weather protection.

After constructing my tuning box, I used my Autek RF-1 analyzer[6] connected to the loop output terminals to set the inductor taps to their design values. I shorted the input

coaxial connector to make these adjustments, using the dc control box in the shack temporarily wired to the tuning box to control the various relays.

Up in the air, each loop should be self-resonant at 3920 kHz. This is with the other loop disconnected from the tuning box. The strong mutual coupling between the two loops resonates the total system at 3800 kHz on phone or 3520 kHz on CW.

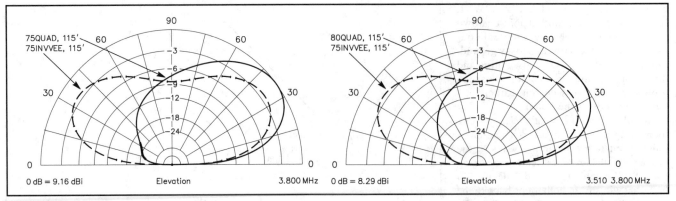

Fig 4—Computed elevation patterns for N6BV quad, compared to response for an inverted-V dipole at the same boom height of 115 feet. At A, response at 3800 kHz is shown. At B, response at 3510 kHz is shown.

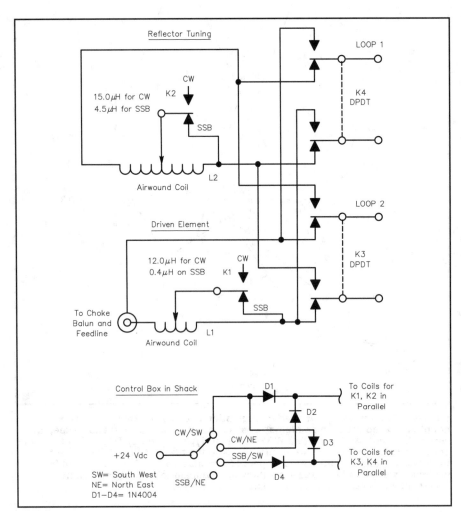

Fig 5—Schematic diagram for tuning/switching box. Relays are surplus open-frame types, with 15-A, 120 V ac contacts and 24 V dc coils. The inductors are air-wound "Air-Dux" coils with 3-inch OD.

Worry, Worry, Worry

During the design phase I was worried that I might have to employ "cut-and-try" techniques to fine-tune the array after installation. Using a remote field-strength meter to tweak a monster wire quad for maximum gain (or F/B), while dangling from the tower at 50 feet (or worse yet, at 115 feet), is an awkward, frustrating and time-consuming task.

Nearby hams are never located where you need them: exactly at the front or back of an antenna. Relying on the observations of nearby hams can also strain neighborly relations. Their enthusiasm for making pattern measurements seems to wane materially somewhere between the third and tenth try! So, if tuning after installation really did prove necessary, putting the tuning/switching components on the tower at the 50-foot point would make life a *lot* easier than if they had been out on the ends of the boom, 17½ feet away!

Deploying the Beast

The following describes how I deployed my quad. You would probably do things differently, depending on your own exact situation. The boom I used for my antenna came from a Hy-Gain 205BA monoband Yagi. The boom is two-inches in diameter, and uses heavy-wall tubing, reinforced with sleeve inserts. It uses a truss from the top to prevent bending when the quad wires are pulled tight. I fitted non-rusting marine pulleys to the ends of the boom. Each pulley held a loop of polyethylene rope taped temporarily at the boom's mounting point.

Down on the ground I had carefully measured and cut each loop in half (making four halves). I measured, tied and securely taped all the ropes associated with each wire. Then I carefully laid all the wires and ropes out on the ground to prevent tangling.

I climbed up and strapped myself in at the 115-foot level of the tower. With help from the ground crew, I pulled the boom up to the 115-foot level on the tower and mounted it to one tower leg using the standard Hy-Gain boom-to-mast clamp. This just fits a Rohn 45 leg. I also secured the boom to a second tower leg, using an angle iron and two U-bolts, one around the boom and one around the leg. The quad boom of course was pointing toward Europe.

Each pulley rope-loop was then carefully untied (Don't let go!) and then retied to make a loop surrounding a tower leg. These loops shuttle the quad-loop wires to each end of the boom. This rather elaborate setup was necessary because the quad wires had to clear everything else already mounted lower on the tower—three TH7DX Yagis and all the guy wires. In my installation, the 90 and 30-foot TH7DXs are fixed to the tower, pointing into Europe. The 60-foot TH7DX is mounted on a sidemount with a rotator. I had pointed the 60-foot antenna at Europe so that all 21 TH7DX elements were in line.

Now I was ready to install each loop. I carefully pulled a half-loop straight up on one side, in-between the driven elements of the three TH7DXs. Then I pulled its mate up on the other side and soldered the two halves of the loop together, using a propane torch with soldering tip attachment. Then I tied the wire to the pulley rope going to the end of the boom. As helpers on the ground carefully pulled each side rope out to move the wires away from the TH7DX booms, elements and tower guy wires, I gently pulled the pulley rope, gradually moving the wire loop out to the end of the boom.

We repeated the procedure for the wire loop on the other side of the boom. The ground crew tied off the two side ropes for each wire loop and I came down the tower to the 50-foot level and belted in. I tossed a weighted piece of string over the reflector element of the 30-foot high TH7DX down to my helpers below. They tied the two wires making up the "tapered transmission line" to the string and tied the 10-foot separator rope to the bottom corners of the wire loop. I fished the two feed wires over the reflector and connected them to the tuning box terminals.

The same procedure was followed for the director end of the TH7DX. Don't forget to tie nylon strings to each of the four wires back to the tower. These "security strings" allow you to disconnect the loop wires and not lose them!

The whole procedure sounds complicated, but it proceeded quite smoothly up in the air. It had helped, of course, that I had cut down virtually every tree, bush and sapling within several hundred feet of the tower using a chainsaw. Not only did this keep trees and branches away from the quad wires, it also helped keep my helpers from tripping while walking with their eyes skyward, watching and listening to me!

Results and Further Thoughts

I had carefully cut my quad wires to the computed dimensions. How did reality compare to the computer simulation? When I tested the resonant frequency of each loop using my Autek RF-1 (with the other loop temporarily disconnected and thus out of the circuit), the resonant frequency was very close to 3920 kHz, as predicted.

Once I reconnected both wire loops, I found to my joy that I needn't have worried about tweaking the quad after installation. It performed just like *NEC/Wires* said it should! I didn't have to touch anything in the tuning box. Isn't science wonderful?!

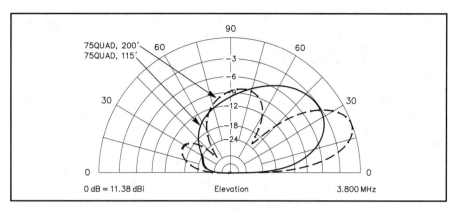

Fig 6—Computed elevation response for N6BV quad at 3800 kHz, with boom mounted at 200 feet, compared with same quad with boom mounted at 115 feet. Height is crucial for excellent performance at low elevation angles! No attempt was made to retune the higher quad for better F/B.

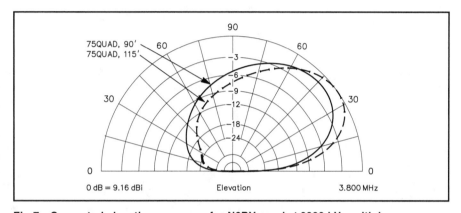

Fig 7—Computed elevation response for N6BV quad at 3800 kHz, with boom mounted at 90 feet, compared with response with boom at 115 feet. The bottom wires are only 9.4 feet off the ground for the lower quad, causing significant detuning. No amount of tuning (by computer) of the reflector tuning coil would improve the pattern significantly, although the SWR didn't change dramatically from that shown by the higher antenna.

In a number of head-to-head tests with reference stations such as KC1XX, I found that I was now within a few dB of the big boys on transmit. Further, the quad receives almost as effectively as my 600-foot long Beverage. This was great news, justifying all the effort needed to put up this antenna.

But these results also implied that I had somehow gained 8 to 13 dB of "gain" over my old inverted-V dipole. That isn't possible, at least not according to the computer model. To prevent interaction, I had removed my old inverted-V dipole before installing the quad, so I can't compare the two antennas on a direct A/B basis.

The computer modeling did predict quite accurately the F/B I experience with this antenna, and the feedpoint SWR is just what the model says it should be. It's just that the performance is better than I expected. My suspicion is that the inverted-V dipole may have been interacting with something else on the tower. Placing the quad loops at the ends of the 35-foot boom apparently moved them away from whatever had been inter-

acting with the inverted-V dipole. Speaking of interaction, I haven't been able to detect any between the quad and the TH7DXs.

I now definitely feel much louder on 75 and 80 meters. My contest results are better than ever before and there are few US pile-ups into Europe and the Middle East that I can't crack quickly. In the 1994 ARRL SS phone contest, I made the Soapbox comment that I had to keep checking the bandswitch, just to make sure I wasn't actually on 40 instead of 75 meters.

What would I do to improve this antenna? The obvious answer, as K1GQ pointed out in his article, is to put it up higher, perhaps at 200 feet. **Fig 6** shows the computed elevation response for the quad at 200 feet, compared with my quad at 115 feet. (In this computation I didn't attempt to retune the higher quad to optimize its pattern.) It has significantly more gain at low elevation angles compared to my quad at its 115 feet boom height. [A set of *.ANT files for *NEC/Wires* are included on the diskette bundled with this book.]

Fig 7 shows what happens to the gain and

pattern when the boom is lowered to 90 feet, where the bottom wires for each loop are only 9 feet off the ground. The quad was badly detuned by ground proximity, which could not be compensated by changing the reflector coil inductance. A 100-foot boom height is about the minimum necessary for good performance with this type of antenna.

The astute reader will observe that this antenna could very easily be scaled to another band, say, 40 meters. The 40QUAD.ANT file on the diskette is a design with a boom at 100 feet. I haven't actually built such an antenna, but I have confidence that this *NEC/Wires* model should work well.

Notes and References

[1] R. D. Straw, N6BV/1, and F. Hopengarten, K1VR, "Stacked Tribanders: A Super Station—Sorta," *QST*, Feb 1994, pp 38-44. See also 17th Edition, *The ARRL Antenna Book* (Newington: 1994), pp 11-24 to 11-30.

[2] B. Myers, K1GQ, "Analyzing 80-Meter Delta Loop Arrays," *Ham Radio*, Sep 1986, pp 10-28. See also article by B. Myers, K1GQ, "The W2PV 80-Meter Quad," *Ham Radio*, May 1986, pp 56-58.

[3] *NEC/Wires*, by Brian Beezley, 3532 Linda Vista Dr, San Marcos, CA 92069. Tele: 619-599-4962, e-mail: k6sti@n2.net.

[4] *The ARRL Antenna Book*, 17th Edition (Newington: 1994), Table 4, p 23-23.

[5] F. Koontz, WA2WVL, "The BIG Wire Beam for 75 Meters," *The ARRL Antenna Compendium, Vol 4* (Newington: 1995), pp 22-26.

[6] Autek RF-1 RF Analyst, Autek Research, Box 302, Dept J, Odessa, FL 33556. Tele: 813-920-5810.

VHF/UHF Antennas

Recycle Those Rabbit Ears!

By Keith Allen, N8QNA
1212 Crawford St
Flint, MI 48507

Photos by Mark Sturtz, N8LIY

N8QNA uses surplus rabbit-ear elements to make a portable Rabbit-Ear Yagi—ideal for fox hunts

In these days of the proliferation of cable TV, the rabbit ears that come with most portable television sets are either thrown away or relegated to the junk box to gather dust. Here is a simple and practical way to recycle those rabbit ears and create a portable Rabbit-Ear Yagi antenna for 144, 222 or 440 MHz. This project makes an ideal fox-hunting antenna because of the collapsibility of the rabbit ears. See **Fig 1** for a photo of the completed Yagi, with the rabbit-ear elements deployed to the correct lengths.

The first step is to disconnect the rabbit-ear elements from the TV cable feed. There are various types of connections, but generally they are fed with 300-Ω twinlead connected with 4-40 hardware. Save the bolts and nuts. They can be used in other projects.

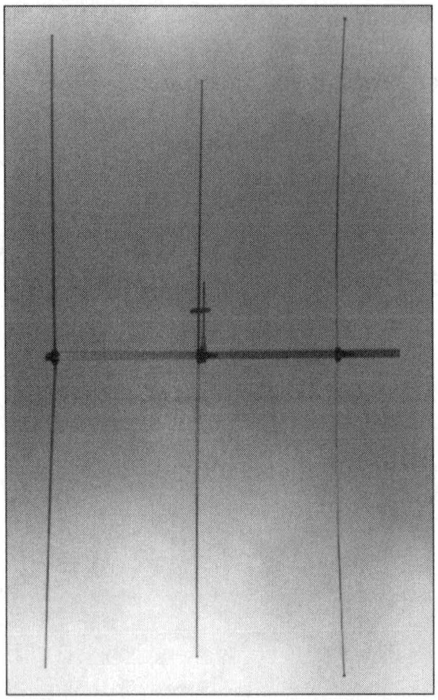

Next, cut the boom to length for the band you want. Use the dimensions shown in **Table 1**. Mark the positions of the elements and the barrel connector and drill a $^5/_8$-inch hole for the connector and three holes for #8 sheet-metal screws for the elements. Use a drill press to ensure accuracy. Drill a clearance hole in each element's mounting bracket for the sheet-metal screws. Mount the elements and install the barrel connector.

Secure a piece of aluminum $^3/_8 \times ^3/_4 \times 1^3/_4$ inch for the gamma-match shorting bar. Construct according to **Fig 2**, using the dimensions in **Table 2**. One hole should be $^1/_4$ inch and the other large enough to clear the driven element's largest diameter. Drill and tap the ends for $^3/_{16}$-inch bolts.

Next slide the center pin into the Radio Shack chassis-mount UHF connector and solder in place. Slide the nylon insulator on and secure with hot glue. Assemble the gamma match to the driven element with the shorting bar. Use an SWR meter to adjust the elements and the gamma match for best SWR.

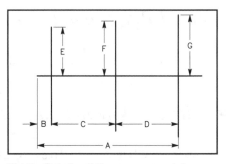

Fig 2—Details of the gamma match assembly.

Fig 1—Photo of completed 2-m Yagi, with elements deployed and telescoped to proper lengths.

Table 1
Yagi Dimensions for Each Frequency Band

Dimension	144 MHz	222 MHz	440 MHz
A	34.80"	23.50"	13.10"
B	1.00"	1.00"	1.00"
C	16.40"	10.75"	5.58"
D	16.40"	10.75"	5.58"
E	18.45"	11.91"	5.90"
F	19.09"	12.50"	6.36"
G	20.00"	13.03"	6.59"
Boom	34.80"	23.50"	13.10"

Table 2
Gamma-Match Dimensions for Each Frequency Band

Item	Size	144 MHz	222 MHz	440 MHz
Rod	0.093" Dia.	6" long	3.8" long	2" long
Tube	*1	6" long	3.8" long	2" long
Sleeve	*2	7" long	4.5" long	2.75" long

*1 Tubing is 0.2" ID, 0.25" OD
*2 Tubing is 0.1" ID, 0.187" OD

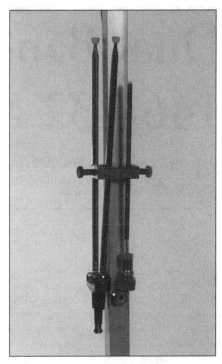

Fig 3—For transport, the elements are telescoped in and the driven-element gamma match is unscrewed from the boom-mounted UHF connector. Then the mounting blocks are twisted so that the elements are parallel to the boom.

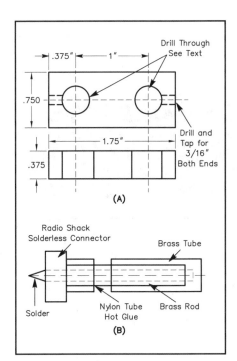

Fig 2—Details of the gamma match assembly.

Final lengths of each element and the position of the shorting bar on the gamma match should be marked so you can quickly return to the same spot each time. Paint pens, permanent markers or an electric etcher are all effective. **Fig 3** shows the driven element folded back along the boom, with the gamma match disconnected for transport.

An auxiliary attenuator attached to this antenna has proven so effective that we have only been consistently beaten by teams using Doppler systems, which cost hundreds of dollars. No matter which band you build the Rabbit-Ear Yagi for, in a pinch it can be used for reception on other bands during a fox hunt. Simply shorten the elements as necessary to attenuate the signal.

In tests conducted versus three other types of popular VHF/UHF antennas, the Rabbit-Ear Yagi shows a significant improvement in gain. The three other types of antennas tested were a standard rubber duckie, a $^5/_8$-λ helical-coil duckie and a $^5/_8$-λ magnetic-mount unit.

The tests were conducted using a linear field-strength meter constructed according to plans in *The ARRL Antenna Book.*[1] When these measurements are done at club meetings and swap meets, raised eyebrows change to incredulous looks and requests for the plans for the Rabbit-Ear Yagi!

[1] *ARRL Antenna Book*, 17th edition, pp 27-39 to 27-40.

Dual-Band Mobile Whip for 146/432 MHz

By Wayde Bartholomew, WA3WMG
RD 3 Box 3769
Pottsville, PA 17901

Here's an easily constructed dual-band mobile antenna you can make in an evening.

I needed a dual-band antenna for my brand-new dual-band handheld. However, I was not really impressed with the cost-to-benefit ratio for several commercial antennas I tried, so I decided to make one myself.

On HF I have enjoyed success using a single radiator with a decoupling stub. On-the-air results were good for DX, indicating to me that I was achieving low angles of radiation. This inspired me to try this design on VHF/UHF.

I used a commercial NMO-style base and magnetic mount. For the radiator and decoupling stub, I used brazing rod, which was coated with a rust inhibitor after all the tuning was done. I started out with a 2-meter radiator that was 20.5 inches long. This is an inch longer than normal so that it could be pruned for best SWR.

I then tacked on the 70-cm decoupling stub, which was 6.5 inches long. I trimmed the length of the 2-meter radiator for best SWR at 146 MHz and then tuned the 70-cm stub on 446 MHz, moving it up and down for best SWR. There was no interaction between the adjustments for either frequency.

See **Fig 1** for the final dimensions I used. The SWR in the repeater portions of both bands was less than 2:1.

I'd like to thank my father Wayne, WA3WLD, who ran all around the local area with this antenna on his minivan to check the performance.

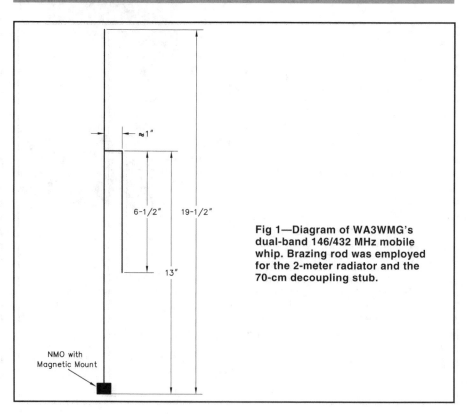

Fig 1—Diagram of WA3WMG's dual-band 146/432 MHz mobile whip. Brazing rod was employed for the 2-meter radiator and the 70-cm decoupling stub.

The K5BO Bi-Square Beam

By Rick Bibby, K5BO
3908 Wood Street
Texarkana, TX 75503

K5BO analyzes beams using Bi-Square elements, using both computer modeling and then field measurements. For the same length boom as a Yagi or conventional quad, the Bi-Square Beam gives more gain.

For many years hams have experimented with parasitic arrays. Usually, designs have been variants of Yagis or quads. The need to have even higher gains for weak signal VHF/UHF work has caused some to seek ways to exceed typical Yagi or quad performance. Since I've never been one to follow the beaten path, I investigated an alternative—the Bi-Square Beam array.

Fig 1 shows a basic Bi-Square loop element. What looks at first to be a 2-wavelength long loop is actually not a loop at all, since the top wires are not connected at the center point. The element actually consists of four $1/2$-wave segments, each carrying current in phase with the segment directly opposite it and 180° out of phase with the other two segments. Published gain figures range from 3.5 to 5 dBd,[1,2] with the lower figure the most accurate.

Adding extra elements to the antenna produces a beam array like the one shown in **Fig 2**. Rather than using a diamond, I prefer a square shape since I find that squares outperform diamonds. Bill Orr, W6SAI, in his book suggested a two-element version of this antenna. However, I found no references of any sort to

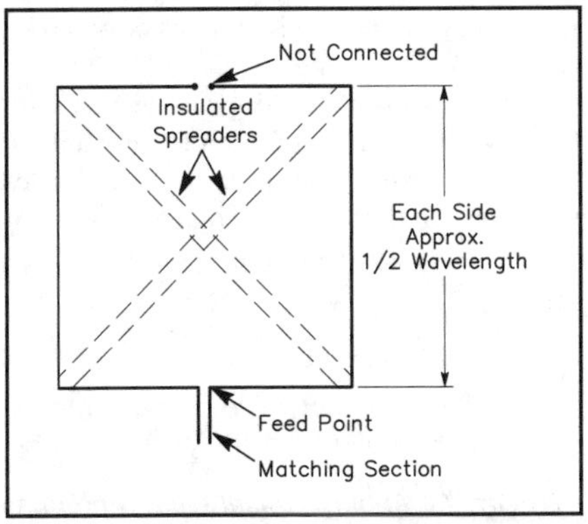

Fig 1—Typical layout of Bi-Square element. Each side is about λ/2 long, depending on whether it is used as a reflector, driven element or director. Since the feed-point impedance is high, the driven element is fed through a λ/4 matching section.

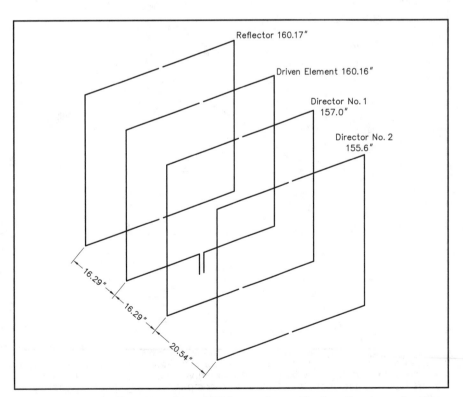

Fig 2—Dimensions for a four-element Bi-Square beam. The lengths shown for the elements are the overall wire length for each. Note that each element is not joined at either the top or bottom.

guide in the construction of a three-element or larger model.

Start with a Computer

Initial development of element size and boom length began entirely on the computer. Three computer modeling programs were used: *AO* (Antenna Optimizer) and *NEC/Wires* (a NEC-2 derivative) for the loop antennas, and *YO* (Yagi Optimizer) for the Yagis.[3] I modeled two, three, four and five-element antennas, comparing results for designs having the same boom length and element spacing. See **Fig 3** for a comparison of the maximum gain at 145 MHz for Yagi, quad and Bi-Square designs.

Antenna modeling must be done using great care to provide unbiased gain, F/B and SWR figures. All models had common boom lengths and element spacings for a given number of elements. Each pass of *AO* and *YO* used three frequencies (143, 144 and 145 MHz). All optimization passes ran to completion. Completed models were then verified using *NEC/Wires*. I must emphasize that my modeling did not produce an ideal antenna that would suit everybody's application. The comparisons between the Bi-Square, conventional quad and Yagi do show, however, that the Bi-Square has more gain than either of the other two antennas, given equal number of elements and the same boom length. The additional gain is not surprising, given that the Bi-Square physically takes up four times the area of a quad with the same number of elements.

Proof in the Pudding

Using the dimensions coming from the computer, I built a four-element Bi-Square Beam using treated pine lumber. I constructed another model using PVC pipe. On-the-air comparisons were made with a four-element Yagi and the four-element Bi-Square Beam, both with the same length booms. At first, I place the antennas on opposite eaves of my house, to minimize interaction between them. However, the presence of very large trees in my yard made this less than a perfect test platform. I needed a way to compare the arrays in unobstructed surroundings.

A small oil-derrick style wood tower provided the simplest solution. Both the tower and the two antennas fit nicely into the back of my half-ton pickup. With a mast that put the array at 14 feet and a high nearby hill I now had a clear test range.

Test results so closely followed expectations that I gained a heightened respect for antenna modeling! The Bi-Square Beam outperformed both the quads and Yagis right from the start. Although the advantage wasn't huge, it definitely was there. The greatest advantage occurred when the signals were weakest, as might be expected.

Construction Notes

Mechanically, the Bi-Square Beam looks much like a conventional square quad array designed for half-frequency, except that the elements aren't connected at the top or bottom. Any method of mechanical construction that is appropriate for a quad will work with the Bi-Square too.

Electrically, the input impedance is very high. I used a quarter-wave matching section to bring the impedance down to where 50-Ω coax can be used to go to the transmitter. Make the matching section a few inches longer than $\lambda/4$ to allow for tuning. Position the antenna so that the directors point straight up in the air while adjusting the matching section, with the reflector propped off the ground on a sawhorse.

Conclusion

From the beginning of this project, my goal was to create an antenna that had better performance than a Yagi or quad on the same length of boom. The Bi-Square Beam certainly meets that criterion, but in no way is it a cure-all. Since the Bi-Square Beam is larger than a quad or Yagi on the same size boom, it's up to you whether the extra gain justifies the bigger size.

If you're into long-distance VHF/UHF work or even EME, then this antenna may be interesting. Four seven-element Bi-Square Beams would probably be adequate for moon-bounce work, given modern low-noise receivers and a high-power transmitter. The computed gain comes out to nearly 19 dBd (21.15 dBi) in free space. This is for individual antennas with booms only 1.67 λ (just 11.33 feet) long!

My thanks to AA5ZT, who helped with the modeling, and to KB5VYD, who helped with the construction. Individual disk files suitable for the *AO* and *NEC/Wires* programs are on the disk that accompanies this book.

Notes and References

[1]R. D. Straw, editor, *The ARRL Antenna Book* (Newington: ARRL, 1994), 17th ed, p 8-43.

[2]W. Orr, *All About Cubical Quad Antennas* (Wilton, CT: Radio Publications, 1971), 2nd ed, pp 53 to 59.

[3]These programs are available from Brian Beezley, K6STI, 3532 Linda Vista Dr, San Marcos, CA 92069, Tel 619-599-4962.

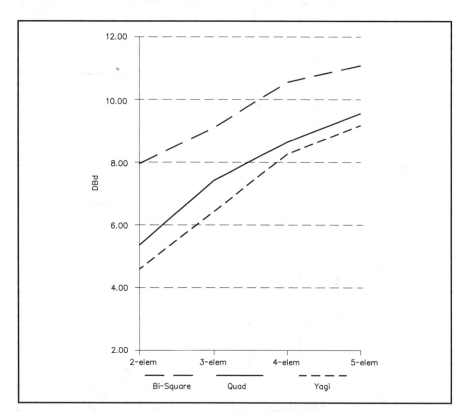

Fig 3—Graph showing gain comparisons between Bi-Square, conventional quad and Yagi, all using the same boom length for a given number of elements. The Bi-Square typically has more than 2 dB of gain compared to the other two types of beams.

Two Portable 6-Meter Antennas

By Markus T. Hansen, VE7CA
mghansen@ansa.com
674 Saint Ives Cres
North Vancouver, BC V7N 2X3

VE7CA provides details on two elegant, practical and portable 6-meter gain antennas—a two-element quad and a three-element Yagi.

Being a ham for over 32 years has provided the time to try many different aspects of this great hobby. Traffic nets, DX, FM mobile, equipment homebrewing and, of course, antenna experimenting have supplied hundreds of hours of enjoyment. Only recently did I become interested in VHF. It began when I found a used Yaesu FT-726R VHF/UHF all-mode transceiver at a reasonable price. With it, I could try my hand at satellite communications and grid collecting on the 144- and 432-MHz bands. Then the opportunity came along to add a 6-meter module and I jumped at it. Several local hams had been telling me about propagation modes I had not experienced: *sporadic-E skip* and *auroral propagation*. As soon as the 6-meter module arrived, up went a dipole.

My first experience working California, Nevada and other distant stations by sporadic E, with only 10 W to a dipole at 20 feet was exhilarating. However, it wasn't too long before I realized that local hams with better antennas were working stations I couldn't even hear. Hearing them work double-hop into Mexico and the East Coast spurred me to action. It was time to build my own gain antenna.

I could have settled for a store-bought Yagi but I have always enjoyed building my own antennas. There is a certain amount of satisfaction saying: "My antenna is home built." So I decided to start with a two-element quad.

After a visit to our local lumber store I came home with a three-foot length of 2×2 clear cedar and four 10-foot lengths of ¹/₂-inch fir dowel. A few evenings in the garage and a two-element quad emerged, ready for tuning. I attached the coax and tuned the reflector for maximum front-to-back ratio, hurrying because the band might just be open.

I noticed a big improvement over the dipole! Despite the improvement, I still found

a local ham using a four-element Yagi working stations I couldn't work. Either I needed more power or antenna gain. Although I could build a 100-W amplifier, I still had to be able to hear them before I could work them, so I decided to make the quad bigger.

I went back to the lumber store, returning with a six-foot length of 2×2 clear cedar and six lengths of ¹/₂-inch fir dowel. I kept the two-element quad up for comparison with my new three-element quad. This time, the director and reflector were hand-tuned for maximum forward gain. The new quad was mounted on a 26-foot pole outside the ham shack and the "armstrong method" was employed to rotate the array. Next, I built a four-element Yagi on a 12-foot boom and put it up on the tower at 52 feet. From my home station I could work just about every station that the locals could and I had a lot of fun on 6 meters.

If you spend enough time hanging around dyed-in-the-wool, hard-core, VHF/UHFers you will find something very peculiar about their behavior. When they talk about contesting they are thinking of high and lofty places—they don't mean the tower in the back yard. They are talking about faraway places on top of the highest mountain peaks man and vehicle are capable of reaching. In other words, if you do not possess a portable gain antenna for your favorite VHF/UHF band, you are not a *true* VHF/UHFer. Not only does an antenna have to be portable, it

Table 1

Gain Comparisons, Computed and Measured

Antenna	Computed* (dBd)	Measured (dBd)
Quad	4.06	4.2
Yagi	5.68	5.8

*Computed using *YO* for Yagi, and *AO* for dipole and quad.

must be easily assembled and disassembled, just in case you have to move quickly to another, supposedly superior, location.

I still liked the idea of a two-element quad, which is easy to assemble, inexpensive to build and just seemed to want to work. The only trouble is that it is a three-dimensional affair. However, it can be made to be quite rugged, and the wire elements are easily repaired in remote locations. As a first attempt at a portable 6-meter gain antenna, I decided to build a quad that would be easy to disassemble and carry in the back of our family van.

A Portable Two-Element 6-Meter Quad

My primary objective was to construct a two-element quad using material found in any small town. I did not want a complicated matching network. The Gamma matches commonly used on quads do not hold up well when you are setting up and taking down these antennas in the field. I planned on adjusting the distance between the driven and the reflector elements so that the intrinsic feed-point impedance was 50 Ω.

When I built my first 6-meter quad, the driven element length was derived from the formula 1005/freq. The reflector was made about 3% longer. Starting with a spacing of 30 inches, I went through the process of moving the elements closer together and adjusting the driven element for minimum SWR and the reflector element for maximum front-to-back ratio with a tuning stub. This took several hours and when I finished I had an element spacing of 24 inches. This produced an SWR under 1.2:1 over a large portion of the 6-meter band. During these adjustments, the antenna was mounted 12 feet high, as this was the height I intended to use for portable operations. The dimensions were later confirmed using a computer modeling program known as *AO*,[1] the Antenna Optimizer, by K6STI.

Fig 1 shows the dimensions for the boom and the boom-to-mast bracket. The boom is made from a 27$\frac{1}{4}$-inch length of 2×2. (The actual dimension of 2×2 is closer to 1$\frac{3}{4}$ by 1$\frac{3}{4}$ inches but it is commonly known in lumber yards as a 2×2.) Use whatever material is available in your area, but lightweight wood is preferred, so clear cedar or pine is ideal. Drill the four $\frac{1}{2}$-inch holes for the spreaders with a wood bit, two at each end, through one of the faces of the 2×2 and the other through the other face. The boom-to-mast bracket is made from $\frac{1}{4}$-inch fir plywood and cut to the dimensions shown in Fig 1.

The spreaders are $\frac{1}{2}$-inch dowel. Our local lumberyard had a good supply of fir dowels but other species of wood are available. The exact material is not important. Maple is stronger but expensive. Fiberglass would be ideal but it is not always available locally. Cut two of the $\frac{1}{2}$-inch dowels to a length of 83$\frac{5}{8}$ inches for the driven element spreaders and two to 88 inches for the reflector spreaders. **Fig 2** is a photo showing

one end of the boom with the two spreaders inserted. The mast was made from two six-foot lengths of 1$\frac{3}{4}$-inch fir dowel. Again, use whatever you may have available. I did not have any aluminum tubing large enough for a mast and wood dowel is a good alternative. Waterproof all wooden parts with at least two coats of exterior varnish.

While you are at the lumberyard or hardware store look for plastic pipe that fits over the end of the $\frac{1}{2}$-inch spreaders. You will need a one-foot length, with some to spare. Cut it into seven equal lengths, approximately one inch long, and one to a length of 1$\frac{1}{2}$ inches. Drill a $\frac{1}{16}$-inch hole through the seven equal lengths $\frac{1}{4}$ inch from the ends, and two holes one above the other $\frac{1}{4}$ inch apart on the 1$\frac{1}{2}$-inch sleeve. I used #14 hard-drawn stranded bare copper wire for the elements. Do not use insulated wire unless you are willing to experimentally determine the element lengths, since the insulation detunes each element slightly.

Cut the reflector element 251 inches long and slip one end of the wire through the holes you drilled in four of the plastic sleeves. Don't attempt to secure the wire to the plastic sleeves at this point. Cross the end of the reflector elements one inch from their respective ends and twist and solder together. The total circumference of the reflector element should be 249 inches when the ends are connected together.

Cut the driven element wire to 241 inches and slip three of the one-inch sleeves onto the wire. Again, don't secure the wires to the sleeves yet. Then the ends are passed through the two holes in the 1$\frac{1}{2}$-inch pipe. Wrap the ends around the pipe and twist them back onto themselves to secure the wire. The coax feed line is attached directly to the two ends at this point. The circumference of the driven-element loop from the points where the coax is attached should be 236$\frac{5}{8}$ inches. Solder the coax feed line to the driven element and waterproof the coax with silicone seal. I used RG-58, as it is lightweight. The length required for a por-

Table 2

Three-Element Yagi Dimensions

Element	Spacing Along Boom (inches)	Center Section Ele. (inches)	Telescoping Length (inches)	Total Length (inches)
Reflector	0	22$\frac{3}{4}$	51$\frac{1}{2}$	125$\frac{3}{4}$
Driven	28	9$\frac{3}{4}$ *	48$\frac{5}{8}$	58$\frac{3}{16}$
Director	63$\frac{3}{8}$	14$\frac{1}{2}$	51$\frac{1}{4}$	117
Hairpin	#14 wire	4 long	1$\frac{5}{8}$ spacing	

*Driven element uses two sections insulated at center.

Fig 2—Photo of one end of the VE7CA quad with the two spreaders inserted.

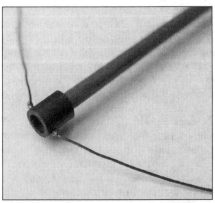

Fig 3—Photo showing one of the plastic sleeves slipped over end of a spreader to provide a mechanical mounting point and support for the wires.

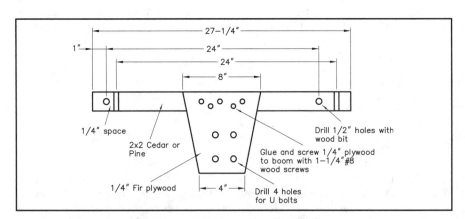

Fig 1—Dimensions for boom and the boom-to-mast bracket for VE7CA's portable two-element 6-meter quad.

table installation is typically not very long, maybe 20 feet, so the loss in the small cable is not excessive. Near the feedpoint, coil the coax into six turns with an inside diameter of two inches. I found this an effective method of choking any RF from flowing on the outside of the coax shield.

Begin assembling the quad by pushing the two reflector spreaders, without wires attached, through one end of the boom and the two shorter driven-element spreaders through the holes in the other end of the boom. Center the spreaders and mark the spreaders with a black felt-tip pen next to the boom. Now insert a 1¹/₂ #8 wood screw or a threaded L-hook into the boom so that it just touches one of the spreaders. Take the screw or L-hook out and file the end flat, then reinsert it so that it is just snug against the spreaders. I only used two L-hooks for the two vertical spreaders; the horizontal

spreaders are held in the proper position by the tension of the wire loops. If you use an L-hook, you can unscrew it with your hands—you won't have to worry about leaving the screwdriver at home.

You are now ready to assemble the wire loops. Take the reflector loop and place the four plastic caps over the ends of the reflector spreaders. Equalize the wire lengths between the spreader so that the loop is square. Now, secure the plastic sleeve pipes by tightly wrapping wire around the sleeve and the wire element and soldering the wire in place. See **Fig 3**, a photo showing one of the plastic sleeves slipped over one of the spreader ends, with the wire element through the hole and fastened in place. Follow the same procedure with the driven element.

Fig 4 is a picture of the quad's boom, with the plywood boom-to-mast bracket fastened with wood screws and glue. Two U-bolts are used to attach to the mast. When the quad is raised, the shape of the loop is commonly known as a diamond configuration. The mast consists of two six-foot lengths of doweling joined together with a two-foot length of PVC plastic pipe, held together with wood screws.

Make a slot the width of a #8 wood screw about one inch deep from the top of the plastic PVC pipe and then put the top mast into the plastic pipe. Insert a one inch #8 wood screw into the bottom of the slot you cut into the top of the pipe and tighten only enough

so that the top mast can be removed without unscrewing it. I drove a nail into the end of the lower mast and left it exposed an inch or more. This end is placed in the ground and the nail holds the pole in place. A strip of wood approximately 1×3 and long enough to cross over the roof rack of our family van is used to hold the center of the antenna mast to the roof rack of the van with small diameter rope. See **Fig 5** for a photo of the quad in action next to the family van.

To disassemble the quad, lay it on its side, slip the plastic sleeves off the ends of the spreaders and roll up the wire loops. Loosen the L-hooks holding the vertical spreaders in place. Push the spreaders out of the boom, loosen the U-bolts and free the mast from the boom.

That is all there is to it. It takes about two minutes to put it up, or take it down. I have found it quite sturdy, surviving several storms with high winds.

A Three-Element Portable 6-Meter Yagi

The idea to build a Yagi antenna resulted when we traded the family van for a compact car. I needed something that would fit into the trunk of the car. At close to seven-feet long, the quad spreaders were too long. After computer modeling a three-element Yagi on a five-foot boom, I found that I could pick up about 1.5 dB gain over the short-boom two-element quad. A five-foot boom fits into the trunk or across the back

Fig 4—Photo of the two-element quad's boom, with the plywood boom-to-mast bracket secured with wood screws and glue.

Fig 5—Ready for action! VE7CA has set up his quad next to the family van.

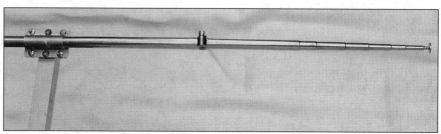

Fig 6—Photo showing a piece of aluminum tubing used as a center section to join the two telescoping tips together.

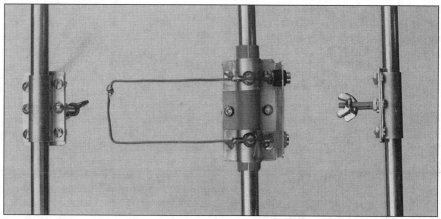

Fig 7—A view of the center sections of the three Yagi elements with their mounting brackets.

Fig 8—Photo detailing attachment of the reflector to the square-section boom, using two #10 bolts and wingnuts.

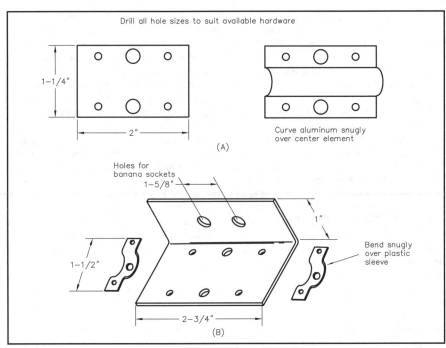

Drill all hole sizes to suit available hardware

1-1/4"

2"

(A)

Curve aluminum snugly over center element

Holes for banana sockets
1-5/8"

1"

1-1/2"

2-3/4"

Bend snugly over plastic sleeve

(B)

Fig 9—At A, details for the reflector and director element-to-boom brackets, made of 1/16-inch plate aluminum. At B, details for the driven-element bracket. These are screwed to the square boom.

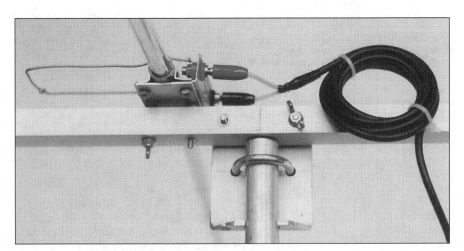

Fig 10—Photo of driven element, complete with hairpin match and the banana plugs used to connect the coax cable to the driven element.

seat of the car, but I had to do something about shortening the nine-foot elements!

Several options were considered. Loaded elements were too complex and tedious to tune. Splitting the elements in the center would work if an easy, reliable method could be devised to quickly fasten and unfasten the elements from the boom. Since my philosophy for portable equipment is to keep it as simple as possible, I continued to look around for good alternatives for elements.

One day I noticed a box of portable-radio telescoping antenna elements at our local radio parts store. To my delight, they were 54 inches long when fully extended. I left the store with six telescoping elements and a big smile on my face. On the way home I stopped by a scrap-metal yard and rummaged around in some bins of aluminum tubing. A 60-inch length of aluminum tubing that fit over the end of my telescoping elements was found and purchased. There are many different sizes of telescoping antenna elements, with different diameters. This is where you will have to use your scrounging skills! **Fig 6** shows how the tubing is used as a center section to join two telescoping elements together. It also serves to extend the total length of each element, since two telescoping elements themselves are not long enough to resonate on 6 meters. Each center section is slotted at both ends with a hacksaw, and stainless-steel hose clamps are used to secure the telescoping elements.

Since the telescoping elements are severely tapered ($7/16$ down to $1/16$ inch), I used the YO^2 Yagi Optimizer modeling program to design the Yagi, taking all tapers into account. After numerous trial runs on YO a satisfactory design was derived on a 65-inch boom. By splitting the driven element in the center and employing a hairpin match, I was able to obtain an SWR below 1.16:1 from 50.05 to 50.2 MHz. **Fig 7** shows the center sections of the three elements with their mounting brackets.

I decided to use a square boom so that I would have one flat surface to work with. **Fig 8** shows how the reflector is

attached to the end of the boom with two $1^1/2$-inch 10-32 bolts and wingnuts. **Fig 9A** provides the dimensions and details for the reflector and director element-to-boom brackets, which are formed from $1/16$-inch plate aluminum. The driven element is split in the center and is insulated from the boom. Fig 9B shows details for the driven-element bracket. **Fig 10** is a photo of the driven element with the hairpin matching wire and the banana plugs used to connect the coax to the driven element. You could use a female PL-259 connector if you wish. I used #14 solid bare copper wire for the hairpin. It is very

durable—even after being severely warped in the car trunk, I can bend it back into shape quickly and easily.

The boom is $3/16$-inch square aluminum, 65 inches long. This material was found at a local hardware store. To detach the elements, just loosen the wing nuts and remove the elements from the boom. I used a similar method to attach the support mast to the boom.

As with the quad design, I decided to use a choke balun, consisting of a coil of coax next to the antenna feedpoint. This chokes off current flowing along the shield of the coax so that the antenna pattern will not be distorted on this balanced feed system. To

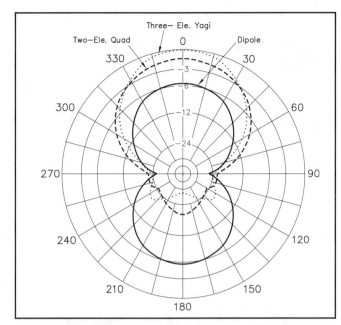

Fig 11—Measured radiation patterns in the azimuth plane for three 6-meter antennas: the portable two-element quad, the portable three-element Yagi and a reference dipole.

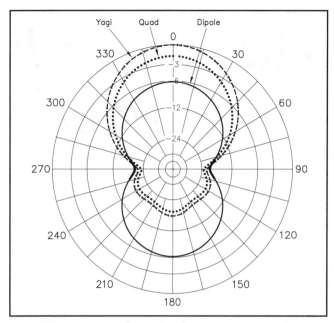

Fig 12—Computed radiation patterns for the same antennas in Fig 11. The computations are remarkably close to the actual measured patterns!

determine the size of the coax coil, I first plotted the antenna pattern feeding the antenna with a straight length of coax. Just like *The ARRL Antenna Book* predicts, the pattern was definitely skewed to one side. I then coiled up the coax for a total of six turns with an inside diameter of 2 inches. Six turns were sufficient.

To tune the hairpin match, assemble the Yagi on its mast and extend the elements. Spray switch contact solution on a cloth and wipe any dirt and grease from the elements. Push the elements together and apart a couple of times so that the contact solution cleans the elements thoroughly. Attach the antenna mast to your vehicle or use whatever method of support you intend to use in the field. Connect an SWR meter and a transmitter to the coax feeding the antenna. I used two alligator clips soldered together to slide along the two hairpin wires to find the position for the lowest SWR. The dimensions computed by computer were correct!

I can take this antenna out of the trunk of the car and assemble it in less than two minutes. One caution: the telescoping elements when fully extended are quite fragile. I have not broken one as yet, but carrying a spare element just in case would be a good idea.

Are These Antennas Effective?

In order to evaluate the two portable antennas, I decided to carry out actual field antenna-pattern measurements. The reference antenna was a 6-meter dipole. A single-transistor, battery-operated crystal oscillator was used as a source. It was placed 200

feet from the antennas under test.

The test location is a large, level grass soccer playing field. The antenna mast was tied to a wooden stepladder in the center of the playing field. A homebrew direct-conversion receiver, step attenuator and a VU meter with a 600 to 8-Ω audio transformer matching the VU meter to the audio output jack of the receiver made up my test equipment.

On a 12-foot mast, I mounted a 6-meter dipole tuned for the 50.13 MHz test frequency. Turning the dipole through 360°, VU meter readings directly in dB were plotted on an ARRL ANTENNA RADIA-TION PATTERN worksheet. Fortunately, the dipole pattern turned out to be text-book symmetrical, indicating that the test site was acceptable for further antenna measurements.

I then removed the dipole and mounted the three-element portable Yagi on the same mast. With the Yagi pointed at the test oscillator and 30 dB of attenuation in line, the receiver audio level was set so that the VU meter read zero. I did not adjust the audio gain control further or move the location of the test oscillator until all antenna measurements were made. Rotating the Yagi through 360°, I stopped every 10° and plotted the VU meter readings on an ARRL ANTENNA RADIATION PATTERN worksheet. As I rotated the Yagi and the VU meter reading decreased to –10 dB, the in-line attenuation was decreased by 10 dB.

After completing the Yagi measurements, I removed it and put up the two-element quad. In a similar manner, I recorded the VU

readings as I rotated the quad through 360°. Finally, the dipole was placed back up on the mast and the antenna pattern plotted on the same plotting sheet.

Fig 11A and **11B** show the results of the actual field measurements. **Fig 12A** and **12B** show computed elevation and azimuth patterns of the three antennas. I was amazed at how close the actual field measurements were to the computer analysis.

Concluding Remarks

Do these antennas perform in the field? Yes, they do. In the 1994 Field Day, the two-element portable quad was instrumental in helping VE7NSR (the North Shore Amateur Radio Club) capture the number one position in the 1A class for Canada.

I have used the three-element portable Yagi on numerous occasions in the field and found it a very effective portable antenna. Side-by-side comparisons between the Yagi and the quad indicate that the Yagi out performs the quad, as expected.

Constructing Amateur Radio antennas is still one area in this great hobby where the results always seem to outweigh the effort. Employing a simple design eases construction, keeps the cost low as well as increasing the likelihood that the design will perform properly. I hope the construction and design techniques outlined in this article will provide readers with the incentive to build their own antennas.

References

[1,2] *YO* and *AO* are antenna modeling programs available from Brian Beezley, K6STI, 3532 Linda Vista Dr, San Marcos, CA 92069.

The Jagi, a 3-Element Yagi with a J-Pole Driven Element

By Michael P. Hood, KD8JB
1441 Burke NE, Apt C
Grand Rapids, MI 49505-5328

> *KD8JB explains that the name of his somewhat offbeat beam can be pronounced "Yagi" if you like, but it's also okay to pronounce it "Jaggy"!*

The all-copper J-Pole antenna I described in *The ARRL Antenna Compendium, Vol 4* is an extremely rugged antenna. It is mounted at 45 feet and gives a good account of itself for omnidirectional work over quite a distance. During good conditions, I have worked the Milwaukee MARS repeater from my QTH in Grand Rapids, a distance of about 125 miles. However, there are times when some directional capability would be nice, such as when conditions are marginal.

There are other reasons, of course. Perhaps that one repeater you need to access is just out of range, or you just happen to be midway between two repeaters and are constantly being bombarded by signals from both. It's rare, to be sure, but it does happen, typically with weaker signals. In this case, a beam antenna of some sort is needed.

By simply adding a reflector and director, you can transform an omnidirectional J-Pole into a vertical directional antenna—a *Jagi*, if you will. The vertical qualities of the antenna as an omnidirectional radiator are only slightly diminished when configured as a Jagi. Because the antenna is at dc ground if bonded to a mast, around a rotator (if used), and subsequently to earth ground, the antenna can be relatively quiet, despite its vertical polarization—a common complaint with vertical antennas. Vertically polarized noise, like ignition shot noise, would still be a problem, however.

The J-Pole antenna described in my previous article is an ideal building block. As it is basically an endfed ³/₄-λ antenna, the top half can be used as the driven element for a directive, vertically polarized Yagi. A small TV-type rotator can easily turn a small 2-meter three-element Jagi. Since the weight is largely balanced, it can be rotated with a larger HF Yagi beneath it with almost no increase in rotator starting torque or braking effort. The feed line is attached in

the usual place for a J-Pole, so there is no need to encumber the boom with any excess weight or stresses resulting from a heavy run of RG-213. Generally, the finished product presents a clean assembly you can be proud to have in your antenna arsenal.

Refer to **Table 1** for a list of parts needed to build this antenna. You'll find that the finished product is quite robust, yet not as cumbersome as you might expect.

Construction

The Jagi's basic structure is the J-Pole antenna shown in **Fig 1**. Index numbers refer to parts in Table 1. If you built the J-Pole as presented in the *Antenna Compendium, Vol. 4*, you'll notice that the upper 19 inches of the ³/₄-λ section is now made of ¹/₂-inch, rather than ³/₄-inch, rigid copper pipe, coupled to the remaining section with a ¹/₂×³/₄ reducer. This provides a bearing surface on which the cedar boom may rest, and eliminates the need for boring a large ⁷/₈-inch hole that may weaken the boom unnecessarily. The cedar used was a full 1-inch thick ("⁵/₄" in lumberyard parlance) and 2¹/₂ inches wide. You may use any wood you're used to working with, but the natural rot-resistant properties of cedar make it an

excellent choice for this application. Oak would work well since it is strong, but it is slightly heavier and will require some sort of preservative if used. Where price is no object, teak or mahogany are excellent choices, too.

Driven Element

Cut the copper pipe to the lengths indicated in Fig 1, from the materials listed in Table 1. Item 10 is a copper nipple cut from the ³/₄-inch copper pipe to a length needed to get from the base of the "J" portion of the assembly to the mast-mating coupler. This nipple should not be longer than 8 inches to minimize bending in locations where strong winds are prevalent.

Assemble all parts together before soldering and carefully recheck the dimensions. Notice that the dimension from the top of the matching stub and the dimension for the ³/₄-λ element is to the top surface of the 1¹/₄-inch nipple—not the center of the nipple diameter. The measurements shown are for 146 MHz. For measurements that are too long, trim the length of the ¹/₂-inch copper pipes a little at a time as necessary. If they are too short, the pipes could be adjusted slightly before soldering, but the bet-

Table 1

Detailed Parts List (See Fig 1)

Item No.	Quantity	Part or Material Name	Item No.	Quantity	Part or Material Name
1	1	³/₄ inch × 10 feet length of rigid copper pipe	13	1	³/₄ × ¹/₂ inch copper reducer
2	1	¹/₂ inch × 10 feet length of rigid copper pipe	14	1	¹/₄ or ³/₈ inch × 40 inch length of aluminum tubing (reflector)
3	2	³/₄ inch copper pipe clamps	15	1	¹/₄ or ³/₈ inch × 36 inch length of aluminum tubing (director)
4	2	¹/₂ inch copper pipe clamps	16	1	⁵/₄ inch × 2¹/₂ × 34 inch length of cedar (boom)
5	1	¹/₂ inch copper elbow			
6	1	³/₄ × ¹/₂ inch copper tee	17	4	8 × 1¹/₂ inch brass wood screws
7	2	¹/₂ × 19 inch copper pipe (make from item 2)	18	6	8-32 × ¹/₂ inch brass machine screws (round, pan, or binder head)
8	2	¹/₂ inch copper end cap	19	6	#8 brass flat washers
9	1	¹/₂ × 1¹/₄ copper nipple (make from item 2)	20	6	8-32 brass hexnuts
10	1	³/₄ inch copper nipple, length as required (make from item 1. See text.)	21	A/R*	Rosin core solder
			22	A/R*	Paste flux
11	1	Your choice of coupling to mast fitting (I used ³/₄ × 1 inch NPT)	23	A/R*	Fine sandpaper, steel wool, or emery cloth
12	1	³/₄ × 37¹/₂ inch copper pipe (make from item 1)	24	A/R*	Solvent to clean away flux after soldering

*Denotes "as required."

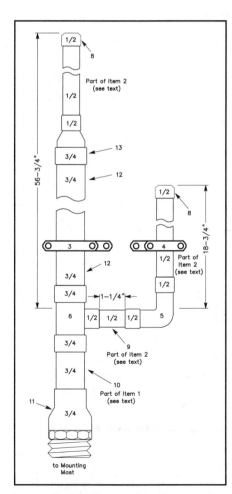

Fig 1—Exploded assembly diagram of Jagi J-Pole driven element. Item numbers refer to parts list in Table 1.

ter choice would be to cut new ones of the correct length. When assembled the ³/₄-λ element should be 56 ³/₄ inches long overall.

Remove burrs from the ends of the pipe after cutting. Clean the mating surfaces with sandpaper, steel wool, or emery cloth. Then apply a very thin coat of flux to the mating elements and assemble the pipe, elbow, tee, endcaps, and stubs. Solder the assembled parts with a torch and rosin core solder. Wipe off excess solder with a damp cloth, being careful not to burn yourself. The copper pipe will hold heat for a long time after you've finished soldering. After soldering, set the assembly aside to cool.

Feedpoint Clamps

Flatten one each of the ¹/₂ and ³/₄-inch pipe clamps. Drill a hole in the flattened clamp as shown in **Fig 2**. Assemble the clamps and cut off the excess metal from the flattened clamp using the unmodified clamp as a template. Disassemble the clamps.

Assemble the ¹/₂-inch clamp around the ¹/₄-λ element and secure with two of the screws, washers, and nuts shown in Fig 2. Do the same with the ³/₄-inch clamp around the ³/₄-λ element. Set the clamps initially to a spot 4 inches above the bottom of the J on their respective elements. Tighten the clamp hardware only finger-tight, since you'll have to move them when tuning.

Boom Assembly

Fig 3 shows the suggested method to make the boom assembly. Note that the wide side of the boom is perpendicular to the elements. Note also that even though the boom isn't large or very heavy to begin with, the ends are tapered slightly to reduce weight. The holes drilled into the cedar are clearance holes for the #8 screws. Drill tap holes into the aluminum tubing and copper reducer at the appropriate places (using the boom as a template) to allow screw thread engagement with the element. **Fig 4** shows views of the tapered boom itself and the placement of screws.

Tuning

Tuning an antenna is simple, although it is more convenient to tune the basic J-Pole antenna without the boom assembly attached. That way you'll get a good feel for how to tune the driven element without the extra hardware over your head. After tuning, install the boom assembly and recheck the tuning to make sure it's correct.

Select the feed line you will ultimately use when the antenna is complete, and attach one side of the feed line to the matching stub clamp assembly, and the other to the ³/₄-λ section clamp assembly. Make sure all the nuts and screws are at least finger-tight. If using coax, it really doesn't matter to which element, ³/₄-λ element or stub, you attach the coaxial center lead. I've done it both ways with no variation in operational effectiveness. Before making any adjustments, mount the antenna vertically, about 5 to 10 feet from the ground. A short TV mast on a tripod works well for this purpose.

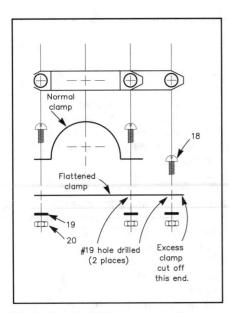

Fig 2—Details of clamp assemblies. Both clamp assemblies are the same.

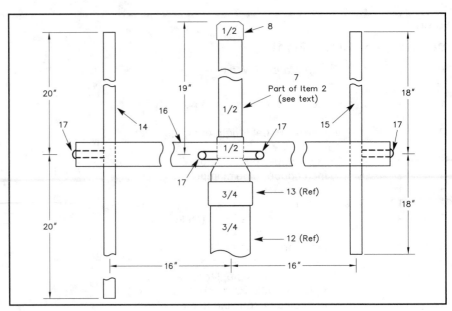

Fig 3—Side view of Jagi showing parasitic reflector and director on wooden boom clamped to J-Pole driven element.

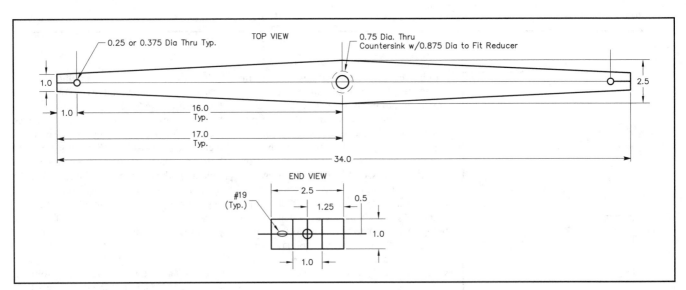

Fig 4—Top view of Jagi boom.

Tuning VHF antennas can be tricky. Seemingly minor mechanical adjustments can have a large effect on antenna performance. Keep in mind that they are extremely sensitive to nearby objects—such as your body. Therefore, you'll need some feed line to separate the antenna and the SWR bridge. Tune as follows:

1. Apply RF at the frequency you want the antenna to perform best.
2. Check the SWR.
3. Turn off the RF.
4. Move the clamps *equally* up from the original position ¹/₂ inch.
5. Reapply RF.
6. Check the SWR again, and turn off the RF.

a. If the SWR went higher, move the clamps 1 inch downward, equally.
b. If the SWR went lower, move the clamps ¹/₂ inch upward, equally.
7. Reapply RF.
8. Check the SWR once more, then turn off the RF. By this time you will be approaching minimum SWR (You may *never* get a 1:1 SWR. Don't sweat it; it's not important enough to worry about.)
9. Adjust the clamps a small amount in the direction of minimum SWR.
10. Repeat steps 7 through 9 until minimum SWR is achieved.
11. Remove RF.

If you elected to tune the antenna before

mounting the boom assembly, install that assembly at this time and recheck the tuning. There may be some interaction, but it should be minimal.

Final Assembly

The final assembly of the antenna will determine its long-term survivability. Perform the following steps with care:

1. After adjusting the clamps for minimum SWR, mark the clamp positions with a pencil.
2. Remove the feed line and clamps.
3. Apply a very thin coating of flux to the inside of the clamp and the corresponding surface of the antenna element

where the clamp attaches.

4. Install the clamps and tighten the clamp screws. Don't attach the feed line yet, and leave off the final washer and nut at the feed line attachment point.

5. Solder the clamps where they are attached to the antenna elements.

6. Apply a small amount of solder around the screw heads and nuts where they contact the clamps. Don't get solder on the screw threads!

7. Clean away excess flux with a non-corrosive solvent.

After final assembly and mounting the antenna in the desired location, attach the feed line, and secure with the remaining washer and nut. It would be a good idea to weather-seal this joint with RTV. Otherwise, you may find yourself repairing the feed line after a couple of years.

Fig 5 shows the computed responses for a J-Pole by itself compared to the Jagi, both mounted 6 feet off the ground. You can see that the Jagi has about 7 dB of gain compared to the J-Pole, a very worthwhile improvement!

References

[1]M. P. Hood, KD8JB, "A True Plumber's Delight for 2 Meters—An All-Copper J-Pole," *The ARRL Antenna Compendium, Vol 4,* 1995, pp 195-197.

[2]J. Brenner, NT4B, "A Portable 2-Meter Beam Antenna," *QST*, Apr 1987, p 57.

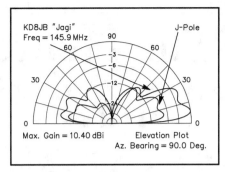

Fig 5—Comparison of computed elevation patterns for J-Pole and Jagi, both mounted 6 feet above ground. Jagi has about 7 dB of peak gain over the J-Pole.

The Hentenna—The Japanese "Miracle" Wire Antenna

By Shirow Kinoshita, JF6DEA/KE1EO
2-19-10-4212, Najima,
Fukuoka-Higashi, 813, Japan

JF6DEA describes an antenna popular in Japan—the "Hentenna." This loop design has gain, a low wind-surface profile and can be configured for either horizontal or vertical polarization.

Perhaps some of you may have heard the name *Hentenna* during QSOs with Japanese operators. The Hentenna is a wire antenna that is very easy to make. Since it first appeared in the Japanese ham literature, the Hentenna has become popular on the HF as well as the VHF/UHF bands.

In Japanese, the word "Hen" means fantastic or miraculous. This is the origin of the name Hentenna, recognized as a "fantastic antenna" by many Japanese amateurs because of its many useful properties.

History

In July 1972, Mr Someya, JE1DEU, the youngest member of the Sagami Club, located near Tokyo, suggested a prototype of this antenna. This was constructed using two quad loops fed in phase. He was just a junior high school student at the time. After that, Mr Tadashi Okubo, JH1FCZ, and a number of other amateurs (JA1RKK, JA1TUT, JH1ECW, JH1HPH, JH1XUQ and JR1SOP) experimented to establish the basic form of the Hentenna. They were encouraged by their measurements, especially since they found the new antenna compared with the gain of a conventional two-element quad. The biggest problem they faced at first was achieving a low SWR.

Five Useful Properties of the Hentenna

The general shape of the Hentenna is shown in **Fig 1**. It is a rectangular loop—λ/6 wide by λ/2 tall. The Hentenna is fed with 50-Ω coaxial cable at the center of a horizontal wire spaced about λ/6 from the bottom.

The azimuth pattern for a horizontally polarized Hentenna is shown in **Fig 2**. The beam pattern is a figure eight, perpendicular to the loop's plane. The basic Hentenna is horizontally polarized. This is the first useful property of the Hentenna—if you should want vertical polarization, the antenna can be rotated 90°, as shown in **Fig 3**.

A second useful property of the Hentenna is that the physical dimensions are not particularly critical. The permitted range of

overall loop length is about ±10%. By mistake, one 50-MHz antenna was constructed with 3.5 m long sides, about 117% of the nominal λ/2 length desired on 6 meters. Nonetheless, even this antenna worked fine, with an SWR of 1.2:1.

A third useful property of the Hentenna is the relative ease by which it may be tuned, which is by sliding the driven-element wire up and down along the vertical element and then soldering it where the SWR is lowest.

A fourth useful property of the Hentenna is that it has gain roughly equal to a two-element Yagi [or a short-boom three-

element tribander.—*Ed.*] Yet it is a lot smaller than a full-size two-element Yagi! JH1FCZ measured a gain of 8 dBd in November 1972. Professor Mushiake at Tohoku University computed the free-space gain at 5 dBi, as announced in the *Journal of the Institute of Electronics, Information and Engineers* in Japan. [No doubt, JH1FCZ measured gain over ground. This would include ground-reflection gain of about 5 dB. Free-space modeling of the Hentenna using *NEC-2* shows a gain of just over 5.1 dBi, just a little less than the gain of a typical full-size two-element Yagi. A gain of 5.1 dBi in

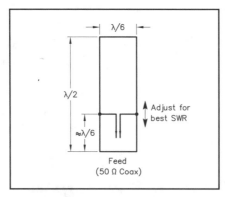

Fig 1—The basic shape of Hentenna, λ/2 high and λ/6 wide. This form of the antenna is horizontally polarized. Exact feed tap points are adjusted experimentally for best SWR.

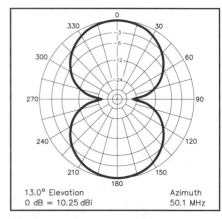

Fig 2—Azimuth pattern for 6-meter Hentenna mounted 20 feet above average ground.

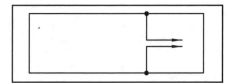

Fig 3—Vertically polarized Hentenna, turned on its side.

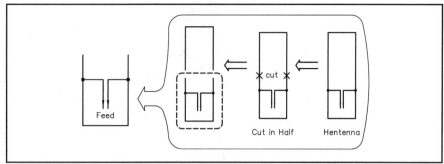

Fig 4—Evolution of Hentenna into "Fork Hentenna" or "Half Hentenna."

free space would result in just over 10.1 dBi of gain—or 8 dBd referenced to a dipole in free space—when the Hentenna is placed over ground with a dielectric constant of 13 and a conductivity of 5 mS/m.—Ed.]

A derivative of the basic Hentenna form results in the so-called *Half Hentenna* shown in **Fig 4**, also called the *Fork Hentenna*. The Hentenna can even be made circular, as shown in **Fig 5**. Here, it is useful as a vertically polarized antenna. The round version of the Hentenna is called the *Hat Hentenna*, because it is similar to a derby hat.

Loop Width and Feed-Point Impedance

The loop width of a typical rectangular Hentenna is λ/6. If you make the width narrower, then the feed-point impedance goes lower. If the width is made wider, the impedance becomes higher. (See **Fig 6**.) A wider loop has a larger SWR bandwidth, implying a lower-Q antenna. When you make a Hentenna, you can adjust the impedance any way you like, so you can make a special Hentenna having the same impedance of your coax or transmitter.

An Example: How to Make a 50-MHz Hentenna

On the 6-meter band in Japan, we usually use horizontal polarization, so the Hentenna is most often constructed in the "tall" configuration. See **Fig 7**. The parts list for this antenna is shown below.

a) Two aluminum spreader tubes, 10 to 12 mm OD (3/8 to 1/2 inches OD), λ/6 = 1 m (3 feet, 3 inches) long
b) Two insulated copper wires, λ/2 = 3 m (9 feet, 9 inches) long
c) #14 or #12 copper wire a little longer than λ/6 = 1 m (3 feet, 3 inches)
d) Two plates with U-bolts to fix the aluminum pipes (a) to the mast
e) A mast, of adequate size to support the antenna

Flatten both ends of the aluminum spreader pipes with a hammer and drill holes to secure lugs crimped onto the vertical copper wires. The two horizontal spreaders attach to the mast λ/2 apart. You don't need to electrically isolate the spreaders from the mast, unless you want vertical polarization, where the Hentenna is turned on its side.

Next, take care to strip off part of the insulation near the lower aluminum spreader.

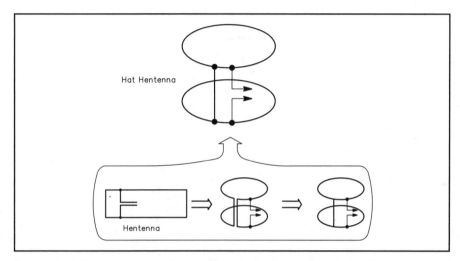

Fig 5—Evolution of Hentenna into circular form—the "Hat Hentenna."

Connect the driven wire to the vertical wires temporarily, perhaps using alligator clips. Adjust the Hentenna by moving the driven-element wire up and down along the vertical elements while measuring SWR. When you find the point with the lowest SWR, solder the connections. Be sure to use conducting paste between the screws and the terminal lugs on the copper wires, and to seal all joints with electrical tape.

Another Example, JA6YBR 6-M Beacon Hentenna

JA6YBR is the station of the Miyazaki University Radio Club, located in Miyazaki, in the Southwest of Japan. Miyazaki is well known as a place which typhoons often visit from the summer through autumn. The -beacon antenna needed to have the following properties:

a) horizontal polarization
b) omnidirectional coverage
c) a small wind-surface area
d) as much gain as possible

The original antenna, installed in 1985, had been a stacked "Squaro," but the gain was inadequate. They next tried a turnstile dipole. They found this to have too large a wind-surface area, considering the typhoon problem. In May 1990, JA6YBR installed a

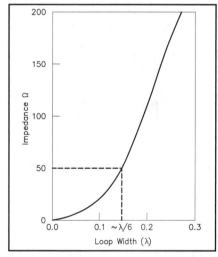

Fig 6—Graph showing the relationship between feed-point impedance and loop width in λ. The impedance is 50 Ω at a width of about λ/6.

two-loop turnstile Hentenna. See photo in **Fig 8**. It has been a good performer, damaged only once during 1993's memorable typhoon.

Throughout Cycle 22, JA6YBR's signal was reported throughout the Pacific, South America and Africa. JA6YBR's 6-meter

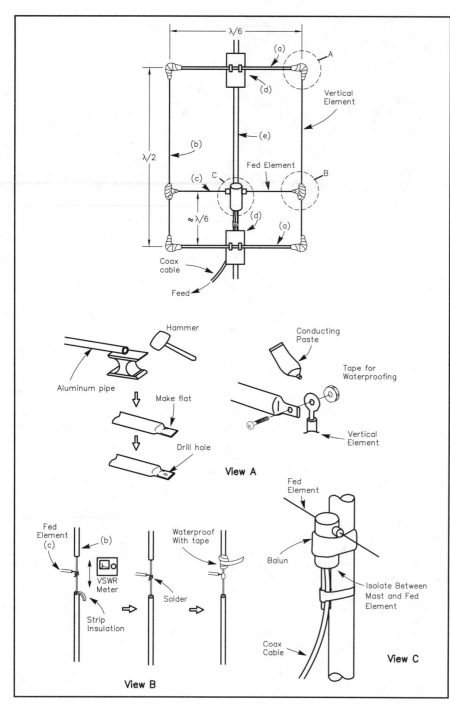

Fig 7—Construction details for 50-MHz Hentenna. See parts list in text of the article.

Fig 8—Photograph of JA6YBR omnidirectional turnstile Hentenna. This antenna has survived numerous typhoons common to the region around Miyazaki.

signal was even received on the US West Coast via multi-hop sporadic E in July 1995. The turnstile Hentenna presently in use at JA6YBR was made by Mr Masayoshi Eguchi, JI6KGZ. A photo of this antenna was published in "The World Above 50 MHz" column in the July 1995 issue of *QST*.

[Note that the gain of the omnidirectional version of the Hentenna will be down about 3 dB compared to the conventional bi-directional version.—*Ed.*]

Hentennas for 75-Meter DX!

The Hentenna can be used on any frequency you like. In Japan, JA1JRK and JM1BPP have used the Hentenna for 75-meter DX and they have had good results. The Hentenna is better than a full-sized dipole, with higher gain and a smaller wind surface area.

Acknowledgments

The Hentenna is a wonderful antenna, small in size and easy to make. Many JA operators have used it as their prime antenna on 6 meters or for field operation. Multi-element Hentennas for VHF/UHF are sold by some antenna manufacturers. These antennas are used for DX and even EME operation.

I want to offer special thanks to Mr Tadashi Okubo, JH1FCZ, for his permission to introduce the Hentenna. Also I want to thank Mr Kenji Tokunaga, JA6QJG, and Mr Shigeru Higasa, JE1BMJ, for their technical advice.

Notes and References
[1] Tadashi Okubo, JH1FCZ, "The Hentenna I," "The Hentenna II" (privately published booklet in Japanese).
[2] Tadashi Okubo, JH1FCZ, "Technology of Hentenna," *Let's HAMing Magazine*, Nov 1995, pp 107-123 (in Japanese).

Broadband Half-Wave 6-Meter Vertical

By Frederick T. Smith, W6DV
4830 Nomad Dr
Woodland Hills, CA 91364

The purpose of this antenna is to provide a vertically polarized antenna that covers both ends of the 6-meter band. I have used this antenna for about four years, to cover repeaters at the top end and local SSB nets at the low end of the band.

> *Want an omnidirectional 6-meter antenna that works at both ends of the band? Here's something W6DV came up with for just that purpose.*

Design

The broad bandwidth is obtained by using a large-diameter, low-Q radiating element that is λ/2 long. This broadbanding technique is discussed in *The ARRL Antenna Book*.[1] An L network is used to match the 50-Ω coax line to the high impedance at the bottom end of the λ/2 element.

I have previously used L networks to feed 12 and 2-meter λ/2 vertical antennas. The design technique is essentially a modification of the J-pole antenna. The λ/4 matching section of the J-pole is similar to a resonant parallel-tuned circuit. Connecting the inner conductor of the coax part way up from the bottom end of the λ/4 matching section corresponds to tapping a parallel circuit inductor several turns up from the bottom to obtain the desired impedance match.

The L network replaces the resonant parallel tuned circuit. A length of RG-58 coax about seven inches long is used to provide the required capacitance in the L network. This eliminates the requirement for a weather-proof container for L-network components.

Formulas for computing the capacitive and inductive reactance values are given in the 1992 edition of *The ARRL Handbook*.[2] The L network consists of the coax capacitor and an inductor connected between the bottom end of the antenna and the coax shield, as shown in **Fig 1**. The output impedance R_2 at the bottom end of the antenna is transformed to $R_1 = 50 \ \Omega$ at the input of the network.

The length of the antenna is computed from Fig 29 in Reference 1, using a coefficient C = 0.85, a frequency of 52 MHz, and multiplying by two for a λ/2 antenna. This yields an antenna length of 96.5 inches. This length is not precise, but small values of capacitive or inductive reactance resulting from the computed length differing slightly from the true resonant length will have a very small effect on the computation of network component values.

From the curves in Reference 3, the resistive component is 710 Ω, and the capacitive reactive component is −100 Ω. Again, these values are approximations. The formulas in Reference 2 assume that the load resistance R_2 is non-reactive. Whatever small reactance is present can be compensated for during final tuning. The equations for inductive and capacitive reactance are:[1]

$$X_L = R_2 \sqrt{\frac{R_1}{R_2 - R_1}}$$

$$X_C = \frac{R_1 R_2}{X_L}$$

The computed values for L and C are:

L = 0.60 µH
C = 16.9 pF

The air-wound inductor was made by winding seven turns of #12 wire, 1¼-inches long, on a ¾-inch diameter form. The form was then removed.

Construction

The construction details of the antenna are shown in Fig 1. The antenna consists of a length of 0.875-inch diameter aluminum tubing. The antenna element is mounted to a piece of 2×2-inch wood support with aluminum clamps fastened to standoff insulators. The standoff insulators are spaced about 12 inches apart on the wooden support. At my installation, the wood support is about 3 feet long with a piece of 1½×1½-inch aluminum angle bolted to the bottom 6 inches of the wood support. The aluminum angle is secured to a 1½-inch pipe mast by two hose clamps.

A 3-inch long by 1¾×¼-inch piece of plastic is mounted on the lower standoff insulator and provides the mounting for one end of the inductor and the end of the coaxial line to the antenna. A pair of solder lugs fastened to the bottom of the antenna provide connections for the shield of the coax capacitor and the other end of the inductor.

A second pair of solder lugs fastened to the lower end of the plastic support connect the end of the inductor to the shield of the coax transmission line feeding the antenna. The inner conductors of the coax capacitor and the coax transmission line are soldered together. The coax capacitor is taped to the

69

antenna with black electrical tape, and the end of the coax is sealed, after adjustments are completed. A 3-inch diameter coaxial choke coil using four turns is located as close as possible to the bottom of the 1/4-inch plastic support and taped to the wood support.

Adjusting for Minimum SWR

The network component values are adjusted by mounting the antenna about six feet above the ground on some form of support. The frequency for minimum SWR is determined. This should be close to 52 MHz. The turns of the inductor are compressed or expanded to minimize the SWR. Next, 1/8 inch is clipped off the coax capacitor, and the coil turns again adjusted for minimum SWR. This process is repeated until a minimum SWR value of 1.1 or less is obtained. If the SWR at each end of the 6-meter band is less than or equal to 1.5, then the adjustment process is completed.

Notes and References

[1]R. D. Straw, Editor, *The ARRL Antenna Book*, 17th Edition (Newington: ARRL, 1994), pp 2-3, 2-38.
[2]*The 1994 ARRL Handbook* (Newington: ARRL, 1993), p 2-55, Fig 87.
[3]Dr H. Jasik, Editor, *Antenna Engineering Handbook*, First Edition (McGraw-Hill Book Co, 1961), pp 3-4 to 3-5.

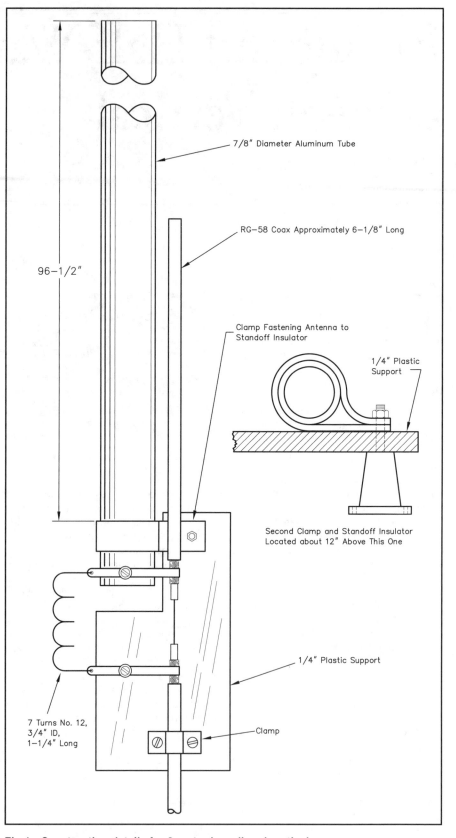

Fig 1—Construction details for 6-meter broadband vertical.

Antenna Modeling

The Effect of Slope on Vertical Radiation Patterns of a Horizontal Antenna

By William Alsup, N6XMW
1120 Ashmount Avenue
Oakland, CA 94610

The earth is not flat around my antenna nor, perhaps, around yours. This presents a problem analyzing our vertical directivity patterns. Those familiar patterns published in *The ARRL Antenna Book* and elsewhere always assume a flat earth. When the actual earth is not flat, the published patterns provide insights into the general relationship between antenna height and the shape of the pattern, but leave us guessing about the effect of sloping earth.

I set out to analyze that effect by considering the trigonometry of the problem and then writing a program in QBasic to draw directivity patterns as a function of different terrain slopes. This article shares my findings. As is usually the case with the published patterns, I assume perfect ground "reflections," ignore diffraction effects and use an individual ray analysis rather than a Fresnel zone approach. The calculations herein all assume a horizontal dipole.

The Single-Slope Case

The trigonometry of the problem is illustrated in **Fig 1A**. A is the antenna, viewed on end; G is the ground vertically under the antenna; and C is the point on the slope at which the reflected ray CH will be parallel to a given upward direct skyray AE. In order for the two to reach the same distant point, although usually with some phase difference, the two rays must be parallel. Because the direct skyray AE and the reflected ray CH have different routes to reach a common destination, they can, and usually

N6XMW examines how real terrain can affect real signals.

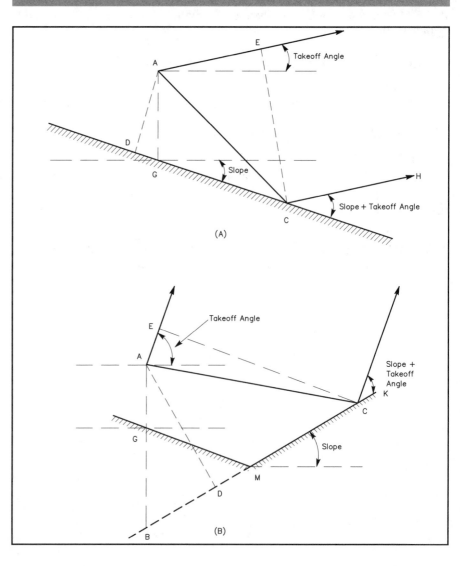

(A)

(B)

Fig 1 — At A, a single-slope terrain, showing definitions for associated angles and distances. At B, a multiple-slope case, showing two different slopes in the foreground, with associated angles and distances.

do, arrive out of phase. Indeed, the slight difference in routes makes a potentially enormous difference in the extent of phase coherence between the two overlapping rays. The two rays, of course, will add or subtract in strength depending on the degree of their phase coherence. At one extreme, when they are exactly in phase, they produce a signal twice the strength of the direct ray alone. At the other extreme, when they are exactly out of phase, they cancel each other and thus produce a null.

The exact degree of phase coherence is simply the difference in path lengths measured in wavelengths times 360°. This path-length difference equals AC minus AE. The rest of their paths are equal in length. (In determining phase coherence, remember that when the downward wave strikes the earth and is "reflected," it undergoes a 180° phase shift.) By the way, to reinforce the idea that a single ground reflection will combine to form a signal that is, at most, twice the skyray strength, all patterns shown are scaled in linear units, not in dB.

Now let's turn to some examples. Note that at the bottom of each pattern shown in this article is a terrain profile representing the ground configuration used to generate the directional pattern. This profile is a convenient way to keep track of the terrain shape under study. Obviously, for a single-slope, flat-earth case, the profile is a flat line.

The flat-earth case is a special case where the ground slope simply equals zero. **Fig 2A** and **2B** compare the patterns resulting from the flat-earth case versus the case in which there is a downward slope of 10°.

The lower lobe resulting from the downward slope, of course, is better for DX. Even a relatively low antenna can achieve a low DX angle when the foreground is sloping downward. (Please recall that the patterns apply to both receive and transmit.)

My own Oakland QTH illustrates the down-slope case. Located about one-third of the way up the East Bay Hills, my horizontal tribander (up 55 feet) looks out over a mile-long downward slope of about 5°. I do well operating over the Pacific. **Figs 3A** and **3B** superimposes the lower lobes (only) for 10, 15 and 20 meters based on an antenna height of 55 feet. In Fig 3A, I assume flat earth and in Fig 3B, a downward slope of 5°.

One of my most memorable QSOs seems to have benefited from this type of factor, although it was during a backcountry hike

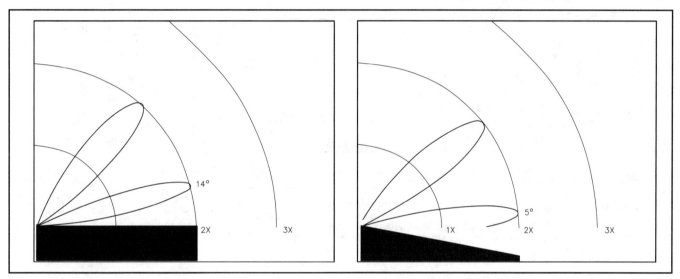

Fig 2 — At A, computed elevation response for a 1-λ high antenna over flat terrain at 14 MHz. The peak response is at about 14°. At B, computed elevation response for a 1-λ high antenna over terrain with long −10° slope. Note that the peak elevation response is now at 5°.

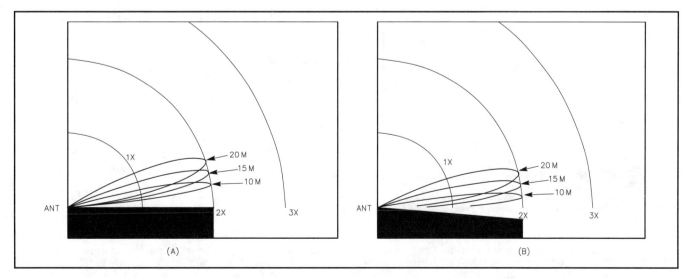

Fig 3 — At A, computed elevation response for an antenna mounted 55 feet above flat ground for 10, 15 and 20 meter bands. At B, elevation responses for a 55-foot high antenna mounted over ground with a −5° slope.

in the High Sierra, not from my home QTH. I took along a QRP rig with a 20-meter dipole. When the time came for an early evening sked with ZL2BIT, we heard each other very clearly. ZL2BIT could hardly believe that I was operating QRP in the Yosemite wilderness. In his direction, my antenna had the benefit of a long stretch of downward sloping foreground at about 5° with a limitless horizon.

Surprisingly perhaps, this advantage can be taken too far. A slope may be so steep that the lowest lobe, the most critical lobe for DX purposes, actually seems to drop *below* the horizon and into the earth. Put differently, if the ground slopes downward, it may actually be a mistake to raise the antenna too high, despite adages to the contrary! In **Fig 4**, the antenna is 2 λ high, over a downward slope of 10°. Compare Fig 4 with Fig 2B, which shows the same terrain but with the antenna at a 1 λ height. At a height of 2 λ, the peak of the lowest lobe has dropped below the horizon. (The computer program truncates all vertical radiation below zero degrees or the visible horizon, whichever is higher.)

When the tilt is upward, the pattern rotates upward. For both upward and downward tilts, note that the pattern shift does not shift exactly degree-for-degree with the tilt. The reason, for both cases, is that when the slope is tilted, the perpendicular height of the antenna above the direction of the slope is reduced. It is measured by AD in Fig 1A, not AG. The more the earth slopes, the more the perpendicular antenna height AD above the direction of the slope is reduced, for both the uphill and downhill slopes. This factor

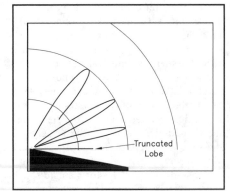

Fig 4 — Computed elevation response for a 28-MHz antenna placed 2 λ over ground with a –10° slope. The main lobe is aimed at the earth, showing that the antenna is too high for effective operating.

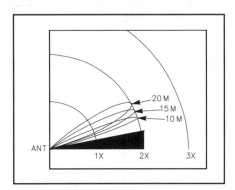

Fig 5 — Elevation response for a triband antenna mounted 55 feet above ground with a constant +10° upward slope. The main lobe has been shifted upward by the effect of the terrain.

tends to raise the lobes for both the uphill and downhill cases. In the downhill case, however, this factor is more than offset by the overall rotation of the overall pattern downward. In the uphill case, on the other hand, both the rotation of the overall patterns and the reduction in effective antenna height conspire to push the lobes higher.

Also note that with an uphill slope, the visible horizon is higher. The earth then acts as a barrier blocking radiation below the visible horizon. In my case, the East Bay Hills above my QTH block any low angles to the northeast. This is especially tough on my QSOs with Europe and Africa. **Fig 5** shows a simple 10° upward slope, where the lowest lobes for 10, 15 and 20 meters for an antenna height of 55 feet are superimposed on the same graph.

The Case of Multiple Slopes

To come closer to real-life terrain, I went one step further to allow for two or more grades in a single terrain. The trigonometry of this case is illustrated in Fig 1B. For the first slope, that is, the slope leading immediately away from the base of the antenna G, the calculations are the same as above. The differences start with the second slope, illustrated by the surface MK. If MK were extended, it would be directly under the antenna at B, thus giving an antenna height of AB relative to the direction of the second slope. As before, it is necessary to find the perpendicular height (AD) above the second slope (BC), along the pretended extension of MK. However, we must be careful to count only those rays that strike the exposed surface, MK.

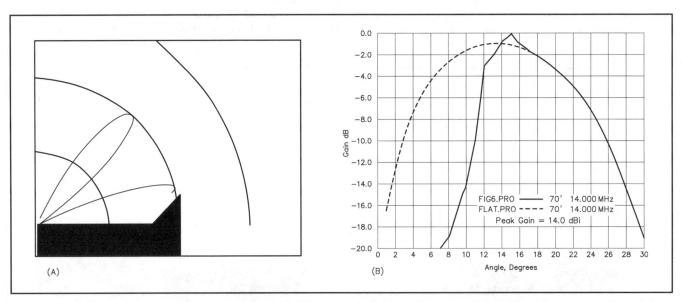

(A)

(B)

Fig 6 — At A, elevation response for a terrain that is flat for a considerable distance (160 λ) in front of the antenna, then rising at a steep 45° slope to block the signal. The 14-MHz antenna is mounted 1 λ above the terrain. [Note that due to diffraction, a tiny signal actually does manage to "sneak over" the top of the obstruction, even though it is very weak at low elevation angles.—Ed.] At B, elevation response for the same terrain, but where both reflection and diffraction has been taken into account for a four-element Yagi. The response is compared to that of an identical antenna over flat ground.

For each takeoff angle, we must check for a parallel reflected ray from each slope in the terrain profile. The phase difference of each such reflection must then be considered and a composite of all parallel rays computed. For each takeoff angle, we must also check for multiple reflections, where a reflected ray eventually leaves the earth, often striking it more than once, parallel to the takeoff angle. We must still remember that each reflection along the way causes a 180° phase shift.

An important multi-slope case is illustrated by **Fig 6**, consisting of a long, flat terrain, followed by a large, distant ridge (in this example, starting 160 λ away, and rising at a steep 45° angle for 40 λ). When the natural horizon is blocked by such a substantial hill, no radiation, whether direct or reflected, can fall below the ridge. [Note: A full reflection analysis including diffraction reveals that a tiny amount of signal does manage to "sneak over" the edge of the ridge. In general, the effects of real-world diffraction tend to "fill in" somewhat the deep nulls and "round off" the extreme peaks computed using reflection-only techniques.—Ed.]

To maximize DX opportunities, this suggests that when a relatively modest ridge blocks the natural horizon but where the foreground is relatively flat, the antenna should be raised so that the peak of the lowest lobe just clears the ridge. Raising the antenna too high, however, will lower the lobe right into the hillside, somewhat as we found in Fig 4.

Another important multi-slope case involves extraordinary ground focusing. The potentially remarkable effect of such focusing is illustrated by **Fig 7**, involving a slope down to a flat plain. In this particular illustration, the antenna is relatively low, only 0.53 λ high. The antenna is on a downward 20° slope sited 3.95 λ (measured horizontally) from the flat plain. Even with this low antenna, a low, broad and powerful (3× plus) lobe peaks at 8° above the horizon while another wide lobe (2× plus) peaks at 22°. This is remarkable because the same antenna at the same height over flat ground would only produce a much higher lobe, peaking at 28°!

In this connection, L. A. Moxon, G6XN, reports that a 4× lobe would be obtained in such a two-slope model as long as

$$AD = \frac{Wavelength}{4 \times \sin(TakeoffAngle + Slope Angle)}$$

$$HS = \frac{Wavelength}{4 \times \sin(TakeoffAngle)}$$

where AD is the AD in Fig 1B and HS is the height above the plane of the point at which

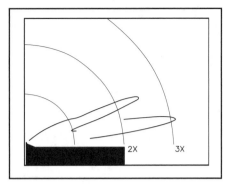

Fig 7 — Elevation response for a 14-MHz antenna mounted at a height of 0.53λ on a small, steep hill facing a wide plain. The terrain has a –20° slope for 4 λ, and flat from then on. This response compares favorably with an antenna mounted at a height of 1 λ over flat ground.

the reflected ray leaves the sloping surface parallel to the takeoff angle.

Why the extra gain? In the 4× case, it is because each surface has a single reflection and one double reflection, and all three are in phase with the skyray. A longer lobe may thus form from the multiple reflections and direct skyray, similar to Fig 7. The lobes so formed, however, are not always broad as in Fig 7.

Consider the terrain in **Fig 8**. The lobes shown are so narrow that they are needle-like. [Again, a full diffraction plus reflection analysis tends to fill in the extreme nulls. See Fig 8B.—Ed.] In this situation, signal strength can vary from very low to maximum strength with only a slight change in angle. Perhaps more importantly, this example leads to an interesting observation—a nearby downward slope can more than compensate for a distant upward slope.

In Fig 8 the terrain drops downward at –5° for 20 λ. Then the ground rises upward at +5° for another 80 λ, such that the peak of the ridge is higher than the antenna. What surprised me is that the resulting profile, despite the ridge, is more favorable for low-angle radiation than the flat-earth case with the same antenna height (see Fig 2A, the flat-terrain case). The ridge does block the angles below the ridge, but the foreground downward slope forms a lower beam than in the flat-earth case, a lower beam whose peak still clears the top of the hill. We saw a similar effect in Figs 6 and 7.

This may help explain why we have all experienced a sudden loss of a strong signal. One or both of the stations in a QSO may project a lobe that peaks in strength right at a nearby ridge line, as in Fig 6. A small shift in the ionosphere may move the relevant angle from the strong tip of the lobe, which clears the ridge, to a slightly lower angle, which does not. A shift in the opposite di-

rection may result in signals seemingly coming in from nowhere to full strength. A similar effect may occur when several slopes contribute to formation of the lobe such that, as mentioned, each lobe is narrow and long with a powerful tip. A small shift in the ionosphere may result in large swings in the signal strength.

From the foregoing, the effect of sloping terrain is clarified in three important respects. The first observation should go without saying, but let's say it anyway for emphasis — little or no radiation will fall below the horizon as seen from the antenna. That is, all radiation by direct and reflected rays will remain above the horizon (again, ignoring diffraction effects).

The second observation is that the slope of the nearby terrain often determines the basic shape of the pattern. In fact, except for very low takeoff angles and multiple reflections of the same ray, precious little far terrain matters in determining the vertical radiation pattern. To combine this with the first observation, the main role of the nearby ground reflection is to determine the shape of the radiation pattern. The main role of the far terrain is to determine how much of that pattern, if any, will be blocked. Again, for DX purposes, we should try to adjust our antenna height or site our antennas such that a substantial lobe just clears a distant ridge or mountain range.

The third observation is that the presence of multiple reflecting grades creates the opportunity for extraordinary gain due to the focusing of multiple ground reflections in the same vertical direction. This can create powerful lobes. The greater the number of contributing reflections in a given takeoff angle, the more needle-like the lobes can exist and the less stable and useful they will be as the ionosphere shifts.

By the way, in thinking about the shape of your own terrain, think on a large scale. Ignore small deformations and consider the general sweep of the landscape. It is possible to put too fine a point on the analysis. Small deformations ought to average out in favor of the general sweep of the landscape.

Flat-Earth Equivalent Antenna Heights for *IONCAP* and *CAPMAN*

Another use of this type of analysis, assuming you have a sloping foreground, is to determine a "flat-earth equivalent" that reasonably fits the lower lobe generated by your actual environment (the lobe that really matters for DX calculations). This is particularly useful for adjusting inputs for antenna heights for propagation programs such as *IONCAP* or *CAPMAN*.

IONCAP and *CAPMAN* (see Dec 1994 *QST*, p 79) predict signal strength at the receiving end of a proposed circuit, such as Oakland to Boston. The user selects a trans-

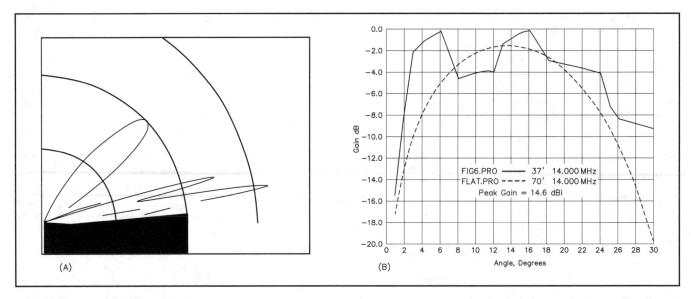

(A)   (B)

Fig 8 — At A, elevation response for a 14-MHz antenna mounted 1 λ over a more complex terrain having two slopes. The first part has a −5° slope for 20 λ, followed by a +5° slope for 100 λ. The response is focused by the shape of the terrain to produce a stronger signal at certain elevation angles. The nulls predicted by this analysis are about 6 dB down on a logarithmic rather than linear scale. [At B, a full diffraction-plus-reflection analysis shows nulls about 4 dB down. — *Ed.*]

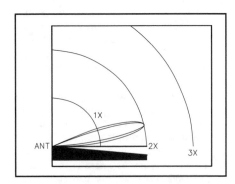

Fig 9 — Elevation patterns of a 55-foot high antenna at N6XMW's QTH toward the West, compared with flat-earth pattern for a 75-foot high (1.14 λ at 14 MHz) antenna. The patterns indicate the equivalence of the terrain/height combinations.

Table 1

Flat-Earth Equivalent Over East Oakland, Toward Pacific (Downhill Case) for Tip of Lower Lobe

Band	Actual Height	Flat-Earth Equivalent
10 meters	55 feet	130 feet
15 meters	55 feet	90 feet
20 meters	55 feet	75 feet

Note that while the actual antenna height in all cases is the same, the flat-earth equivalents differ for each band. The 10-meter case illustrates a limitation of this analysis. The tip of the "actual height" and "flat-earth" lobes are equal but the overall shape of the "flat earth" lobe is narrower. This could lead to substantial error for angles much off the peak.

Table 2

Flat-Earth Equivalent for East Oakland Toward Boston (Uphill Case) for Tip of Lower Lobe

Band	Actual Height	Flat-Earth Equivalent
10 meters	55 feet	36 feet
15 meters	55 feet	39 feet
20 meters	55 feet	43 feet

mitting and a receiving antenna, such as a horizontal dipole or Yagi. For each, the user specifies a height — above flat ground. The program then calculates the transmitter gain based on the particular path and the vertical directivity patterns associated with each antenna. Obviously, a circuit using tips of the lobes has more gain than one that is, say, half-way down the lobe, all other things being equal.

For example, as stated, my horizontal tribander is 55 feet above ground. However,

my QTH enjoys a long 5° slope downward toward the West. This slope will produce a lower lobe peaking at 13° on 20 meters. The flat-earth equivalent for the lower lobe is a horizontal tribander at a height of 1.14 λ, which translates to 75 feet for that band. **Fig 9** shows the two lobes overlapping. The fit is imperfect but reasonable.

Since the lower lobe is of the most impor-

tance for DX, substituting a flat-earth equivalent height for the actual height may give more realistic results. Table 1 gives the equivalents for the 10, 15 and 20-meter bands for QTHs like the one described above; that is, a long downward slope of about 5°.

In addition, if the horizon is blocked by a ridge or mountain, then it is necessary in such signal tracing programs to adjust for the obstacle. For *CAPMAN* and *IONCAP*, the user may eliminate all paths that fall below a specified angle (up to 10°).

Notes and References

L. A. Moxon, *HF Antennas For All Locations*, 2nd edition (RSGB: 1991), p 140.

The OpenPF Plot-File Standard

By Brian Beezley, K6STI
3532 Linda Vista Dr
San Marcos, CA 92069

Ever wrestle with trying to interpret a software standard while writing a program? K6STI gives details and practical examples about how to use the OpenPF Plot-File Standard in your own programs.

In the fall of 1994, at the urging of Dean Straw, N6BV (editor of this compendium), I began looking into modeling antennas over irregular terrain. Available antenna-modeling algorithms either assumed perfectly flat ground (often unrealistic) or ground with step-changes in elevation (unrealistic and ignores diffraction and shadowing). I wanted to generate a free-space elevation pattern with an antenna-analysis program and then feed the pattern to a terrain-analysis program. The terrain program would combine the pattern with topographic data to calculate realistic elevation patterns for the antenna as installed at an actual site.[1]

I quickly realized that free-space pattern phase as well as magnitude was needed because irregular terrain can combine antenna rays emitted at different angles into one ray in the far field. The magnitude of the combined ray depends directly on the relative phase of its constituents.[2] However, plot files generated by existing antenna-analysis programs did not accommodate phase data. Rather than modify an existing file format, I decided to create a new one that not only provided for phase data but was more general in many other ways. I hoped that other software authors might adopt this file format so that radiation patterns could be transported easily among antenna-analysis, terrain-analysis, data-recording, and pattern-plotting programs. The result was the OpenPF Plot-File Standard.[3]

I wanted OpenPF to handle the most complex patterns produced by *NEC*, *MININEC*, and Yagi programs. But I also wanted the standard to be easy to implement for simple data. For example, if a ham rotated his Yagi and recorded the approximate pattern using his receiver's S-meter, I wanted it to be easy for him to write a simple BASIC program to generate a plot file from the measurements.[4] He could then use any OpenPF plotting program to display the measured pattern and

perhaps compare it with others.

This article contains examples, pointers, and tips for writing software that uses OpenPF. Except as background, I don't think the information presented here will be of much interest to someone who simply uses software that implements the standard!

The text of version 1.0 of the OpenPF standard is reproduced at the end of this article. You'll also find it in OPENPF.DOC on the disk that accompanies this compendium. (OPENPF.DOC is formatted for printing. You can print the file with the DOS PRINT command by typing PRINT OPENPF.DOC. You can examine it on your screen with the *V.EXE* file viewer by typing V OPENPF.DOC.) Future OpenPF versions will be posted on the ARRL BBS and at the ARRL Internet ftp mirror sites. Look for file names like OPENPF11.ZIP, OPENPF20.ZIP, etc.

Data Organization

OpenPF provides for both simple and complex data by using data blocks. All data in an OpenPF file is organized into blocks. Each block begins with a data-type identi-

fier. When generating an OpenPF file, you need only write blocks that describe your data. You don't need to write extraneous data blocks that don't apply. Similarly, when reading an OpenPF file, a program only must read data blocks of interest. It can ignore blocks it doesn't need or isn't programmed to understand. These include data blocks defined in future revisions of OpenPF.

OpenPF data blocks provide another nice feature. The length of each block is given right after the block type. Programs that use this length instead of an assumed length to find the file location of the next data block are very likely to remain compatible with future OpenPF versions. Future versions may define additional information fields at the end of existing data blocks. However, the block length will reflect the new block size. Therefore, programs that use this length will automatically skip over newly defined fields. Provided that the new fields contain only ancillary information, older programs should be able to read files written with newer versions of OpenPF.

The OpenPF standard allows for nonstandard data. The upper half of the block-identifier number space is reserved for non-standard blocks. Software authors can use these block numbers to define data blocks not intended to be read by all programs. This allows OpenPF to be customized. (For widest interoperability, you should avoid doing this whenever possible.) If you do make use of this feature, consider embedding a signature in your private data block. The nonstandard block identifier you pick may be used by another program. A four-byte signature will provide virtual certainty that the nonstandard block you read is really yours.

Finally, the block organization allows for future extensions to the standard without the overhead of reserved data fields. New fields simply can be defined at the end of a block and the block length adjusted accordingly.

Example: A Simple OpenPF File

At this point I'm going to assume that you've read the text of the OpenPF Plot-File Standard. Here's a recipe for writing a simple OpenPF file that contains the pattern of a Yagi measured every 10°:

First, let's create a simple file header. For OpenPF version 1.0, the version byte should be 16. We won't annotate this simple file, so set the header length to 8 and the source, title, environment, and notes lengths to 0. Note that the notes length is a word, not a byte. This allows for really long-winded comments!

Now, let's create a single data block for our measurements. We'll pick block type 1 (total magnitude data). Total field implies that each data point was taken using the transmit or receive polarization (possibly elliptical) that maximized the signal. Although it's very unlikely that we actually did this, block type 1 is the most generic polarization category. It can represent any antenna polarization. Quick-and-dirty plotting programs may respond to just this single block type. So, although it's not strictly accurate to say that our data represents the total field, let's use block type 1 anyway for maximum generality.

We'll come back to block length later after we determine its value. We've decided not to annotate our data, so set the title, environment, and notes lengths to 0 (again, notes length is a word, not a byte). Next, write the floating-point frequency in MHz. The next byte is 0 for azimuth data. Write a floating-point 90 for the plane angle. (We'll assume that the response was measured at 0° elevation angle or 90° zenith angle.) Let's assume that we've taken data over 360° and not made any symmetry assumptions, so set the symmetry byte to 0. Next, we have 36 measurements if they're 10° apart, so write a word containing 36.

Let's arbitrarily align the X axis with the main beam and begin writing our measurements from there. OpenPF defines 0° to be along the X axis, so write a floating-point 0 for the first angle. The angular increment should be a floating-point 10. Finally, we write 36 floating-point values that represent our measurements beginning at north and proceeding counterclockwise. (We don't go clockwise because OpenPF uses a standard coordinate system that defines azimuth angle as increasing in the counterclockwise direction. We could supply our measurements for increasing compass angle by specifying –10 for the angular increment.) We've elected not to annotate the file so the title, environment, and notes strings don't exist. Finally, add up the byte length of everything in the block (you should get 171) and write this number in the block-length field.

That's all there is to creating a simple OpenPF file. Any general-purpose OpenPF plotting program, like *PLOT.EXE* on the included disk, should be able to read and display the azimuth pattern we've generated.[5]

Not Specified

Nowhere does OpenPF define the orientation of the XY axes with respect to compass direction or screen presentation. It only tells you that the Y axis is counterclockwise from the X axis. Although you may want to assign the +X direction as north, your plotting program may display this direction to the right on the screen, like typical X-Y graphs in mathematics. In this case radiation patterns for fixed antennas will be difficult to correlate with maps that have true north at the top. If your plotting program won't reassign the display axes as *PLOT.EXE* does, you can redefine the angle for the first data point as 90° instead of 0°. (**PLOT.EXE** places +X toward the top of the screen.)

An Ambiguity

The standard states that header lengths less than 3 are invalid. This might reasonably lead you to expect that any length greater than 2 is valid. However, the source, title, environment, and notes length fields aren't described as optional like the strings themselves, so the minimum header length is 8. I regard this as an ambiguity in the standard. To avoid confusion, the standard either should state that header lengths less than 8 are invalid or that the string-length fields are optional. Robust OpenPF file parsers can accommodate the ambiguity by accepting all header lengths greater than 2 and assuming that missing string lengths are 0. Conversely, to be safe you should generate OpenPF files with an explicit entry for each of these lengths.

A Numerical Detail

A note in the OpenPF text in the context of

ellipticity data states that the floating-point code for minus infinity is valid. This value can occur at points where an elliptically polarized field collapses to a linear field and the ellipticity becomes zero (minus-infinity dB). The note doesn't mention that this situation may also occur for magnitude data. For example, an antenna-analysis program may generate a value of minus-infinity dB for the response of a free-space dipole off its ends. Since this value is defined in IEEE-754-1985, the floating-point standard that OpenPF uses, this is a valid data point. A general OpenPF file parser should be prepared to handle this value. However, some computer languages may generate a floating-point exception when infinity is encountered. This can be a nuisance. When generating an OpenPF file, you may wish to substitute a value like –200 for minus-infinity dB. This will make life easier for quick-and-dirty file parsers without compromising the data in any practical way.

Another Numerical Detail

It's possible that the angular increment can't be exactly represented as a binary floating-point number. Repeated use of an inexact number can result in accumulated round-off error. For example, let's say you generate 3600 data points with 0.1-degree resolution (it's a government project and you need to use up this year's budget). You write 0.1 in the angular-increment field. You also write 0.1 for the first angle. If the plotting program accumulates the floating-point representation for 0.1 3599 times, it may get 359.9128° as the final angle, not 359.9°. It's better to multiply the angular increment by the index of each data point. This will ensure that round-off errors don't accumulate.

Polarization—Complete Field Descriptions

The list of far-field block types in OpenPF might suggest that 16 blocks are needed for a complete field description that includes all polarization components and metrics. Actually, all of this data can be derived from just four blocks: Horizontal and vertical magnitude and phase. Plotting programs that do the derivations themselves can greatly reduce the size of polarization-complete OpenPF files.

Character Set

If you'd like to get fancy and use characters in OpenPF strings beyond those provided in the 128-character ASCII code, note that OpenPF explicitly uses the PC-8 symbol set. This is the 256-character code provided by default DOS code page 437. This set includes some foreign characters, a few currency symbols, a couple of fractions, typographic marks, area-fill characters, line-drawing characters, Greek math symbols, and a few other odds and ends. If you're generating OpenPF files with a Windows program, note that the default Windows character set is not PC-8.

Absolute Fields

OpenPF provides for absolute fields as well as relative fields. Absolute fields are defined at specific locations in three-dimensional space and the field values depend on input power. To accommodate existing *MININEC* and *NEC* output, field locations can be specified in rectangular, cylindrical, or spherical coordinates. Data for block types 64 through 69 can be derived by a smart plotting program when given data for blocks 70 through 81. Note that absolute-field symmetry is defined somewhat differently than symmetry for relative fields.

OpenPF closely follows *NEC* conventions for absolute fields but fails to explicitly specify whether fields in volts/meter or amps/meter use RMS or peak values. To date, all OpenPF files containing absolute-field data have been generated by *NEC*, so you can safely assume peak values. This omission will be corrected in the next revision of OpenPF.

It's not easy to display four-dimensional data (three coordinates plus a data value) in two dimensions. OpenPF doesn't specify how data is to be displayed. *PLOT.EXE* employs a very general solution to this problem for absolute-field data—it ignores it.

Acknowledgment

I'd like to thank Roy Lewallen, W7EL, and Jim Breakall, WA3FET, for reviewing early drafts of OpenPF and making valuable suggestions for improvement.

Notes and References

[1] The result was the **TA** 1.0 Terrain Analyzer.

[2] It turns out that the radiation phase for nearly all antennas varies little in the forward lobe. At low angles over most terrain, very little error is introduced by assuming constant phase. However, this is not true at higher angles. Therefore, you can safely ignore antenna phase if your primary interest is low-angle DX. But for short-hop, NVIS applications, antenna phase (and ground quality) are quite important.

[3] The name OpenPF was inspired by that of OpenGL, a graphics-language standard. Both are open, nonproprietary standards intended to be widely used.

[4] It's no longer necessary to write such a program. GENPF.EXE on the disk that accompanies this compendium generates an OpenPF plot file from measured data you supply.

[5] The first OpenPF file parsers implemented in **AO** 6.5, **YO** 6.5, **NEC/Wires** 2.0, and **NEC/Yagis** 2.5 were restricted to files with data resolution of 1 or 2 degrees because these programs only generated such data.

OpenPF Plot-File Standard
Version 1.0   June 1, 1995

OpenPF is an open, nonproprietary standard for computer
files that contain electromagnetic-field data. OpenPF is
intended for files created by antenna-analysis programs and for
files containing measurements. OpenPF files can be used to
archive or transport data. Programs can use the files to plot
fields and radiation patterns on a screen, printer, or plotter.

The OpenPF file format is extensible. Data is organized
into blocks that begin with a block-type identifier followed by
block length. This structure allows a file parser to skip and
ignore blocks it's not designed to handle, including block types
defined in future revisions of OpenPF.

The blocks defined below provide for 2-D relative far-
field data (azimuth and elevation radiation patterns) and 3-D
absolute near- and far-field data. 3-D radiation patterns can
be represented by a set of 2-D data blocks.

The file format enables a single file to contain near-
and far-field data, absolute and relative data, azimuth and
elevation data, low- and high-resolution data, partial- and
full-data cuts, data for multiple frequencies, and unidentified
data.

The file format provides for new fields in future OpenPF
revisions without the overhead of reserved bytes. New fields
can be appended to any file block. File parsers that use block
length to advance to the next block will skip over new fields.

OpenPF uses standard coordinate systems. For
rectangular coordinates, X and Y are in the horizontal plane and
Z is height. For spherical coordinates, R is radial distance,
phi is azimuth angle, and theta is zenith angle. For
cylindrical coordinates, rho is horizontal distance, phi is
azimuth angle, and z is height. Azimuth angle is 0 along the X
axis and increases in the XY plane toward the Y axis
(counterclockwise). Zenith angle is 0 along the Z axis and
increases in the Z-phi plane toward the phi ray. Units are
meters and degrees.

OpenPF files use the Intel-processor "little-endian"
data convention (least-significant byte first for multibyte
data). Bit number 0 is the least-significant bit. Bits marked
as reserved are undefined and should not be assumed to be 0.
OpenPF files use the file extension ".PF".

In the descriptions that follow, word means a 16-bit
unsigned integer and FP means a 32-bit, single-precision,
floating-point number that conforms to the IEEE-754-1985
standard. ASCII means a character string that uses the
extended-ASCII PC-8 symbol set.

Header

OpenPF files begin with a block of the following format:


```
            Version               byte
            Header length         word
            Source length         byte
            Title length          byte
            Environment length    byte
            Notes length          word
            Source                ASCII
            Title                 ASCII
            Environment           ASCII
            Notes                 ASCII
```


Version

        This byte identifies the OpenPF version used to generate
the file. The version number consists of a major revision (an
integer) and a minor revision (an integer between 0 and 9)
separated by a decimal point. The upper nibble (4 bits)
contains the major revision and the lower nibble the minor.

Header Length

        This field is the total length of the header in bytes.
File parsers should use this length to advance past the header
instead of assuming that data blocks begin immediately after the
notes string. Future OpenPF revisions may define additional
fields after the notes string. Header lengths less than 3 are
invalid.

Source, Title, Environment, and Notes Lengths

        These fields specify the lengths in bytes of the ASCII
strings. Zero is valid.

Source String

        This string identifies the data source. This may be an
antenna-analysis program, an antenna-test range, a laboratory,
etc. No bytes are allocated for a zero-length string.

Title, Environment, and Notes Strings

        The title string describes the antenna. The environment
string describes the antenna environment (free space, antenna
height, ground constants, etc.). The notes string contains
auxiliary information that might be displayed only upon request.
Data blocks that follow may contain strings that override these
global strings. No bytes are allocated for zero-length strings.

Data following the header is organized into blocks. Each data block begins with block type (a byte) followed by block length (a word). Block types 0 through 127 are standard and reserved. Programs may use types 128 through 255 for special purposes, but block length must always follow block type so that file parsers can skip over the block. Blocks may appear in any order. The following standard blocks are defined:

NOP Block

Block type	byte (0)
Block length	word
Rest of block	

Block type 0 is a no-operation block. Its purpose is to provide a standard ignore-me block. Normally this block isn't used, but it may be convenient when patching a file. Block length is the total length of the block in bytes. Block lengths less than 3 are invalid.

Relative Far-Field Blocks

Block type	byte (1-16)
Block length	word
Title length	byte
Environment length	byte
Notes length	word
Frequency	FP
Plane	byte
Plane angle	FP
Symmetry	byte
Number of points	word
First angle	FP
Angular increment	FP
Data points	each FP
Title	ASCII
Environment	ASCII
Notes	ASCII

Block Type

Block types 1-16 contain planar cuts of relative far-field data as follows:

Block	Data	Units
1	Total magnitude	dBi
2	Horizontal magnitude	dBi
3	Vertical magnitude	dBi
4	Right-circular magnitude	dBic
5	Left-circular magnitude	dBic
6	Major-axis magnitude	dBi
7	Minor-axis magnitude	dBi

8	Ellipticity	dB
9	Total phase	degrees
10	Horizontal phase	degrees
11	Vertical phase	degrees
12	Right-circular phase	degrees
13	Left-circular phase	degrees
14	Major-axis phase	degrees
15	Minor-axis phase	degrees
16	Polarization tilt	degrees

Ellipticity is the length of the minor axis of the polarization ellipse divided by the length of the major axis expressed in dB (the IEEE-754-1985 floating-point code for -infinity is valid). Polarization tilt is the counterclockwise angle between the major axis and the E(theta) direction.

Block Length

This field is the total length of the block in bytes. Block lengths less than 3 are invalid.

Title, Environment, and Notes Lengths

These fields specify the lengths of the ASCII strings in bytes. Zero is valid.

Frequency

Frequency is in MHz.

Plane

This byte is 0 for azimuth data (constant zenith angle) and 1 for elevation data (constant azimuth angle). Other values are reserved.

Plane Angle

This field is the zenith angle for azimuth data and the azimuth angle for elevation data.

Symmetry

The bits of this byte have the following meaning when set:

Bit	Azimuth Data	Elevation Data
0	X symmetry	XY symmetry
1	Y symmetry	Z symmetry

Symmetry means that the data has the same value when reflected about the line or plane indicated. For azimuth data, bits 0 and 1 refer to the X and Y axes. For elevation data, they refer to the XY plane and the Z axis. Both bits may be set simultaneously. Bits 2-7 are reserved.

Number of Points

This field is the number of data points that follow. 0 is valid.

First Angle

This field is the angle of the first data point.

Angular Increment

This field defines the sign and magnitude of the angular difference between successive data points.

Data Points

The data set begins at the first angle. The angles of successive data samples differ by the angular increment.

Title, Environment, and Notes Strings

When string length is nonzero, the string overrides the corresponding header string for this block. When zero, the header string applies for this block and no string space is allocated.

Absolute Near- and Far-Field Blocks

Field	Type
Block type	byte (64-81 & 96-101)
Block length	word
Title length	byte
Environment length	byte
Notes length	word
Frequency	FP
Power	FP
Coordinate system	byte
Symmetry	byte
Number of A points	word
First A	FP
A increment	FP
Number of B points	word
First B	FP
B increment	FP
Number of C points	word
First C	FP
C increment	FP
Data points	each FP
Title	ASCII
Environment	ASCII
Notes	ASCII

Block Type

Block types 64-81 contain absolute fields. Block types 67-81 are field components resolved in rectangular coordinates. These blocks are defined as follows:

Block	Data	Units
64	Power density	watts/square-meter
65	Peak E magnitude	volts/meter
66	Peak H magnitude	amps/meter
67	Px Poynting vector	watts/square-meter
68	Py Poynting vector	watts/square-meter
69	Pz Poynting vector	watts/square-meter
70	Ex magnitude	volts/meter
71	Ey magnitude	volts/meter
72	Ez magnitude	volts/meter
73	Hx magnitude	amps/meter
74	Hy magnitude	amps/meter
75	Hz magnitude	amps/meter
76	Ex phase	degrees
77	Ey phase	degrees
78	Ez phase	degrees
79	Hx phase	degrees
80	Hy phase	degrees
81	Hz phase	degrees

Block types 96-101 contain absolute electric-field components resolved in spherical coordinates as follows:

Block	Data	Units
96	E(R) magnitude	volts/meter
97	E(phi) magnitude	volts/meter
98	E(theta) magnitude	volts/meter
99	E(R) phase	degrees
100	E(phi) phase	degrees
101	E(theta) phase	degrees

Use block type 65 for peak E magnitude.

Block Length, String Lengths, and Frequency

These fields are the same as for relative far-field blocks.

Power

This field is antenna input power in watts.

Coordinate System

This byte specifies the coordinate system used to locate data points (it does not specify the coordinate system used to resolve field components). The byte is 0 for rectangular coordinates, 1 for spherical coordinates, and 2 for cylindrical coordinates. Other values are reserved. Coordinates are defined as follows:

Coordinate System		A	B	C
0	Rectangular	X	Y	Z
1	Spherical	R	phi	theta
2	Cylindrical	rho	phi	z

Symmetry

The low-order bits of this byte have the following meaning when set:

Bit	Meaning
0	A-coordinate symmetry
1	B-coordinate symmetry
2	C-coordinate symmetry

Here, symmetry means that the data has the same value when the coordinate is negated (this definition is not the same as that for relative fields). Any of the bits may be set simultaneously. Bits 3-7 are reserved.

Number of A, B, or C points

These fields specify the number of points for each coordinate. 0 is invalid. The total number of data points is the product of A, B, and C. For rectangular coordinates, the field is computed over a point, line, rectangular surface, or rectangular volume. For spherical coordinates, the field is computed over a point, line, circle, disc, spherical surface, spheroid, or portion thereof. For cylindrical coordinates, the field is computed over a point, line, circle, disc, cylindrical surface, cylindrical volume, or portion thereof.

First A, B, and C

These fields specify the starting values of each coordinate.

A, B, and C Increments

These fields define the sign and magnitude of the difference between successive coordinate values.

Data Points

Points are ordered as follows: A cycles from first to last, then B, and finally C. The coordinate values of successive data samples differ by the A, B, or C increments.

Title, Environment, and Notes Strings

These fields are the same as for relative far-field blocks.

Notes

The Gain of an End-Fire Array

By Darrel Emerson, AA7FV, G3SYS
3555 E. Thimble Peak
Tucson, AZ 85718
e-mail: aa7fv@amsat.org
(CompuServe): 74010,2230

Everyone knows that (usually) the bigger an antenna the higher its gain. This article discusses the maximum theoretical antenna gain of an antenna of a given size or of a given number of discrete elements, and whether that gain is ever likely to be achieved in practice. Although the emphasis is on *end-fire* arrays, which include Yagi and helical designs, much of the discussion is more generally applicable. The theory is compared with some standard, practical designs.

> *AA7FV takes you through a lively and informative discussion on the theoretical and practical limits of gain from end-fire arrays.*

Introduction

Radio amateurs are always interested in generating the biggest signal possible within the limitations of their equipment and their license. The antenna is probably the most critical part of the entire amateur station. A higher-gain antenna may enable lower power to be used from the transmitter on a given communication link. The correspondingly narrower beam concentrates the energy where it's most useful, at the intended receiving station. This reduces the likelihood of the transmission causing unintended interference to other stations, gives a stronger signal to the receiver, and reduces interference that might be received from unwanted signals on the same frequency but from different directions. Everybody wins.

For optimum received signal-to-noise ratio (S/N) the sidelobe response of an antenna may be even more important than its forward gain. This is particularly true in receiving weak signals from space at UHF and higher frequencies. It is possible, even common, for an antenna to have relatively high forward gain, and yet for perhaps half the total radiated power to go into the sidelobes. It's the sum of the power in *all* the sidelobes that counts here, not just the peak in any one sidelobe. If such an antenna tracks a satellite in the direction of cold sky, many of the sidelobes are likely to pick up the relatively strong ground radiation, which may dominate the total noise at the receiver. An antenna with somewhat less forward gain, but with much lower *average* sidelobe response, may give a much higher S/N. EME operators are well aware of these considerations.

A well-designed antenna will take all these factors into account. There is usually a trade-off between optimizing for antenna gain versus antenna sidelobe response. Antenna bandwidth, efficiency, tolerances and ease of construction are all important parameters for a practical antenna.

However, this article considers only the antenna gain, and gives some examples of what can theoretically be achieved when a design is optimized to give the maximum possible gain, disregarding other factors. This article does not aim to give any practical solutions or recommendations, but looks only at what could in principle be achieved. Practical details, like how to feed the antennas, and the effect of the ground or of real losses in the antenna, are ignored. Although the emphasis is on end-fire arrays, much of the discussion is more generally applicable.

Ordinary Antenna Gain

While this article concentrates on the end-fire array, it is worth summarizing the characteristics of some other antenna arrays, to serve as a basis for comparison.

A Random Array of Phased Dipoles

The equations introduced in this section may all be found in standard antenna texts.[1,2] The normal approach to designing an antenna is to arrange all components of the antenna to generate radiation that adds constructively in the desired direction. Because of the physical separation of the components of the antenna, the radiation will usually only add completely constructively in one direction. In other directions, the components of radiation from different parts of the structure may add destructively, giving minima in the far field radiation pattern. The larger the antenna, the narrower is the angle over which the radiation adds more or less constructively, and so the narrower the width of the main beam and the higher the forward gain of the antenna.

For antennas made of a discrete number of elements, such as parallel half-wave dipoles, the coupling between the elements has to be considered—the mutual impedances are important in considering the resultant antenna gain. However, if the parallel dipoles are far enough apart that mutual impedances may be neglected, and if all dipoles are fed with equal currents phased to add constructively in one direction, then the power gain of this phased array over one element alone is exactly equal to the number of elements.

The reason is simply that, for n elements carrying equal currents, the in-phase far field will be n times greater than that from a single element. Since the power density in the far field goes as the field squared, the far field power density has increased by n^2. The power radiated from each dipole is the same, so the total power supplied by the transmitter has only increased by a factor of n in feeding the n dipoles. The normal power gain of an n-element array, with respect to a single element, is then simply n.

Mutual impedances become important with element spacings of about a half-wavelength ($\lambda/2$) or less. If the elements are packed more closely than this, even with the radiation from each element still adding in phase in the chosen direction, the gain may be very different from n.

A Broadside, Linear Array

This could be a straight-line array of parallel half-wave dipoles, perhaps much closer to each other than a half wavelength, all fed with equal currents exactly in phase. Because of the mutual impedances, we cannot simply use the total number of elements n to calculate the gain. Standard antenna texts often consider a slightly simplified case, where the individual elements are not half-wave dipoles, but hypothetical isotropic elements. This makes the mathematics slightly easier, and sometimes gives more insight into the operation of the antenna. The standard texts (eg, Reference 2, p 3-7) show that the directive gain of a linear phased array of isotropic elements, radiating in the broadside direction, is given by:

$$Gb = \frac{(k \times d)n^2}{(n \times k \times d) + 2x \sum_{m=1}^{n-1} \frac{(n-m)}{m} x \sin(m \times k \times d)}$$

where

Gb is the broadside gain of the antenna
$k = \dfrac{2\pi}{\lambda}$
n = the number of individual elements
d = the spacing between the elements
L = (n–1)×d is the total length of the array

This expression simplifies dramatically for two important cases:

(a) for the case where the element spacing d is any precise multiple of $\lambda/2$, or where d >> λ, then Gb = n. Note that this is equal to the expression given in the section above for the gain of n widely spaced but phased elements.

(b) for the case where the elements are very close together, so that the array may be considered as a continuous distribution rather than one made of discrete elements, and where the total array length L >> λ, then

$$Gb = 2\frac{L}{\lambda}$$

In this case the gain is independent of the number of elements, but depends just on the total length L of the array. The two expressions in (a) and (b) become equivalent when the element spacing d = $\lambda/2$.

An Ordinary End-Fire, Uniform Linear Array

This array might be the same straight line array of closely spaced elements. This time, they are fed with equal currents but with a relative phase delay so that the fields from all elements add constructively along the line of the array. For example, with dipoles spaced, say, $\lambda/4$, then each dipole would be fed with current that is progressively delayed by a quarter-period (90°) with respect to the dipole to its left. Then by the time the radiation from the dipole on the immediate left arrives at the next element, it will be exactly in phase with the field now being generated by this adjacent dipole. Radiation traveling to the right along the line of the array will be progressively reinforced.

The corresponding expression[3] (see also p 3-29 of Reference 2) for the gain Ge of an equally spaced array of N isotropic elements phased to give gain in the end-fire direction is:

$$Ge = \frac{(k \times d)n^2}{(n \times k \times d) + \sum_{m=1}^{n-1} \frac{(n-m)}{m} \sin(2m \times k \times d)}$$

As with the broadside array, this expression simplifies dramatically for these special cases:

(a) Where the element spacing is any whole multiple of a quarter wavelength, the end-fire gain Ge becomes exactly Ge = n.

(b) Where the elements are very close together, so that the array may be considered as a continuous distribution rather than one made of discrete elements, and where the total array length L >> λ, then

$$Ge = 4\frac{L}{\lambda}$$

These two expressions become equivalent when the spacing d is $\lambda/4$. Note that for a given number of elements n, the ordinary uniform end-fire antenna will have a gain identical to the broadside array if its element spacing d, and hence its total length, is one half that of the broadside array. An end-fire array of the same length as a broadside array will have twice the gain.

The difference in directive gains between the broadside and ordinary end-fire arrays can be explained by considering the respective radiation patterns. The broadside linear array gives directivity only in the direction at right angles to its length—in azimuth for a horizontal linear array. Its radiation pattern is a flat, circular pancake. The end-fire array on the other hand gives directivity simultaneously in both directions (that is, both in azimuth and in elevation) more like a pencil beam. This probably at least in part accounts for the popularity of end-fire an-

tennas like the Yagi. The beamwidth of the ordinary end-fire array in either azimuth or in elevation is greater than that of the azimuth beam of the broadside array of the same length, but the total angular area of its beam is much less.

Note that Kraus, in Table 4-3 of Reference 1, gives a value of $2\pi L$ for the gain of an ordinary uniform end-fire, rather than the 4L quoted above, which is taken from References 2 and 3. However, Kraus derived his gains indirectly, using just the antenna beamwidths, and clearly states his results to be approximate. A similar comment applies to the different quoted gains of the end-fire array with increased directivity, discussed below.

Fig 1 shows the gain of the ordinary, uniform end-fire array as a function of length. The solid line is for a continuous array of isotropic elements rather than dipoles. Discrete points mark the gains of arrays of from 2 to 10 isotropic elements, with uniform spacings of $\lambda/8$, $\lambda/4$ and $3\lambda/8$. The arrays with discrete elements have very similar gains to the continuous distribution, although for spacings > 0.4 λ the gain begins to fall quite sharply. Here the length of an n-element array of elements with regular spacing d is defined to be n×d. Strictly speaking, the boom length of an n-element array is (n–1)×d rather than n×d, but the end-fire array of discrete elements performs as if it had the longer length.

The linear arrays plotted as the solid line in Fig 1 are made up of hypothetical isotropic elements, rather than practical dipoles. If real dipoles are substituted, parallel to each other and at right angles to the length of the array, the total array gains will be a little higher—but not necessarily by the full gain of a dipole. This is illustrated by comparing the solid and dashed lines of Fig 1. The dashed line is for a continuous end-fire array of dipoles—in effect, where the individual dipoles are packed very closely together. The gain of a dipole comes from the fact that radiation off its ends is suppressed, thereby concentrating the energy at right angles to its length.

With the end-fire array, radiation is already suppressed to a significant degree at right angles to the array. The polar diagram of individual dipoles in an array may only result in a very minor modification of the total array response. The dashed line in Fig 1 was calculated for an array of short dipoles, following the method given on p 3-13 of Reference 2. For arrays 0.5 λ or less in length the extra ≈2 dB expected from the additional gain of a dipole over an isotropic antenna is obtained. However, for an array 2 λ long the extra gain is reduced to only 0.7 dB, and for an array 10 λ long only an additional 0.2 dB is obtained.

This demonstrates that for reasonably

long ($\geq 2 \lambda$) end-fire arrays such as Yagis or loop arrays, the precise nature of the individual elements is of little consequence in determining the total gain. Other desirable properties of the antenna, such as its Q or its bandwidth, may depend strongly on the details of the individual elements, but not the total gain. Some of the following discussion is strictly limited to arrays consisting only of isotropic elements. Many textbooks consider only isotropic elements. This makes the calculations much easier, and for all but the shortest antennas the results are nearly indistinguishable from dipole arrays.

Capture Area

If a continuous aperture is fed in phase to give broadside radiation, with uniform excitation over the entire aperture, the antenna gain is directly proportional to its area A. Such an antenna might be a field of closely spaced dipoles over a reflecting screen, or a parabolic dish uniformly illuminated from its feed. The gain G of such an antenna is simply given by $G = 4\pi A/\lambda^2$. In practice the gain may be a little less. For example, parabolic dish illumination has to be tapered in some way towards the edge of the dish to avoid some of the radiation spilling off beyond the dish surface. This may halve the gain from the above formula. However, this expression gives the maximum gain that can be achieved in a given area with uniform illumination and phasing across the surface.

Any antenna has a certain *capture area* A. For some antennas, such as the uniformly illuminated surface, the capture area may equal the physical area. For a given flux density S of incoming radiation from a distant source, the total received power available at the antenna feed is simply SA. The capture area of many antennas bears little relationship to the physical area of the antenna. The effective capture area of any antenna with isotropic gain G at wavelength λ is given by $A = G\lambda^2/4\pi$. This is really the same equation as given above, for the gain of a continuous aperture.

It is shown below that the gain of an end-fire antenna can be enhanced very considerably with very careful adjustment of the relative amplitudes and phases of the currents in its elements. As the gain G of an antenna increases, so does its capture area A. Anything intruding into the capture area, such as other antennas or masts, will almost certainly detract from the gain. The corresponding capture areas give an indication, for example, of how far apart phased Yagi antennas should be spaced. The individual capture areas should not overlap.

End-Fire Arrays With Increased Directivity: the Hansen-Woodyard Array

The gain Ge of an ordinary uniform end-fire array of length L, for reasonably long

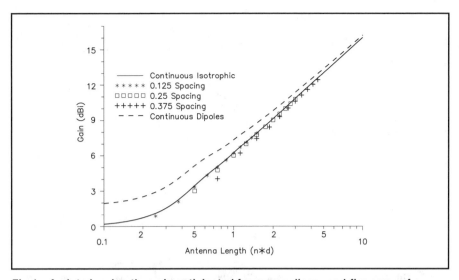

Fig 1—A plot showing the gain anticipated from an ordinary end-fire array of n isotropic elements spaced uniformly by d, as a function of antenna length given by n×d in wavelengths. Although the boom length of such an antenna is (N−1)×d, the antenna behaves as if its electrical length is just N×d. Symbols are plotted for uniform inter-element spacings of 0.125, 0.25 and 0.375 λ. The solid line is for a continuous antenna—the limit of the elements being spaced very close to each other. Discrete elements spaced by λ/8, or even by λ/4, are little different from the continuous antenna. The upper, dashed line shows the end-fire gain if the antenna is comprised of dipoles, rather than theoretical isotropic elements. Although the extra gain approaches that of an isolated dipole for very short antennas, the difference between arrays of isotropic elements and arrays of dipoles soon becomes insignificant as the antenna is lengthened.

antennas, is Ge = 4 L/λ. In this antenna, all the very closely spaced elements have equal currents, with the phase of the currents changing progressively, element to element, by an amount equal to the free-space propagation delay between the elements. The excitation of the elements in effect creates a traveling wave along the antenna, traveling towards the direction of maximum radiation from the antenna. In the ordinary end-fire array, the traveling wave in the antenna travels at the same speed as the radio wave in free space.

In 1938 W. W. Hansen and J. R. Woodyard[4] investigated the end-fire array to see whether greater antenna gain could be obtained with some other wave velocity in the antenna. With some very elegant mathematical manipulation they showed that significantly higher antenna gain could be obtained if the phase velocity in the antenna were slightly less than that of free space. Maximum gain occurs, for antennas at least a few λ in length, when the total change in phase from one end of the antenna to the other becomes 2.9215 radians (167°) more than the free-space propagation delay in length L. (Many antenna textbooks, such as reference [1], approximate this extra phase delay to π radians.) Hansen and Woodyard showed that if this condition were satisfied, then the forward gain of the antenna, for antennas several λ or more in length, is given by G = 7.214 L/λ. This is

nearly twice the gain of the ordinary end-fire array. Even more gain per length is available for shorter antennas.

Fig 2A shows the antenna gain of a Hansen-Woodyard end-fire array of isotropic elements, with the gain of the ordinary end-fire antenna shown for comparison. Also shown is the maximum gain obtainable with a small number of discrete elements, but phased according to the Hansen-Woodyard optimization. The upper solid curve in Fig A shows the gain of the Hansen-Woodyard array allowing for the small extra directivity using dipoles, rather than isotropic elements. As with the ordinary end-fire array, this correction is rather small for all but the shortest antennas.

With modern computer hardware and software, it is trivial to calculate this constant in a more general way than was possible for Hansen and Woodyard in 1938. **Fig 2B** shows the excess phase in radians required along the length of a Hansen-Woodyard array in order to get the optimum antenna gain. The upper solid curve is for a continuous distribution, effectively where elements of an array are infinitely close together. The excess phase is plotted against antenna length, or n×d for an array of n elements with the elements separated by d λ. The dashed horizontal line shows the 2.92 radians of excess phase calculated by Hansen & Woodyard, which is strictly only valid for an antenna of many wavelengths; however, the more exact

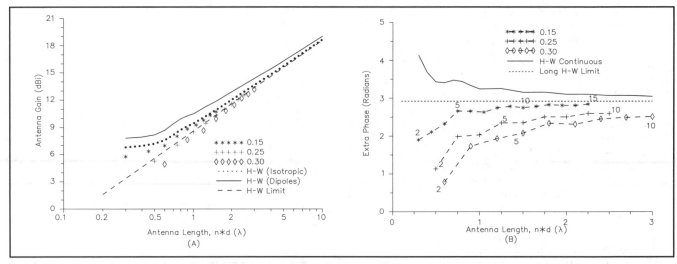

Fig 2—Characteristics of the Hansen-Woodyard end-fire array with enhanced directivity. At A, the gain of the H-W array versus antenna length N×d. Discrete symbols are shown for arrays from 2 to 10 elements, with element spacings of 0.15, 0.25 and 0.30 λ. The lower, dashed curve shows the gain using the original H-W constant derived in 1938. This is strictly only valid for long antennas. The lower solid curve shows the gain of a continuous H-W array of isotropic elements, with the upper solid line showing the gain of a continuous H-W array of dipoles. As in Fig 1, the characteristics of the individual elements are relatively unimportant. The array of discrete elements separated by 0.15 λ performs nearly as well as the continuous distribution. Even with a spacing of 0.25 λ the difference is little more than 1 dB for very short antennas, becoming insignificant for N.d ≥ 2 λ. At B, is shown how the elements of a H-W array must be phased to yield the maximum gain for this antenna. The quantity plotted is the excess phase in radians, accumulated along the length of the antenna by the driven elements, compared to the same distance in free space.

The horizontal, dashed line is the excess phase using the 1938 H-W constant, which is strictly only valid for long antennas. The upper, solid line shows the excess phase for a continuous distribution. For an array of discrete elements, the optimum value of excess phase is plotted for arrays from 2 to 10 elements (spaced 0.3 and 0.25 λ) and for arrays from 2 to 15 elements where the element spacing is 0.15 λ .

To calculate the optimum phasing for each discrete element, read off the extra phase φ from Fig 2B for the corresponding number n of elements and element spacing d. The extra phase per element is then f/(n−1). This then must be added to the free-space propagation delay (2πd/λ) between elements, to give the relative phase angle of each element with respect to its neighbor. See text for an example.

values calculated from numerical integration become very close to this H-W value for antennas even 3 λ long.

Points are also plotted in Fig 2B for arrays of discrete elements, with constant element spacings of 0.15, 0.25 and 0.30 λ. Especially for the shorter arrays, the optimum value of excess phase along the array becomes significantly lower than that for a continuous feed distribution. For example, for a five-element array spaced λ/4 between elements, Fig 2B shows the necessary excess phase to be 2.35 radians, or 134.6°. The extra phase increment between each of the n elements (n = 5) is then 134.6/(n−1) or 33.7°. Adding this to the free space propagation delay between elements (90 degrees for λ/4 spacing) we arrive at an optimum inter-element phasing of 123.7 degrees.

This should be a progressive phase lag from one element to the next, in the direction of the maximum end-fire radiation. With this phasing and with equal currents in all five elements, the antenna gain becomes 9.44 dBi rather than the 7.0 dBi that would be obtained with phasing as an ordinary end-fire array. Table 1 shows a detailed comparison between a uniform five-element ordinary end-fire, the Hansen-Woodyard end-fire and the superdirective array described

below in the section on supergain.

The curves in Fig 2 were computed with the help of the software tool *Mathcad*,[5] using numerical integration to find the optimum phase velocity and hence maximum possible gain of the antennas of different lengths, following exactly the methods given by Ma in References 2 and 3. The *Mathcad* results agree exactly with examples from these and other references, and also with the gains of Hansen-Woodyard arrays of discrete elements calculated in Reference 6.

Surface Wave Antennas

The Hansen-Woodyard array is a special case of the *Surface-Wave antenna*. With the H-W array, there is a traveling wave along the antenna, with a phase velocity somewhat less than that of free space, and with equal amplitude along the length of the array. In a more general surface-wave antenna, the energy is launched into the antenna structure at one end, and travels to the far end of the antenna but gradually decays in amplitude. The antenna structure could be a leaky waveguide, or a dielectric slab or rod. The Yagi antenna, with its closely coupled elements, is often considered a surface-wave antenna. The velocity of the wave traveling down the length of the antenna is deter-

mined by the spacing and tuning of the directors along the antenna.

Just as for the H-W array, for a given surface wave antenna there is an optimum wave velocity which will give maximum antenna gain; however the optimum wave velocity is dependent on details of the feed and of the amplitude distribution along the array. The optimum wave velocity for a given antenna length cannot easily be calculated. There is a fairly detailed discussion of this phenomenon in Chapter 12 of Reference 2.

As stated above, the maximum gain G of a H-W array of isotropic elements tends to G = 7.214 L/λ. For the surface wave, the maximum gain of an optimized antenna with 3 λ < L < 8 λ is close to G = 10 L/λ, a little higher than H-W (see p 12-12, Reference 2). The continuous lines in **Fig 3** show, as a function of antenna length, the gain of the ordinary end-fire array and the Hansen-Woodyard array (both corrected for dipole rather than isotropic elements) and the maximum gains reported in the literature (from Reference 2, Fig 12-9) for optimized surface wave antennas. Also shown in Fig 3 are the maximum gains available from the axial-mode helix antenna,[7] the gains of optimized Yagi antennas taken from Lawson[8] and from *The ARRL Antenna Book.*[9]

At least for $L > 2\lambda$, the Yagi antenna gains are in excellent agreement with the optimized surface wave antenna gains. The gain for the Hansen-Woodyard array of dipoles is consistently about 0.5 to 1 dB lower than the surface-wave or Yagi antenna gains. The main conclusion to be drawn here is that the modern optimized Yagi designs apparently already give a gain performance very close to the theoretical optimum for this type of antenna. The helical antenna has many desirable properties, but for $L > 2\lambda$ its gain is not competitive with a good Yagi design.

Supergain

Ordinarily, the bigger an antenna, the higher the expected antenna gain. However, this isn't always true. John Kraus, W8JK, noted in 1937 an article by Brown,[10] which showed that theoretically the gain from two elements spaced $\lambda/8$ and fed with equal currents exactly in antiphase, should exceed the gain from two similar elements spaced $\lambda/4$ and fed in the conventional way, 90° out of phase.

Kraus (Reference 1, p 454) recounts that Brown had tried to publish this result as early as 1932, but that the concept was so revolutionary at the time that it was not accepted for publication in a professional journal until five years later. This theoretical result led Kraus to develop the W8JK flat-top array.[11,12] Kraus confirmed that, contrary to intuition and to the conventional wisdom of the time, the gain indeed increased as the spacing between the two elements decreased. Kraus found that a spacing even as small $\lambda/16$ was workable,[13] but that with spacings less than about $\lambda/8$ the antenna Q became very high, with very high currents in each element and excessively low radiation resistance. These concepts are now well established and can be found in the standard antenna textbooks. The W8JK antenna is an early example of superdirectivity, or supergain.

With the ordinary end-fire array described earlier, each element is phased to add constructively in the chosen direction. There is an alternative design philosophy. Rather than having fields *add* constructively in the *wanted* direction, design the antenna such that the fields from the different elements together *cancel* each other as far as possible in the *unwanted* directions. This is the philosophy of the W8JK beam, and of supergain antennas in general. The approach leads to significantly higher antenna gains, but at a high price.

There are very high currents in the conductors that have to be precisely balanced in amplitude and in phase so that the distant fields in the unwanted directions cancel. This implies very high Q in the antenna. (Q is nearly proportional to current squared,

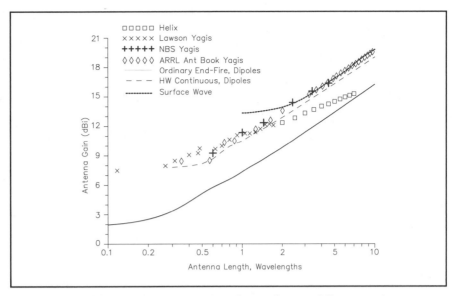

Fig 3—A comparison of the antenna gains of an ordinary end-fire array, the continuous Hansen-Woodyard array (corrected for an array of dipoles rather than isotropic elements), the optimized surface wave antenna, various Yagi designs taken from References 8 and 9, and the optimized gain of a helix, taken from Reference 7. The gains of the Yagi designs are on average about 0.5 dB higher than the H-W antenna, but are well represented by surface-wave antennas. The gain of the optimized helix is several dB inferior to the H-W or Yagi antennas for lengths $\geq 2\lambda$.

with a given power supplied to the antenna. See p 457 of Reference 1.) This results in very narrow bandwidths (~$1/Q$), extraordinarily demanding precision in construction and feed of the antenna elements, and potentially very high losses from conductor resistance and dielectric losses.

There is a most surprising rule, which runs directly counter to intuition. R. C. Hansen states it clearly: "Thus the important theorem: a fixed aperture size can achieve (in theory) any desired directivity value."[14]

The snag is that as the gain increases beyond that predicted by "ordinary gain," the Q increases exponentially. Hansen (Reference 14, p 177) gives an example for a particular array $\lambda/2$ in length, which has been optimized for maximum gain subject to the constraint of sidelobes being no greater than –10 dB. With two isotropic elements spaced $\lambda/2$ the gain is 3 dB. Increasing the number of isotropic elements to seven within the same $\lambda/2$ length, giving an element separation of $\lambda/12$, it is theoretically possible to raise the gain to 6.4 dB. However, the antenna Q has risen to $> 10^8$.

The fractional antenna bandwidth is $1/Q$; if it were possible to build this antenna for 144 MHz, the extra 3.4 dB of (superdirective) gain would reduce the antenna bandwidth *to less than 1 Hz*! If nine elements are used within the same half-wavelength, giving a spacing of $\lambda/16$, the extra gain from superdirectivity increases to 4.3 dB, but the antenna Q becomes $> 10^{14}$! This is a good demonstration of the practical limitations to

superdirectivity or supergain. The laws of physics place no limit on the gain of an antenna of given size, but the laws of engineering in practice may limit the achievable gain to little more than the "ordinary" arrays.

Although there is no limit, according to the laws of physics, to the maximum theoretical antenna possible with an antenna of a given size, there is a well-defined limit in terms of the complexity of an antenna. Uzkov[15] (quoted on p 110 of Reference 3) showed in 1946 that the limiting gain of an antenna consisting of n isotropic elements is n^2, and that this limiting gain is approached as the element spacing tends to zero. Note the surprising result that the potential gain actually *increases* to this limit of n^2 as the physical extent of the n-element array *decreases* toward zero.

This is the ultimate limit of array gain set by the laws of physics and mathematics. However, problems of high Q, high circulating currents, resistance of conductors and necessary precision of construction and of supplying the very precise relative amplitudes and phases in the elements, mean that practical antennas can rarely if ever begin to approach the potential array gain of n^2.

Ma (section 2.9 of Reference 3) gives a detailed mathematical procedure for calculating exactly the optimum amplitudes and phases required in a given n-element configuration, in order to achieve the maximum theoretically possible antenna gain. Although the mathematics is a little complicated, it is fairly easy to implement the procedure in a program such as *Mathcad* or

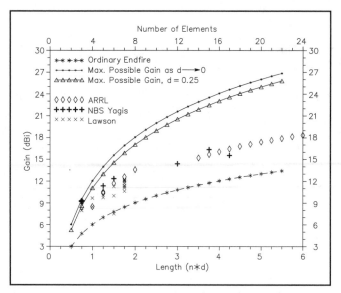

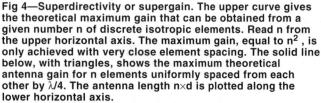

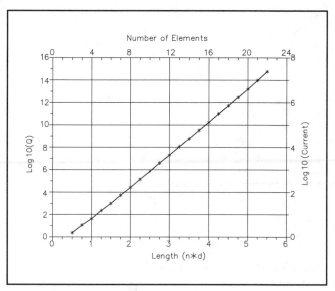

Fig 4—Superdirectivity or supergain. The upper curve gives the theoretical maximum gain that can be obtained from a given number n of discrete isotropic elements. Read n from the upper horizontal axis. The maximum gain, equal to n^2, is only achieved with very close element spacing. The solid line below, with triangles, shows the maximum theoretical antenna gain for n elements uniformly spaced from each other by $\lambda/4$. The antenna length n×d is plotted along the lower horizontal axis.

Discrete points are plotted for the gains of Yagi antennas, taken from References 8 and 9. Note that the Yagi gains benefit from the additional directivity of the dipole elements, compared to the maximum gains of arrays of isotropic elements given by the solid curves. The Yagi gains are plotted against number of elements along the top axis, rather than according to antenna length. The lower, dashed curve shows the gain of an ordinary end-fire antenna for the corresponding antenna lengths.

Fig 5—The approximate Q of a maximum-gain superdirective antenna of n elements (upper horizontal axis) spaced from each other by d. Note that the $\log_{10}$ (Q) is plotted on the left vertical axis. For a 22-element array, Q becomes $\sim 10^{15}$! The right vertical axis shows the amplitude of current which has to flow in the antenna elements, compared to an ordinary end-fire antenna with the same total radiated power. The high currents relative to an ordinary end-fire ($\sim 10^7$ greater for a 21-element array) are indicative of the impractically low radiation resistance of the supergain antenna.

Table 1
Optimized End-Fire Arrays: Five Elements, $\lambda/4$ Spacing, L=λ

	Ordinary End-Fire		Hansen-Woodyard		Maximum Supergain	
Ele. Nr	Current (Amps)	Phase (Degrees)	Current (Amps)	Phase (Degrees)	Current (Amps)	Phase (Degrees)
1	1	0°	1.93	0°	4.6	0.0°
2	1	−90°	1.93	−123.7°	11.6	−169.6°
3	1	−180°	1.93	−247.3°	15.1	−340.7°
4	1	−270°	1.93	−371.0°	11.6	−151.8°
5	1	−360°	1.93	−494.6°	4.6	−321.4°

Element Type	Gain (dBi)	Gain (dBi)	Gain (dBi)
Isotropic	7.0	9.4	13.0
Dipole	8.3	10.7	13.5

Mathematica. **Fig 4** displays the maximum possible gain calculated for end-fire arrays of from 2 to 22 elements, where the inter-element spacing is fixed at $\lambda/4$. The calculated points for these hypothetical arrays fall quite close to the theoretical maximum gain n^2, plotted by the solid line.

If the element spacing is reduced below $\lambda/4$, the theoretical gains are increased and approach the theoretical limit of n^2 even more closely. Also shown on Fig 4 are the gains for the same Yagi designs plotted in Fig 3. The Yagis are plotted against number of elements (top horizontal axis) rather than length. Note that these Yagi gains benefit from the small additional directivity of dipole elements, while the solid curves in Fig 4 are for arrays of isotropic elements. For all but the smallest antennas, the Yagi gains, while significantly higher than the ordinary end-fire arrays, are well below the theoretical maximum gain for a given number of elements.

Fig 5 shows on the left vertical scale the approximate Q of the same superdirective antennas, relative to the ordinary end-fire antenna with the same numbers of elements and the $\lambda/4$ element spacings. Note the logarithm of the Q is plotted; for a hypothetical 22-element array the Q becomes $\sim 10^{15}$! The values are obviously quite impractical; real antennas with this degree of supergain could never be built. The right vertical scale of Fig 5 plots the current that would flow in a central element, with respect to an ordinary end-fire antenna with the same power input.

Table 1 shows a comparison of element currents and gain of a fairly unambitious five-element array with element spacing of $\lambda/4$, and total length λ. The relative element currents and phasing angles are shown for the ordinary end-fire, the Hansen-Woodyard array, and the optimized superdirective array that gives the maximum possible gain for this particular configuration of (isotropic) elements. The current amplitudes for each antenna are normalized for the same total radiated power.

The Hansen-Woodyard and superdirective parameters were calculated independently using *Mathcad*, but agree perfectly with values given in Table 2.6 of Reference 3. Note that even with this modest five-element array and λ/4 spacing, the superdirective array requires 15 times the antenna current of the ordinary end-fire array, for the same total radiated power. The radiation resistance is extremely low. The relative antenna gains are 7.0, 9.4 and 13.0 dBi. The theoretical maximum possible directive gain from any five-element array of isotropic elements, which would be obtained with smaller element separations, is 5^2 or 14.0 dBi. Further calculations using the same current ratios and phasing, but allowing for the extra directivity from dipole rather than isotropic array elements, give gains of 8.3, 10.7 and 13.5 dBi for the ordinary, H-W and superdirective arrays.

Superconductive Arrays

The high Q of a supergain array, with its high circulating currents, implies very low radiation resistance, and hence potentially very low antenna efficiency. With superconducting antennas the conductor losses can be removed and so the efficiency may, at least in principle, be increased. However, the matching difficulties remain, and so does the impractically high Q. In a review article on superconducting antennas, R. C. Hansen states: "Size reduction of antennas will generally not be aided by superconductors."[16]

There are some examples in the literature of the successful application of superconductivity to the construction of supergain antennas. Ivrissimtzis, Lancaster, and Alford[17] were able to demonstrate a supergain enhancement of about 1.8 or 2.5 dB, increasing the gain of a particular 1-GHz array of four closely spaced dipoles from 4.6 to 7.1 dB. Element separations from 0.02 to 0.2 λ were investigated. The experimenters commented that it was found essential for the element matching networks to be superconducting, not just the antenna elements.

Summary of the Properties of Superdirective Arrays

To summarize the properties of superdirectivity or supergain:
1) The Q becomes astronomical, reducing the bandwidth to an impractical value.
2) There are huge circulating currents in the elements.
3) The construction, current balance and phasing become unmanageably critical.
4) The feed impedance becomes almost impossible to match.
5) The space around the antenna must be kept free of other conductors or dielectrics, to an extent comparable to the electrical capture area, or to the size of an ordinary antenna with comparable gain.

Given these factors, even if room-temperature superconductors become readily available, it is difficult to imagine many circumstances where significant amounts of supergain could be exploited. Although the

Table 2
Summary of End-Fire Antenna Gains

Broadside Array	Ordinary End-Fire	Hansen-Woodyard Enhanced Directivity	Surface-Wave Yagi	Supergain Maximum Theoretical
$2L/\lambda$	$4L/\lambda$	$7.28\ L/\lambda$	$10\ L/\lambda$	no limit
n at d = λ/2	n at d = λ/4	1.82 n at d = λ/4		n^2 as $d \Rightarrow 0$

laws of physics would allow a 100% efficient 20-dB gain antenna for 20 meters to be carried in your back pocket, it will never happen in practice!

Example: Field Distributions

Fig 6A and **6B** illustrate snapshots of the computed free-space electric field surrounding a 13-element antenna, with element spacing λ/4 and a total length of 3 λ. The axis of the antenna is horizontal, and in both cases it is phased for maximum radiation to the right. The calculations are for any plane containing the array. The antenna is circularly symmetric about its axis (the elements are assumed isotropic), so the complete 3-D field picture would be obtained by spinning the images in Fig 6 around the axis of the antenna.

In Fig 6A is the field of an ordinary end-fire antenna, where all elements have equal currents, and there is a progressive phase delay of 90° from one element to the next in the direction of maximum radiation. The gain of this antenna (assuming isotropic elements) is 11.14 dBi.

In Fig 6B, the amplitude and phase in each element is adjusted to give the maximum possible end-fire gain available from such a configuration, using the mathematical procedure described in Reference 3. The theoretical gain is 21.24 dBi. The field intensities in Figs 6A and 6B have been scaled to correspond to the same power applied to each antenna.

For the same total power radiated, the current in the center element of the superdirective array has to be a factor of ~10^4 higher than the ordinary end-fire array. The Q has increased by a factor of approximately 10^8. Were it possible to build and feed such an antenna, the bandwidth would be about 1 Hz for a 2-meter antenna! If the relative current in the center element is either too high or too low by as little as 0.1%, the gain is reduced by 12 dB. Although not contravening any physical laws, this is a purely fictional antenna whose performance could never be approached in practice, other than as a computer model. However, the theoretical field distributions around these different antennas give an interesting illus-

tration of how the antennas work.

A detailed comparison of Fig 6A and 6B shows:

1) For the ordinary array in Fig 6A, except very close to the driven elements, there are no exceptionally high field values surrounding the antenna. The near-field region merges smoothly into the far field. Because the energy density close to the antenna is not abnormally high, the Q of the antenna is fairly low, and hence its bandwidth is fairly high. On the other hand, for the supergain array in Fig 6B, the immediate field surrounding the antenna is extremely high— literally thousands of times higher than for the ordinary array, for equal total radiated powers.

2) For the supergain antenna in Fig 6B, not only is the field in the neighborhood of the antenna extremely high, but the high-field regime extends over a considerable volume of space. With the theoretical supergain value of 21.24 dBi, the antenna capture area is 10.6 square wavelengths. This capture area about equals the cross-sectional area of the high-field region obvious in Fig 6B. Although the physical extent of the antenna is relatively small, its effective electrical extent is very significantly greater. Any other unintended conductors or dielectrics within this high-field region will almost certainly compromise the performance of the antenna.

3) Along the ordinary end-fire array in Fig 6A, the distance between peaks of instantaneous field is equal to the free-space wavelength. This is seen easily by looking at the distance between wave crests within the antenna and some distance from it: they are the same. With the supergain array in Fig 6B, however, the apparent wavelength close to the antenna is much smaller than the free-space wavelength. This reflects the much slower equivalent-wave velocity within the confines of the antenna, and the fact that adjacent elements of the supergain antenna are nearly in antiphase.

4) As the near field merges into the far field for the ordinary end-fire case, the wavefront is nearly spherical. The wave crests to the right of the antenna in Fig 6A, in the direction of maximum radiation, are roughly segments of

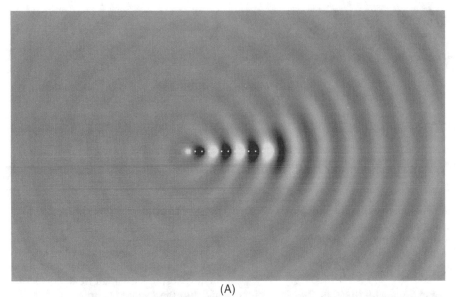

(A)

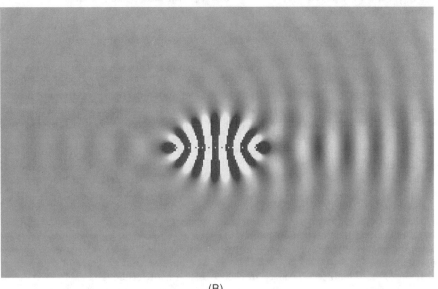

(B)

Fig 6—At A, a representation of the field surrounding an ordinary end-fire array of 13 elements. At B, the same elements fed to give the maximum possible gain from this configuration. In both cases the elements are uniformly spaced, separated from their neighbors by λ/4. The direction of maximum radiation is to the right of each figure. The field intensity scale is normalized to the same total radiated power from each antenna.

There are obvious differences between the two cases in terms of the peak central field, the extent of the high field region, and the emerging wavefronts being nearly parallel in the supergain case, rather than nearly spherical as for the ordinary array. The ordinary end-fire antenna at A has a gain of 11.14 dBi, and the superdirective array at B theoretically has a gain of 21.24 dBi. See text for details. Although the computer model of antenna at B is useful to demonstrate the differences between the two types of array, a real antenna with this magnitude of superdirectivity is quite impractical!

a circle, centered on the antenna. In the supergain case in Fig 6B, however, there is a focusing effect, rather like an optical lens. A short distance to the right of the antenna, in the direction of maximum radiation, the wavefronts actually become parallel. The extent over which the wavefront is parallel also approximates the electrical capture area of the

antenna. The divergence of the parallel wavefronts is obviously much less than for nearly spherical wavefronts, concentrating much more energy in the wanted direction.

Even if this antenna could be fed in the correct way without excessive losses, it is difficult to see any real advantage over a more conventional antenna. The space sur-

rounding the antenna has to be free of any conductors or dielectrics, to an extent comparable to the size of a conventional array with equivalent gain. So, instead of striving for precision of 1 part in 10^5 to feed the antenna, and instead of trying to achieve and live with an impossible Q of $> 10^8$, why not fill the high-field volume with more elements? This would reduce the Q and relax all the tolerances. Although it makes a fascinating illustration, it is difficult to imagine any real motivation for trying to achieve antenna gain with even this much superdirectivity.

Conclusions

Different types of end-fire designs have been presented:

a) The ordinary end-fire array, where the amplitude of current in each element is identical, but the phase of the current is arranged to cause the distant fields to add constructively in the desired direction. The phase velocity along the array is the same as that of free space. The gain available from an end-fire antenna of given length is twice that expected from a broadside antenna of the same length.

b) The Hansen-Woodyard end-fire array with improved directivity, where the phase velocity along the array is constant, but is somewhat less than free space.

c) The surface-wave antenna, which is a modification of the Hansen-Woodyard array in that the current in the antenna decays along its length. Slightly higher gains are available, although the antenna is more difficult to analyze. Yagi arrays can be described very well as surface-wave antennas.

d) The supergain or superdirective array, where the relative amplitude and phase of the current in each element is optimized to give the maximum possible antenna gain with any given configuration. The maximum gains theoretically possible are unlikely to be achieved in practice, because of high Q, narrow bandwidth, high circulating currents and the necessary precision in adjusting the relative amplitudes and phases in the individual elements. In theory, a superdirective antenna of any desired gain can be contained in any finite size. In practice, the additional gain from superdirectivity may be limited to a few extra dB.

The theoretical performances of the different arrays are summarized in Table 2. The gains are approximately true for antennas at least a few wavelengths in length. The effective length of a discrete array of n elements spaced d apart is n×d, rather than the boom length of (n–1)×d. Somewhat higher gain per length than shown in the table below is available for shorter antennas. The table is strictly for isotropic elements, but except for short (≤ 2 λ) antennas the additional gain computed by replacing theoretical isotropic elements by dipoles is negligible. The first line in the table gives the gain in terms of the length of the array, assuming that discrete elements are sufficiently ($\leq \lambda/4$) close together. The second line gives the gain in terms of the number of elements n, assuming an element spacing of $\lambda/2$ for the broadside array and of $\lambda/4$ for the endfire arrays.

The Hansen-Woodyard antenna (corrected for dipole rather than isotropic elements) predicts a gain within < 1 dB of that obtained from optimized Yagi models, at least up to 14 λ total antenna length. The surface wave antenna gives even better agreement with optimized Yagi designs. It seems that modern Yagi designs have approached the theoretical limit of gain for this type of antenna.

There remains the tantalizing fact that, according to the laws of physics, there is no limit to the theoretical gain available from an antenna of given size. For an array of n elements, in theory an array gain of up to n^2 is possible. In practice, even using superconducting antennas, existing designs may already be close to the limit of gain that can realistically be achieved in a given size of antenna.

References

[1] John D. Kraus, W8JK, *Antennas*, 2nd ed. (McGraw-Hill Book Company, 1988).

[2] *Antenna Engineering Handbook*, 2nd ed. Editors Richard C. Johnson & Henry Jasik (McGraw-Hill, 1984).

[3] M. T. Ma, *Theory and Application of Antenna Arrays* (John Wiley & Sons, Inc, 1974).

[4] W. W. Hansen & J. R. Woodyard, "A New Principle in Directional Antenna Design," *Proc IRE*, Vol 26, No 3, Mar 1938, pp 333-345.

[5] *Mathcad for Windows 5.0*, published in 1994 by MathSoft, Inc, 101 Main St, Cambridge, MA 02142. See W. Sabin, "*Mathcad 6.0*: A Tool for the Amateur Experimenter," *QST*, Apr 1996, pp 44-47.

[6] R. C. Hansen, "Hansen-Woodyard Arrays with Few Elements," *Microwave & Optical Technology Letters*, Vol 5, No 1, Jan 1992, pp 44-46. [Note that the vertical scale in Figure 2 of this article is mislabelled as running from 0 to 10, instead of the correct 5 to 10.]

[7] Darrel Emerson, "The Gain of an Axial-Mode Helix Antenna," *The ARRL Antenna Compendium, Vol 4* (Newington: ARRL, 1995), pp 64-68.

[8] James L. Lawson, W2PV, *Yagi Antenna Design* (Newington: ARRL, 1986).

[9] R. Dean Straw, N6BV, ed, *The ARRL Antenna Book*, 17th ed. (Newington: ARRL, 1994).

[10] George H. Brown, "Directional Antennas," *Proc IRE*, Vol 25, Jan 1937, pp 78-145.

[11] John D. Kraus, "A Small but Effective Flat Top Beam Antenna," *Radio*, No 213, Mar 1937, pp 56-58.

[12] John D. Kraus, W8JK, "Directional Antennas with Closely-Spaced Elements," *QST*, Jan 1938, pp 21-23.

[13] John D. Kraus, W8JK, "By Popular Demand," *Radio*, No 220, Jun 1937, pp 10-16.

[14] R. C. Hansen, "Fundamental Limitations in Antennas," *Proceedings of the IEEE*, Vol 69, No 2, Feb 1981, pp 170-182. Quotation cited is on p 173.

[15] A. I. Uzkov, "An Approach to the Problem of Optimum Directive Antenna Design," *Compt Rend Dokl Acad Sci URSS*, Vol 3, 1946, p 35.

[16] R. C. Hansen, "Superconducting Antennas," *IEEE Transactions on Aerospace and Electronic Systems*, Vol 26, No 2, Mar 1990, pp 345-355.

[17] L. P. Ivrissimtzis, M. J. Lancaster, M. McN. Alford, "Supergain Printed Arrays of Closely Spaced Dipoles Made of Thick Film High-Tc Superconductors," *IEE Proc Microw Antennas Propag*, Vol 142, No 1, Feb 1995, pp 26-34.

Ground Parameters for Antenna Analysis

R. P. Haviland, W4MB
1035 Green Acres Circle North
Daytona Beach, FL 32119

W4MB reviews techniques for determining the exact characteristics of the ground under your antenna.

I n the past, amateurs paid very little attention to the characteristics of the earth (ground) associated with their antennas. There are two reasons for this. First, these characteristics are not easy to measure—even with the best equipment, extreme care is needed. Second, the "so what" factor is important! Almost all hams have to put up with what they have—there are very few who can afford to move because their location has poor ground conditions. Further, the ground is not a dominant factor in the most popular antennas—a triband Yagi at 40 feet or higher, or a 2-meter vertical at roof height.

Even so, there has been a desire and even a need for ground data, and for ways to use it. It is very important for vertically polarized antennas. Ground data is useful for antennas mounted at low heights generally, and for such specialized ones as Beverages. The performance of such antennas changes a lot as the ground changes. And now the analysis has become easier. *MININEC* started this and *NEC* is now readily available for amateur use. These programs make accurate analysis of the effect of the ground possible, even simple. Now the problem becomes: what are the ground parameters?

Importance of Ground Conditions

To see why ground conditions can be important, let us look at some values. For a frequency of 10 MHz, *CCIR Recommendation 368*,[1] gives the distance at which the signal is calculated to drop 10 dB below its free-space level as:

Conductivity (mS/meter)	Distance for 10 dB Drop (km)
5000	100
30	15
3	0.3

The high-conductivity condition is for sea-water. Inter-island work in the Caribbean on 40 and 80 meters is easy, whereas 40-meter ground-wave contact is difficult for much of the USA, because of much lower ground conductivity. On the other hand, the Beverage works because of poor ground conductivity.

Fig 1 shows a typical set of expected propagation curves for a range of frequencies. This data is also from *CCIR Recommendation 368* for relatively poor ground, with a dielectric constant of 4 and a conductivity of 3 mS/m (1 milliSiemens/meter is 0.001 mho/meter). The same data is available in the *Radio Propagation Handbook*.[2] There are equivalent FCC curves, found in the book *Reference Data for Radio Engineers*,[3] but only the ones near 160 meters are useful. Here in Florida, we have difficulty hearing stations across town on ground wave, an indication of the poor soil conditions: reflected sky-wave signals are often stronger.

Securing Ground Data

It seems that there are only two basic ways

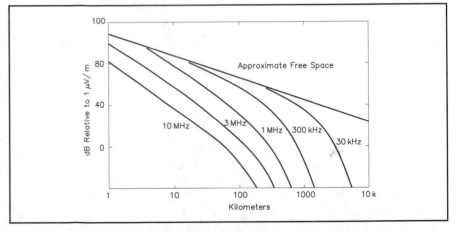

Fig 1—Variation of field strength with distance. Typical field strengths for several frequencies are shown. This is from CCIR data for fairly poor soil, with dielectric constant of 4 and conductivity of 3 mS/m. The curves for good soil are closer to the free-space line, and those for seawater are much closer to the free-space line.

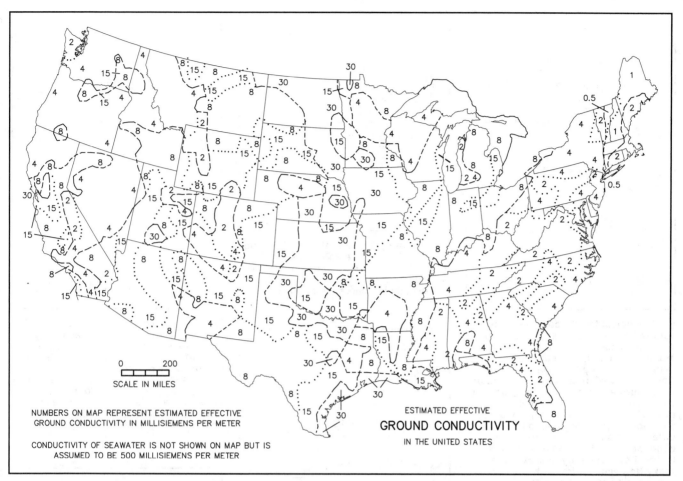

Fig 2—Estimated effective ground conductivity in the United States. FCC map prepared for the Broadcast Service, showing typical conductivity for continental USA. Values are for the band 500 to 1500 kHz. Values are for flat, open spaces and often will not hold for other types of commonly found terrain, such as seashores, river beds, etc.

to approach this matter of ground data. One is to use "generic" ground data typical to the area around the antenna. The second is to make measurements, which haven't really gotten easier. For most amateurs, the best approach seems to be a combination of these—use some simple measurements, and then use the generic data to "close-in" on a better estimate. Because of equipment costs and measurement difficulties, none of these will be highly accurate for most hams. But they will be much better than simply taking some condition preset into an analysis program. Having a good set of values to plug into an analysis can mean abandoning that pet idea as not worthwhile, before a lot of construction time is spent.

Generic Data

In connection with its licensing procedure for broadcast stations, the FCC has published generic data for the entire country.[4] This is reproduced in **Fig 2**, a chart showing the "estimated effective ground conductivity in the United States." A range of 30:1 is shown, from 1 to 30 mS/m. An equivalent chart for Canada has been prepared, originally by DOT, now DOC.

Of course, some judgment is needed when trying to use this data for your location. Broadcast stations are likely to be in open areas, so the data should not be assumed to apply to the center city. And a low site near the sea is likely to have better conductivity than the generic chart for, say, the coast of Oregon. Other than such factors, this chart gives a good first value, and a useful cross-check if some other method is used.

Still another FCC-induced data source is the license application of your local broadcast station. This includes calculated and measured coverage data. This may include specific ground data, or comparison of the coverage curves with the CCIR or FCC data to give the estimated ground conductivity. In the past, it was likely that the larger stations had measuring equipment, but the lawyers who now run the FCC do not think that such capability is necessary.

Another set of curves for ground conditions are those prepared by SRI.[5] These give the conductivity and dielectric constant versus frequency for typical terrain conditions. These are reproduced as **Fig 3** and **Fig 4**. By inspecting your own site, you may select the curve most appropriate to your terrain. The curves are based on measurements at a number of sites across the USA, and are averages of the measured values.

Figs 5 through 7 are data derived from these measurements. **Fig 5** gives the ground-dissipation factor. Sea water has low loss (a high dissipation factor), while soil in the desert or in the city is very lossy, with a low dissipation factor. **Fig 6** gives the skin depth, the distance for the signal to decease to 63% of its value at the surface. Penetration is low in high-conductivity areas and deep in low-conductivity soil. Finally, **Fig 7** shows the wavelength in the earth. For example, at 10 meters (30 MHz), the wavelength in sea water is less than 3 meters. Even in desert, the wavelength has been reduced to about 6 meters at this frequency. This is one reason why buried antennas have peculiar properties.

Lacking other data, it is suggested that the values of Figs 3 and 4 be used in *MININEC* and *NEC* analysis.

Measuring Ground Conditions

W2FNQ developed a simple technique to

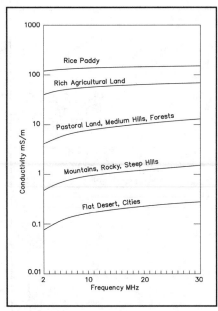

Fig 3—Typical terrain conductivities versus frequency for five types of soils. This was measured by SRI. Units are mS/m. Conductivity of seawater is usually taken as 5000 mS/m. Conductivity of fresh water depends on the impurities present, and may be very low. To extrapolate conductivity values (for 500 to 1500 kHz) shown in Fig 2 for a particular geographic area to a different frequency, move from the conductivity at the left edge of Fig 3 to the desired frequency. For example, in rocky New Hampshire, with a conductivity of 1 mS/m at BC frequencies, the effective conductivity at 14 MHz would be approximately 4 mS/m.

measure low-frequency earth conductivity (see Reference 4), which has been used by W2FMI. The test setup is drawn in **Fig 8**, and uses a very old technique of 4-terminal resistivity measurements. For probes of ⁵/₈-inch diameter, spaced 18 inches and penetrating 12 inches into the earth, the conductivity is:

$$C = 21\ V_1/V_2 \quad mS/m.$$

The voltages are conveniently measured by a digital voltmeter, to an accuracy of about 2%. In soil suitable for farming, the probes can be copper or aluminum. The strength of iron or copperweld may be needed in hard soils. A piece of 2×4 or 4×4 with guide holes drilled through it will help maintain proper spacing and vertical alignment of the probes. Use care—there is a shock hazard. An isolating transformer with a 24-V secondary instead of 115 V will reduce the danger.

Ground conditions vary quite widely over even small areas. It is best to make a number of measurements around the area of the antenna, and average the measured values.

While this measurement gives only the low-frequency conductivity, it can be used to select curves in Fig 3 to give an estimate

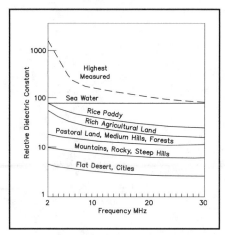

Fig 4—Typical terrain relative dielectric constant for the five soil types of Fig 3, plus seawater. The dashed curve shows the highest measured values reported, and usually indicates mineralization.

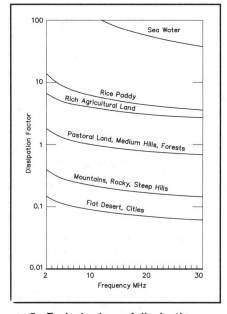

g 5—Typical values of dissipation .ctor. The soil behaves as a leaky dielectric. These curves show the dimensionless dissipation factor versus frequency for various types of soils and for seawater. The dissipation factor is inversely related to soil conductivity. Among other things, a high dissipation factor indicates that a signal penetrating the soil or water will decrease in strength rapidly with depth.

of the conductivity for the common ham bands. Assume that the 60-Hz value is valid at 2 MHz, and find the correct value on the left axis. Move parallel to the curves on the figure to develop the estimated curve for other soil conditions.

A small additional refinement is possible. If the dielectric constant from Fig 4 is plotted against the conductivity from Fig 3 for a given

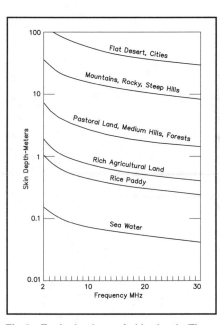

Fig 6—Typical values of skin depth. The skin depth is the depth at which a signal will have decreased to 1/e of its value at the surface (to about 30%). The effective height above ground is essentially the same as the physical height for seawater, but may be much greater for the desert. For practical antennas, this may increase low-angle radiation, but at the same time will increase ground losses.

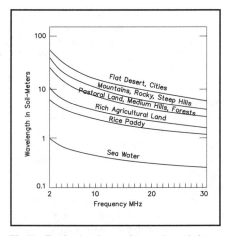

Fig 7—Typical values of wavelength in soil. Because of its dielectric constant, the wavelength in soils and water will be shorter than that for a wave traveling in air. This can be important, since in a Method of Moment program the accuracy is affected by the number of analysis segments per wavelength. Depending on the program being used, adjust the number of segments for antennas wholly or partly in the earth, for ground rods, and for antennas very close to earth.

frequency, a scatter plot develops, showing a trend to higher dielectric constant as conductivity increases. At 14 MHz, the relation is:

$$k = \sqrt{10000/C}$$

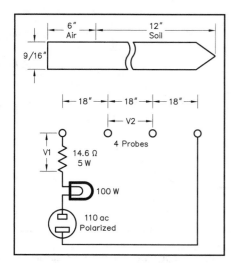

Fig 8—Low-frequency conductivity measurement system. A 60-Hz measuring system devised by W2FNQ and used by W2FMI. The basic system is widely used in geophysics. Use care to be certain that the plug connection is correct. A better system would use a lower voltage and an isolation transformer. Measure the value of V_2 with no power applied—there may be stray ground currents present, especially if there is a power station or an electric railway close.

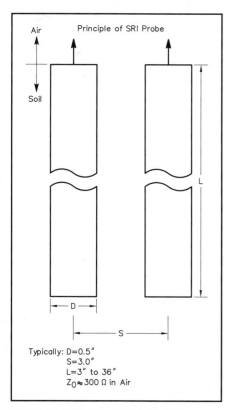

Fig 9—High-frequency conductivity/ dielectric constant measurement system. System for measuring ground conditions at frequencies up to about 100 MHz, devised by SRI and used to obtain the data in Figs 3 through 7. Basically, this is a section of transmission line with soil as the dielectric. Requires measurement of high impedances to good accuracy.

where k is the dielectric constant and C is the conductivity, as measured. Using these values in *MININEC* or *NEC* calculations should give better estimates than country-wide average values.

Direct Measurement of Ground Properties

For really good values, both the conductivity and dielectric constant should be measured at the operating frequency. One way of doing this is the two-probe technique of Reference 5. This was the technique used is securing the data for Figs 3 through 7. The principle is sketched in **Fig 9**. In essence, the two probes form a short, open-circuited, two-wire transmission line. As shown by the equations for such lines, the input impedance is a function of the conductivity and dielectric constant of the medium. A single measurement is difficult to calculate, since the end effect of the two probes must be determined, a complex task if they are pointed for easy driving. The calculation is greatly simplified if a set of measurements is made with several sets of probes that vary in length by a fixed ratio, since the measured difference is largely due to the increased two-wire length, with some change due to the change in soil moisture with depth.

The impedance to be measured is high because of the short line length, so impedance bridges are not really suitable. An RF vector impedance meter, such as the HP-4193A, is probably the best instrument to use, with a RF susceptance bridge, such as the GR-821A, next best. With care, a Q-meter can be substituted, see Reference 3.

Because of the rarity of these instruments among amateurs, this method of measurement is not explored further here. Consult Reference 5 for further information.

Indirect Measurement

Since the terminal impedance and resonant frequency of an antenna change as the antenna approaches earth, measurement of an antenna at one or more heights permits an analysis of the ground characteristics. The technique is to calculate the antenna drive impedance for an assumed ground

condition, and compare this with measured values. If not the same, another set of ground conditions is assumed, and the process is repeated. It is best to have a regular plan to guide the assumptions.

In connection with his studies of transmission lines, Walt Maxwell, W2DU, made such measurements on 20, 40 and 80 meters. Some of the data was included in his book *Reflections.*[6] The following example is based on his 80-meter data. Data came from his Table 20-1, for a 66-foot, 2-inch dipole of #14 wire at 40 feet above ground. His table gives an antenna impedance of $72.59 + j\,1.38\ \Omega$ at 7.15 MHz.

Table 1 shows calculated antenna impedances for ground conductivities of three different ground conductivities: 10, 1 and 0.1 mS/m, and for dielectric constants of 3, 15 and 80. The nearest value to the measured drive impedance is for a conductivity of 0.1 mS/m and a dielectric constant of 3. Figs 3 and 4 indicate that these are typical of flat desert and city land. The effect on antenna performance is shown in **Fig 10**. The maximum lobe gain for soil typical of a city is over 2 dB lower than that for the high-conductivity, high-dielectric constant value.

The ground at the W2DU QTH is a suburban Florida lot, covered with low, native vegetation. The ground is very sandy (a fossil sand dune), and is some 60 to 70 feet above sea-level. Measurements were made near the end of the Florida dry season. The water table is estimated to be 20 to 30 feet below the surface. Thus the calculated and measured values are reasonably consistent.

In principle, a further analysis, using values around 0.1 mS/m conductivity and 3 for dielectric constant, will give a better ground parameter estimate. However, the results should be taken with a grain of salt, because the opportunities for error in the computer modeling must be considered. The antenna should have no sag, and its length and height should be accurate. The measurement must be with accurate equipment, free from strays, such as current on the outer conductor of the coax. The feed-point gap effect must be estimated. [Further, the ground itself under the antenna must have constant

Table 1

Calculated values of drive resistance, in ohms, for a 40-meter dipole at 40-feet elevation versus conductivity and dielectric constant.

Conductivity (mS/m)	Dielectric Constant		
	3	15	50
10	$89.78 - j\,12.12$	$88.53 - j\,10.69$	$88.38 - j\,7.59$
1	$80.05 - j\,17.54$	$83.72 - j\,10.23$	$87.33 - j\,6.98$
0.1	$76.44 - j\,15.69$	$83.18 - j\,9.85$	$97.30 - j\,6.46$

The value measured by W2DU was $72.59 - j\,1.28\ \Omega$, and compares closest to the poor soil condition of dielectric constant of 3 and conductivity of 0.1 mS/m.

characteristics for modeling to be completely accurate.—*Ed.*]

Finally, the feed-line length and velocity constant of the transmission line must be accurately measured for transfer of the measured values at the feeding end of the transmission line to the antenna itself. Because of all the possibilities for error, most attempts at precision should be based on measured values at two or three frequencies, and preferably at two or three heights. Orienting the antenna to right angles for another set of measurements may be useful. Obviously, this can involve a lot of detailed work!

I have not been able to find any guidelines for the best height or frequency. The data in the article *Exact Image Method for Impedance Computation of Antennas Above the Ground*[7] suggests that a height of 0.3 λ will give good sensitivity to ground conditions. Very low heights may give confusing results, since several combinations of ground parameters can give nearly the same drive impedance. Both this data and experience suggest that sensitivity to ground for heights above 0.75 λ is small or negligible.

If an overall conclusion about ground characteristics is needed, we can just restate from the first paragraph—it is not greatly important for the most common horizontally polarized antenna installations. But it's worth taking a look when you need to depart from typical situations, or when the performance of a vertically polarized antenna is contemplated. Then the techniques outlined here can be helpful.

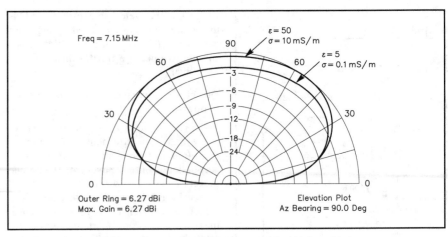

Fig 10—Plot showing computed elevation patterns for 40-foot high, 40-meter dipole for two different ground conditions: poor ground, with dielectric constant of 3 and conductivity of 0.1 mS/m, and good ground, with dielectric constant of 50 and conductivity of 10 mS/m. Note that for a low horizontal antenna, high-angle radiation is most affected by poor ground, with low-angle radiation least affected by ground characteristics.

Notes and References
[1] *CCIR Recommendation 368*, Documents of the CCIR XI Plenary assembly, ITU, Geneva, 1967.
[2] Peter N. Saveskie, *Radio Propagation Handbook* (Blue Ridge Summit, PA: TAB Books, 1960). Data from Reference 1 is partially included in this handbook.
[3] *Reference Data for Radio Engineers*, 5th edition (Indianapolis: Howard W. Sams, 1968), Chapter 28.
[4] *The ARRL Antenna Book*, 17th edition, p 3-3.
[5] George H. Hagn, SRI, "HF Ground Measurements at the Lawrence Livermore National Laboratory (LLNL) Field Site," *Applied Computational Electromagnetics Society Journal and Newsletter*, Vol 3, Number 2, Fall 1988. (A comprehensive set of references is included.)
[6] M. Walter Maxwell, W2DU, *Reflections* (ARRL: Newington, CT, 1990), p 20-3. Out of print.
[7] Ismo Lindell, Esko Alanen, Kari Mannerslo, "Exact Image Method for Impedance Computation of Antennas Above the Ground," *IEEE Transactions on Antennas and Propagation*, AP-33, Sep 1985.

An Introduction to "Patch" Antennas

R. P. Haviland, W4MB
1035 Green Acres Circle North
Daytona Beach, FL 32119

T his article is intended to be an introduction, or overview, of a widely
used antenna not often found in the
Amateur Service. This antenna is formed
with a metal patch, and is sometimes called
a *microstrip* antenna.

The antenna is a sheet of metal a short
distance above a ground plane. In its most
common form, the half-wave patch, the
sheet will have a length L just less than a
λ/2, and a width W a little less than twice
this, as shown in **Fig 1A**. The cross-sectional view shows feed from the center conductor of a coax, with the outer conductor
connected to the ground plane. For this feed,
the connection is at the edge of a long side,
at its center. The patch is about λ/100 wavelength above the ground plane. Insulating
material between the parts is shown here,
but air insulation is often used at the lower
useful frequencies.

The wavelength λ referred to that *in the
dielectric* under the plate, and the length L
is given by:

$$L = 0.49\, \lambda_d = 0.49 \frac{\lambda_o}{\sqrt{E_r}} \qquad \text{(Eq 1)}$$

where
λ_d = wavelength in the dielectric
λ_0 = wavelength in free space
E_r = relative dielectric constant. Note that
in air, $\lambda_d = \lambda_0$.

The field fringes at the ends of the plate
as is sketched in Fig 1B. Because of this
fringe presence, the plate edge radiates
very nearly as a gap or slot antenna, of
width equal to the plate spacing, best kept
at about λ/100 wavelength. The radiated
field produced by one edge is uniform
throughout the 180° above an infinite
ground plane. Since there are two radiating
edges separated by a distance, and in phase,
the net field is maximum at right angles to
the plate.

The E-plane field strength with an infinite ground plane is given by:

W4MB gives the lowdown on the Patch, an antenna that deserves more recognition in the amateur community.

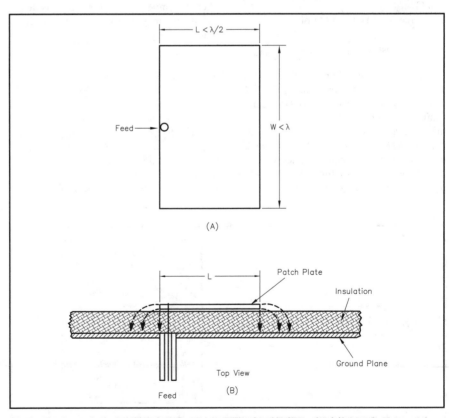

Fig 1—At A, a plan view, and at B, a cross-sectional view of a λ/2 patch antenna.
Dimensions are discussed in the text. This drawing and the ones following are not
drawn to scale. This design shows a coax feed, but other feed types can be used.
The dielectric can be air, with appropriate small insulators for mechanical support,
or it can be other materials.

101

$$E_0 = k \cos\left(\frac{\pi L}{\lambda_0 \cos\theta}\right) \qquad \text{(Eq 2)}$$

which is close to the pattern of a dipole. The H-plane pattern is that resulting from a line source of uniform intensity:

$$E_h = k \tan\theta \sin\left(\frac{\pi W}{\lambda_0 \cos\theta}\right) \qquad \text{(Eq 3)}$$

If the ground plane is not infinite, there will be some radiation to the rear of the plate, and the pattern will show some signal strength variations with angle. For example, with a 6-λ ground plane, and a dielectric constant of 2.45, the E-plane radiation to the rear is about 10 dB down along the line of the plate, and about 20 dB down at 20° below the plate. The forward lobe may show variations of a few dB at intervals of 360° divided by the ground-plane dimension in wavelengths. While lower in strength than the forward lobe, the variations in rear lobe strength can be greater. These effects are due to coupling between the top and rear of the ground plate along its edges.

The feed impedance is determined by the width W of the plate, and by the distance the feed is inset from the edge. At the edge:

$$R_0 = \frac{60\,\lambda_0}{W} \qquad \text{(Eq 4)}$$

so if W is just less than $\lambda/2$ wavelength wide, the drive-point impedance will be about 130 Ω.

For feeds away from the plate edge:

$$R_{in} = R_0 \cos^2\left(\frac{\pi d}{W}\right) \qquad \text{(Eq 5)}$$

where d is the distance from the edge. The feed resistance can be set to any value up to 130 Ω. This ability to change feed impedance can be useful in setting up arrays of plates. Note that the drive resistance goes to zero when D is equal to W/2. If a grounded plate is needed, a bridge to ground at the plate center will have negligible effect on performance. A rivet or plated through-hole can be used.

One way of looking at a plate antenna is that it is composed of a pair of one-turn coils connected in parallel, shunted by the capacitance of the plate to the ground plane. As is common in high-C tank circuits, the bandwidth will be narrow. To fair accuracy, the bandwidth is:

$$BW = 4f^2\left(\frac{t}{1/32}\right) \qquad \text{(Eq 6)}$$

where BW is in MHz, f is in GHz and the insulation thickness t is in inches. The reason for this form is that dielectric material is usually available in steps of $1/32$ inch. For example, for a frequency of 2 GHz, and a thickness of $1/32$ inch, the bandwidth will be about 1.6%. Some common substrates are Rexolite, $E_0 = 2.6$, and Duroid, $E_0 = 2.32$. An expensive one is an alumina block, $E_0 = 9.8$. Teflon-glass is used often. Epoxy-glass is often used for practice designs, although its loss is rather high.

At the higher frequencies, size of coax connectors becomes a problem, and microstrip transmission line is much easier to use. The principle of such feed is shown in **Fig 2**. The typical source connection is to 50 Ω line Z_S. This is matched to the $Z_0 = 130$-Ω point of the edge of the plate by a quarter-wave transformer Z_1. The impedance for this is:

$$Z_1 = \sqrt{Z_0 Z_S} \qquad \text{(Eq 7)}$$

where the impedance of the line Z_1 is determined primarily by its width W. Approximately:

$$Z_1 = \frac{\dfrac{377}{\sqrt{E_r}}}{\dfrac{W}{t} + 2} \qquad \text{(Eq 8)}$$

The 2 in the denominator accounts for field fringing along the edges of the line. More accurate relations include the thickness of the line conductor and its loss as factors. Further, the dielectric constant E_r is the effective value, which is affected by the

geometry. For line widths greater than about 50 times the thickness, there is negligible effect. For smaller line widths, the reduction in dielectric constant may reach 20%.

A variation in construction produces the $\lambda/4$ plate, shown in **Fig 3**, essentially half of the above type, with a ground along the edge opposite to the feed point (that is, the ground passes through the point, which can be grounded on the half-wave plate). The ground may be a series of plated through-holes or rivets. Sometimes a narrow slot is cut through the insulation substrate, with a piece of foil that passes through this and is soldered to the plate and to the underside of the ground plane.

Since this type is essentially half of the $\lambda/2$ type of plate, the impedance at the plate-center drive point is twice as great, or 260 Ω. The 50-Ω point is approximately $2/3$ of the distance from the edge, toward the grounded side.

Matching can be by a $\lambda/4$ section, as in Fig 2 above. An alternate is to cut a slot at the center of the plate, connecting the feed line to the bottom of this, as shown in **Fig 4**. The notch depth is essentially that given by

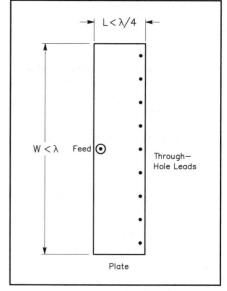

Fig 3—Design of a $\lambda/4$ antenna, essentially half of Fig 1, grounded along the line of symmetry. See text.

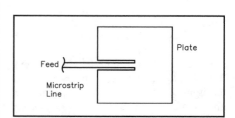

Fig 4—Another matching method produced by cutting a slot into the plate at the feed point, and feeding the bottom of the slot. The impedance is calculated using Eq 5.

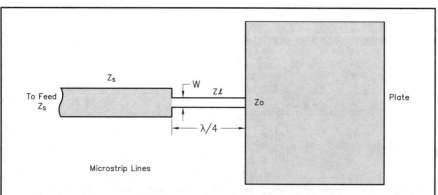

Fig 2—An alternative feed system using microstrip transmission line. The ground plane is present as in Fig 1. The wide section of line is designed for 50 Ω, and the narrow $\lambda/4$-matching section for 80 Ω, the square-root mean of 50 Ω and the 130 Ω patch impedance.

Eq 5, with Z_0 set to the edge-feed impedance of a λ/4 plate, or approximately 260 Ω.

Other shapes and patch dimensions are also possible. Some shapes that have been successfully used are triangles, pentagons, disks, and rings. In the full-wave mode, the plate area will be approximately equal to that of the rectangular plate above. Just as with conventional dipoles that are 1, 2, 3, or more half-waves in length, higher modes are possible with patch antennas. These take the form of larger plates. Such modes, however, are not often found, because of the complex patterns associated with them.

It is not necessary for the ground plane and plate to be a flat surface. Cylinders are often used for the ground "plane," as well as such curved surfaces, such as the skin of an aircraft.

A useful variation on Fig 1 is to make the plate exactly square, of dimension L. With a feed to two adjacent sides, the plate will radiate a signal that is both horizontally and vertically polarized. If the sides are fed 90° out of phase, the signal will be circularly polarized. Two possible feeds are shown in **Fig 5**. The more complex feed of **Fig 6** allows choice of right- or left-hand circular polarization by reversing the position of the feed line and the load resistor.

As with conventional antenna types, several patches can be arranged and fed as an array. **Fig 7** shows a typical technique. The equal line lengths mean that the maximum radiation is broadside to the ground plane. Drive-point to plate matching is secured by selecting the widths of the feed line appropriately. The maximum gain that can be developed is:

$$G_{max} = 10 \log \left(\frac{4\pi A}{\lambda_0 / \lambda_0} \right) - ATT \times (D_1 + D_2)/2$$

(Eq 9)

where

$D_1 = (n+1) \times H_{spacing}$
$D_2 = (m+1) \times V_{spacing}$
$A = D_1 \times D_2$
ATT = attenuation per unit length of line

and

n, m are the number of plates in the horizontal and vertical directions.

Differing line lengths can be used to feed one set of plates with a phase difference compared to another set. As with conventional antennas, lobe switching or even lobe scanning is possible. A simple switch is shown in **Fig 8**. Reversing the bias on the PIN diodes causes the line length to change from L to (L+ dL). Four such devices in cascade with the dLs arranged binomially would allow 16 beam positions.

In the figures above, the lines have been shown as joining at right angles, and making right-angle bends. This is not the best way, since each introduces a discontinuity, increasing the reflection and the line SWR.

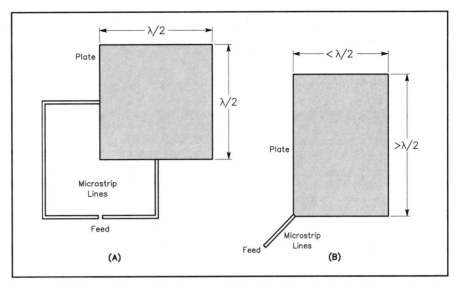

Fig 5—Feeds to produce a circularly polarized signal. At A, the needed 90° phase shift is produced by line sections differing by λ/4. At B, phase shift is produced by making the plate slightly rectangular, to have equal reactance of opposite signs for the two plate directions. The conductance and susceptance terms must be equal in magnitude, so this is a very narrow bandwidth design.

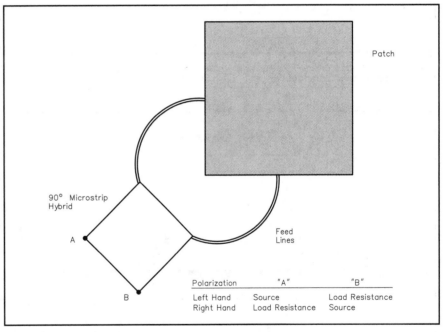

Polarization	"A"	"B"
Left Hand	Source	Load Resistance
Right Hand	Load Resistance	Source

Fig 6—Circularly polarized antenna using a ring divider to give either +90° or −90° phase shift, and right or left-hand circular polarization, by interchanging the input and load resistance.

Gradual curves are best. Fig 7 shows one possible arrangement. As an added refinement, this also shows a short section of tapered line, where the two thin lines join into a wider one, again to reduce reflection. There is another feed variation, series-feed from one plate to another. The two methods are analogous to feed systems used with dipoles, for broadside and end-fire arrays.

There is ample room for ingenuity in the layout of large and special-purpose arrays.

There are also other methods of feeding, and of array types, including power tapering for side-lobe control and traveling-wave structures. See the references for examples of these.

As amateur use of the centimeter wave bands increases, patch antennas will become more common. In the range 200 to 900 MHz, air-insulated designs can be used. From about 500 to 20 GHz, the insulation substrate type seems far better. For

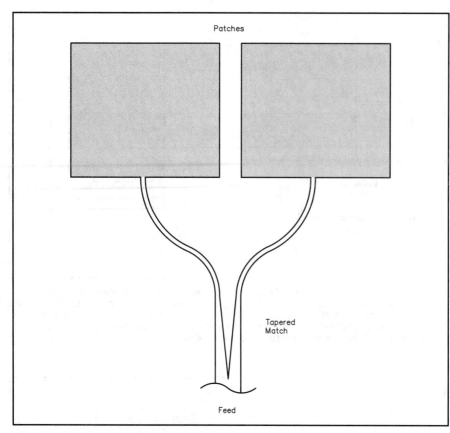

Patches

Tapered
Match

Feed

Fig 7—Two patches phased for increased gain. A tapered match is used to feed the two patches in this broadside array.

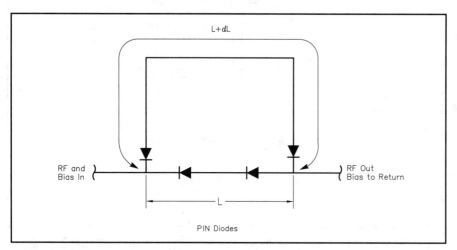

L+dL

RF and
Bias In

RF Out
Bias to Return

L

PIN Diodes

Fig 8—PIN diode switch to change length of a transmission line. This technique is often used for lobe switching. A number of such switches in cascade may be used for stepped-beam position scanning.

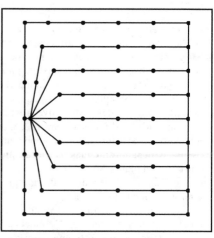

Fig 9—Possible wire and segment arrangement for simulating a patch antenna as a wire grid. This can be analyzed by a method-of-moments program such as *MININEC* or *NEC*. *NEC* is recommended, since it gives better accuracy with close-spaced wires.

H- and E-plane pattern. Another is for several types of transmission line, including microstrip. Programs for calculating patterns of arrays are included. Any of a number of commercial or public domain array programs may also be used.

Modeling of patch antennas by programs such as *MININEC* or *NEC* is possible. *NEC* includes specific provisions for patches and patch radiation, but no provision for a dielectric above a ground plane. In principle, either program can model a plate antenna by using a wire grid approximation. A major point in setting these up is that there must be a direct path from the feedpoint to all high current areas.

Fig 9 shows a possible wire and segment arrangement for a model of a half-wave patch. Usually these models find that the far-field radiation pattern is reasonably accurate, but that the drive impedance is often considerably in error. Nevertheless, such modeling is useful if unusual shapes must be designed.

Notes and References
[1] John D. Kraus, *Antennas*, 2nd edition (New York: McGraw-Hill, 1988).
[2] Richard C. Johnson and Henry Jasik, *Antenna Engineering Handbook*, 2nd edition (New York: McGraw-Hill, 1984).
[3] I. J. Bahl and P. Bhartia, *Microstrip Antennas* (Dedham, MA: Artech House, 1982).
[4] David Pozar, *Antenna Design Using Personal Computers* (Dedham, MA: Artech House, 1985).
[5] Chuan-Li Chi and Nicolaos G. Alexopoulos, "An Efficient Approach for Modeling Microstrip-Type Antennas," *IEEE Trans. Ant. Prop.*, Vol 38, No 9, Sep 1990.
[6] James R. Fisk, "Microwave Transmission Line," *Ham Radio*, Jan 1978.
[7] Technical Correspondence, *QST*, Jul 1992.

example, $^1/_8$-inch thick fiberglass-Teflon or even fiberglass-epoxy is good at 1200 MHz. $^1/_{32}$-inch glass-Teflon would be suitable for 20 GHz. Anyone with experience in PC-board construction will be at home with these designs.

The relations given here are of sufficient accuracy for fabrication of experimental antennas, but trimming may be required to get the exact design frequency or feed impedance. With care, this can be done with a motorized grinding wheel to reduce size. Plating with solder will increase size somewhat, especially if thick foil board is used. Soldering small trim tabs at high voltage points has been used also.

For serious work, the more accurate expressions found in the Notes and References should be used in design. Reference 4 is most useful, since it includes BASIC programs for design. One is for half-wave patches, and includes calculation of the

Linear-Loaded Short Wire Antennas

By John Stanford, NNØF
1327 Clark Ave
Ames, IA 50010
e-mail: stanford@iastate.edu

NNØF measures and then models linear-loaded antennas to take some of the mystery out of this very useful method of reducing antenna size without compromising efficiency.

*T*he *ARRL Antenna Book* says commercial antenna manufacturers use "linear loading" but that amateurs generally have not taken much advantage of these techniques. This stimulated my interest. What *is* linear loading anyway? I think I have found out something about it, and decided to share what I learned.

One thing I found is that linear loading can significantly reduce the required length for resonant antennas. For example, it is easy to make a resonant antenna that is as much as 30 to 40% shorter than an ordinary dipole for a given band. The shorter overall lengths come from bending back some of the wire. The increased self-coupling lowers the resonance frequency. These ideas are applicable to short antennas for restricted space or portable use.

First I'll discuss experiments I made with linear-loaded antennas. At the end I will also tell about computer modeling of these systems. The calculations nicely support the measurements I made, and allow a simple rule to be given to aid in the design of short linear-loaded wire antennas.

Experiments

My experiments began after I happened across a short article that (ever so sketchily) described some experiments with linear loading. It was from a paper "A New Class of Wire Antennas" presented at an antenna symposium some years ago by J. Rashed and C. Tai.[1] The article unfortunately did not explain things as completely as I would have liked, but it stimulated me to try some experiments of my own. Here's a summary of what I learned so far.

I first did a number of experiments with linear loaded monopoles, working over my old 48-radial ground system of electric fence wire. The nine-year old wires are now rusting, but are still evidently a reasonably good ground system. My measurements in-

clude ground losses, but these are evidently low, since an ordinary quarter-wave vertical measures within an ohm or so of the theoretical, perfect-ground value of 36-Ω. I used an MFJ-204B Antenna Bridge and Optoelectronics frequency counter.

The results of my measurements are shown in **Fig 1** and are also consistent with values given by Rashed and Tai. This shows several simple wire antenna configurations, with resonant frequencies and impedance (radiation resistance). The reference dipole has a resonant frequency f_0 and resistance R = 72 Ω. The f/f_0 values give the effective reduced frequency obtained with the linear loading in each case. For example, the two-wire linear-loaded dipole has resonance frequency lowered to about 0.67 to 0.70 that of the simple reference dipole of the same length.

The three-wire linear-loaded dipole has its frequency reduced to 0.55 to 0.60 of the simple dipole of the same length. As you will see later, these values will vary with conductor diameter and spacing.

The two-wire linear-loaded dipole (Fig 1B) looks almost like a folded dipole but, unlike a folded dipole, it is open in the

middle of the side opposite where the feedline is attached. My measurements show that this antenna structure has a resonant frequency lowered to about two-thirds that

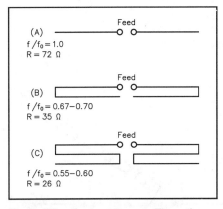

Fig 1—Wire dipole antennas. The ratio f/f$_0$ is the measured resonant frequency divided by frequency f$_0$ of a standard dipole of same length. R is radiation resistance in ohms. At A, standard single-wire dipole. At B, two-wire linear-loaded dipole, similar to folded dipole except that side opposite feed line is open. At C, three-wire linear-loaded dipole.

of the reference dipole, and R equal to about 35 Ω. A three-wire linear-loaded dipole (Fig 1C) has even lower resonance frequency and R about 25 to 30 Ω.

Linear-loaded monopoles (one half of the dipoles in Fig 1) working against a radial ground plane have similar resonant frequencies, but with only half the radiation resistance shown for the dipoles.

A Ladder-Line Linear-Loaded Dipole

Based on these results, I next constructed a linear-loaded dipole as in Fig 1B, using 24 feet of 1-inch ladder line (the black, 450-Ω plastic kind widely available) for the dipole length. I hung the system from a tree using nylon fishing line, about 4 feet from the tree at the top, and about 8 feet from the ground on the bottom end. It is slanted at about a 60° angle to the ground. This antenna resonated at 12.8 MHz and had a measured resistance of about 35 Ω. After the resonance measurements, I fed it with 1-inch ladder open-wire line (a total of about 100 feet to my shack).

For brevity, I will call this a vertical *LLSD* (linear-loaded short dipole). A tuner resonated the system nicely on 20 and 30 meters. On these bands the performance of the vertical LLSD seemed comparable to my 120-foot long, horizontal center-fed Zepp, 30 feet above ground. In some directions where my horizontal, all-band Zepp has nulls, such as towards Siberia, the vertical LLSD was definitely superior. This system also resonates on 17 and 40 meters. However, from listening to various signals, I have the impression that this length LLSD is not as good on 17 and 40 meters as the horizontal 120-foot antenna.

Using Capacitive "End Hats"

I also experimented with an even shorter resonant length by trying a LLSD with capacitive "end-hats." The hats, as expected, increased the radiation resistance and lowered the resonant frequency. Six-foot long, single-wire hats were used on each end of the previous 24-foot LLSD, as shown in **Fig 2**. The end wires were self-supporting, springy copperclad-steel wire from old telephone lines. This antenna was supported in the same way as the previous vertical dipole, but the bottom-end hat wire was only inches from the grass. This system resonated at 10.6 MHz with a measured resistance of 50 Ω.

If the dipole section were lengthened slightly, by a foot or so, to about 25 feet, it should hit the 10.1-MHz band and be a good match for 50-Ω coax. It would be suitable for a restricted space, shortened 30-meter antenna. Note that this antenna is only about half the length of a conventional 30-meter dipole, needs no tuner, and has no losses due to traps. It does have the loss of the extra wire, but this is essentially negligible.

Any of the linear-loaded dipole antennas can be mounted either horizontally or vertically. The vertical version can be used for longer skip contacts—beyond 600 miles or so—unless you have rather tall supports for horizontal antennas to give a low elevation angle. Finally, I will mention that using dif-ferent diameter conductors in linear-loaded antenna configurations yields different results depending on whether the larger or small diameter conductor is fed. I experimented with a vertical ground plane antenna using a 10-foot piece of electrical conduit pipe (5/8 inch OD) and #12 copper house wire.

Fig 3 shows the configuration. The ground radial system was buried a couple of inches under the soil and is not shown. Note that this is not a folded monopole, which would have either A or B grounded.

The two conductors were separated by 2 inches, using plastic spreaders held onto the pipe by stainless-steel hose clamps obtained from my local hardware store. Hose clamps intertwined at right angles were also used to clamp the pipe on electric fence stand-off insulators on a short 2×4 post set vertically in the ground.

The two different diameter conductors make the antenna characteristics change, depending on how they are configured. With the antenna bridge connected to the larger diameter conductor (point A in Fig 3), and point B unconnected, the system resonated at 16.8 MHz and had R = 35 Ω. With the bridge at B (the smaller conductor), and point A left unconnected, the resonance lowered to 12.4 MHz and R was found to be about 24 Ω.

The resonant frequency of the system in Fig 3 can be adjusted by changing the over-all height, or for increasing the frequency, by reducing the length of the wire. Note that a 3.8-MHz resonant ground plane can be made with height only about half that of the usual 67 feet required, if the smaller conductor is fed (point B in Fig 3). In this case, the pipe would be left unconnected electrically. The lengths I give above can be scaled to determine a first-try attempt for your favor-

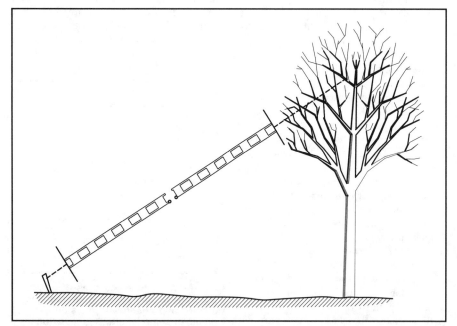

Fig 2—Two-wire linear-loaded dipole with capacitive end hats. Main dipole length was constructed from 24 feet of "windowed" ladder line. The end-hat elements were stiff wires 6 feet long. The antenna was strung at about a 60° angle from a tree limb using monofilament fishing line. Measured resonant frequency and radiation resistance were 10.6 MHz and 50 Ω.

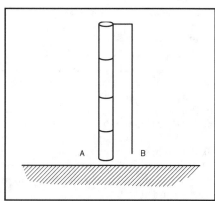

Fig 3—Vertical ground-plane antenna with a 10-foot pipe and #12 wire as the linear-loaded element. Resonant frequency and radiation resistance depend on which side (A or B) is fed. The other side (B or A) is *not* grounded. See text for details.

ite band. Resonant lengths will, however, depend on the conductor diameters and spacing.

The same ideas hold for a dipole, except that the lengths should be doubled from those of the ground plane in Fig 3. The resistance will be twice that of the ground plane. Say, how about a shortened 40-meter horizontal beam to enhance your signal?!

Modeling Linear-Loaded Dipoles

I later performed modeling calculations for the linear-loaded systems I measured. I purchased a copy of W7EL's *EZNEC* antenna modeling software.[2] It is an excellent program, user friendly and fun to experiment with. I modeled the system shown in Fig 1B, using #18 wire (to approximate my ladder line wires) spaced 1 inch apart, 30 feet above ground. I ran the model with a real ground with similar conductivity and dielectric constant as that in my area (very good ground). Due to the small spacing between the wires, a large number of seg-

ments are needed to obtain convergence in the calculations. After tests, I used 499 wire segments for reliable results.

The model calculations very closely matched my experimental measurements. For example, choosing a length of 30 feet for the linear-loaded dipole, the resonant frequency was predicted to be 10.15 MHz, corresponding to $f/f_0 = 0.698$, and radiation resistance of 34 Ω. These values are very close to my observations (see Fig 1B) of $f/f_0 = 0.66$ to 0.70 and R = 35 Ω. The calculations were for bare wires. Using windowed plastic ladder line should be roughly 5% lower f/f_0, near the 0.66 value measured due to the loading effect of the insulation.

From these calculations, one can predict a resonant length of a two-wire linear-loaded dipole of Fig 1B using the approximate formula:

L = 310/f (using windowed ladder line)

or

L = 327/f (using bare #18 wires)

where L will be in feet, if f is in MHz.

The calculated bandwidth at SWR = 2:1 is approximately 2.5% for the two-wire linear-loaded dipole (Fig 1B). If you like, you may match the 35-Ω impedance of the two-wire loaded dipole to 50 Ω with a simple gamma match. Since it is only about two-thirds as long as a standard dipole, this configuration makes a nice antenna for reduced-space or portable situations.

In summary, I've enjoyed experimenting with linear-loaded antennas. They offer interesting possibilities for reduced size antennas. I hope you will be stimulated to try a variation or two yourself. Have some fun experimenting and let the rest of us know what you find out!

Notes and References

[1] J. Rashed and C. Tai, "A New Class of Wire Antennas," 1982 International Symposium Digest, Antennas and Propagation, Vol 2, IEEE.
[2] *EZNEC*, available from Roy Lewallen, W7EL, PO Box 6658, Beaverton, OR 97007.

Multiband Antennas

The Coupled-Resonator Principle: A Flexible Method for Multiband Antennas

By Gary Breed, K9AY
7318 S Birch Street
Littleton, CO 80122

K9AY describes the evolution and refinement of the Coupled-Resonator Principle.

In 1995, *QST* published two antenna designs that use an interesting technique to get multiband coverage in one antenna. Rudy Severns, N6LF, described a wideband 80 and 75-meter dipole using this technique, and Robert Wilson, AL7KK, showed us how to make a three-band vertical. Both of these antennas achieve multi-frequency operation by placing resonant conductors very close to a driven dipole or vertical—with no physical connection. Since the principle that makes these antennas possible is not well known in the amateur community, I wanted to offer some background.

The Coupled-Resonator Principle

As we all know, nearby conductors can interact with an antenna. Our dipoles, verticals and beams can be affected by nearby power lines, rain gutters, guy wires and other metallic materials. The antennas designed by Severns and Wilson use this interaction intentionally, to combine the resonances of several conductors at a single feedpoint. While other names have been used, I call the behavior that makes these antennas work the *coupled-resonator* (C-R) principle.

Take a look at **Fig 1**, which illustrates the general idea. Each figure shows the SWR at the feedpoint of a dipole, over a range of frequencies. When this dipole is all alone, it will have a very low SWR at its half-wave resonant frequency (Fig 1A). Next, if we take another wire or tubing conductor and start bringing it close to the dipole, we will see a "bump" in the dipole's SWR at the resonant frequency of this new wire. See

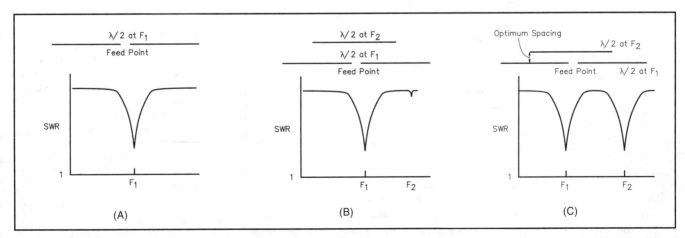

Fig 1—At A, the SWR of a dipole over a wide frequency range. At B, a nearby conductor is just close enough to interact with the dipole. At C, when the second conductor is at the optimum spacing, the combination is matched at both frequencies.

Fig 1B. We are beginning to see the effects of interaction between the two conductors. As we bring this new conductor closer, we reach a point where the SWR "bump" has grown to a very deep dip—a low SWR. We now have a good match at both the original dipole's resonant frequency and the frequency of the new conductor, as illustrated in Fig 1C.

We can repeat this process for several more conductors at other frequencies to get a dipole with three, four, five, six, or more resonant frequencies. The principle also applies to verticals, so any reference to a dipole can be considered to be valid for a vertical, as well.

We can write a definition of the C-R principle this way:

Given a dipole (or vertical) at one frequency and an additional conductor resonant at another frequency, there is an optimum distance between them that results in the resonance of the additional conductor being imposed upon the original dipole, resulting in a low SWR at both resonant frequencies.

Some History

In the late 1940s, the *coaxial sleeve* antenna was developed (**Fig 2A**), covering two frequencies by surrounding a dipole or monopole with a cylindrical tube resonant at the higher of the desired frequencies. In the 1950s, Gonset briefly marketed a two-band antenna based on this design. Other experimenters soon determined that two conductors at the second frequency, placed on either side of the main dipole or monopole, would make a skeleton representation of a cylinder (Fig 2B). This is called the *open-*

sleeve antenna. The Hy-Gain Explorer tribander uses this method in its driven element to obtain resonance in the 10-meter band. Later on, a few antenna developers finally figured out that these extra conductors did not need to be added in pairs, and that a single conductor at each frequency could add the extra resonances (Fig 2C). This is the method used by Force 12 in some of their multiband antennas.

This is a perfect example of how science works. A specific idea is discovered, with later developments leading to an underlying general principle. The original coaxial-sleeve configuration is the most specific, being limited to two frequencies and requiring a particular construction method. The open-sleeve antenna is an intermediate step, showing that the sleeve idea is not limited to one configuration.

Finally, we have the coupled-resonator concept, which is the general principle, applicable in many different antenna configurations, for many different frequency combinations. Severns' antenna uses it with a folded dipole, and Wilson uses it with a main vertical that is off-center fed. I've used it with conventional dipoles and quarter-wave verticals. Other designers have used the principle more subtly, like putting the first director in a Yagi very close to the driven element, broadening the SWR bandwidth the same way Severns' design does with a dipole.

In the past, most antennas built with this single-conductor technique have also been called *open-sleeve* (or *multiple-open-sleeve*) antennas, a term taken from the history of their development. However, the term *sleeve* implies that one conductor must surround another. This is not really a physical or elec-

trical description of the antenna's operation, therefore, I suggest using the term *coupled-resonator,* which I believe is the most accurate description of the general principle.

A Little Math

The interaction that makes the C-R principle work is not random. It behaves in a predictable, regular manner. I have derived an equation that shows the relationship between the driven element and the additional resonators for ordinary dipoles and verticals:

$$\frac{\log d}{\log (D/4)} = 0.54 \qquad \text{(Eq 1)}$$

where

  d = distance between conductors, *measured in wavelengths* at the frequency of the chosen additional resonator
  D = the diameter of the conductors, also in wavelengths at the frequency of the additional resonator.

Eq 1 assumes they are both the same diameter and that the feedpoint impedance at both frequencies is the same as a dipole in free space (72 Ω) or a quarter-wave monopole over perfect ground (36 Ω).

The equation only describes the impedance due to the additional resonator. The main dipole element is always part of the antenna, and it may have a fairly low impedance at the additional frequency. This is the case when the frequencies are close together, or when the main element is operating at its third harmonic. At these frequencies, the spacing distance must be adjusted so that the parallel combination of dipole and resonator results in the desired feedpoint impedance.

I worked out two correction factors, one to cover a range of impedances and another for frequencies close together. These can be included in the basic equation, which is rearranged below to solve for the distance between the conductors:

$$d = 10^{0.54 \log(D/4)} \times \frac{Z_0 + 35.5}{109} \times$$
$$\left[1 + e^{[((((F_2/F_1)-1.1)\times 11.3)+0.1)]}\right] \qquad \text{(Eq 2)}$$

where

  d and D are the same as above.
  Z_0 = the desired feed-point impedance at the frequency of the additional resonator (between 20 and 120 Ω). For a vertical, multiply the desired impedance by two to get Z_0 if you want a 50-Ω feed, use 100 for Z_0.
  F_1 = the resonant frequency of the main dipole or vertical.
  F_2 = the resonant frequency of the additional conductor. The ratio F_2/F_1 is more than 1.1.

Eq 2 does not directly allow for conductors of unequal diameters, but it can be used as a starting point if you use the diameter of the driven dipole or vertical element for D

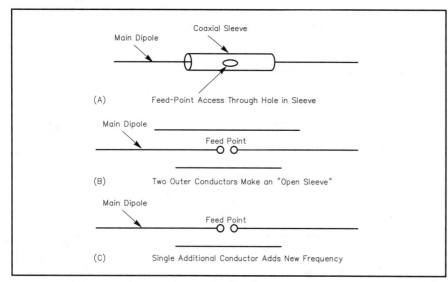

Fig 2—Evolution of coupled-resonator antennas: At A, the *coaxial-sleeve* dipole; at B, the *open-sleeve* dipole; and at C, a *coupled-resonator* dipole, the most universal configuration.

in the equation. Work is underway to refine the equation so it applies to nearly any combination of conductors and frequencies.

Characteristics of C-R Antennas

Here's the important stuff—what's different about C-R antennas, what are they good for and what are their drawbacks? The key points are:

- Multiband operation without traps, stubs or tuners
- Flexible impedance matching at each frequency
- Independent fine-tuning at each frequency (little interaction)
- Easily modeled using *MININEC* or *NEC*-based programs
- Pruning process same as a simple dipole
- Can accommodate many frequencies (seven or more)
- Virtually lossless coupling (high efficiency)
- Requires a separate wire or tubing conductor at each frequency
- Mechanical assembly requires a number of insulated supports
- Narrower bandwidth than equivalent dipole
- Capacitance requires slight lengthening of conductors

To begin with, the most obvious characteristic is that this principle can be used to add multiple resonant frequencies to an ordinary dipole or vertical, using additional conductors that are *not physically connected*. This gives us three variable factors: (1) the diameter of the conductor, (2) its length, and (3) its position relative to the main element.

Having the freedom to control these factors gives us the advantage of *flexibility*; we have a wide range of control over the impedance at each added frequency. Another advantage is that the behavior at each frequency is quite *independent*, once the basic design is in place. In other words, making fine-tuning adjustments at one frequency doesn't change the resonance or impedance at the other frequencies. A final advantage is *efficiency*. With conductors close together, and with a resonant target conductor, coupling is very efficient. Traps, stubs, and compensating networks found on other multiband antennas all introduce lossy reactive components.

There are two main disadvantages of C-R antennas. The first is the relative *complexity* of construction. Several conductors are needed, installed with some type of insulating spacers. Other multiband antennas have their complexities as well (such as traps that need to be mounted and tuned), but C-R antennas will usually be bulkier. The larger size generally means greater windload, which is a disadvantage to some hams.

The other significant disadvantage is *narrower bandwidth*, particularly at the highest of the operating frequencies. We can partially overcome this problem with large conductors that are naturally broad in bandwidth, and in some cases we might even use an extra conductor to put two resonances in one band. It is interesting to note that the pattern is opposite that of trapped antennas. The C-R antenna gets narrower at the highest frequencies of operation, while trap antennas generally have narrowest bandwidth at their lowest frequencies.

There are two special situations that should be noted. First, when the antenna has a resonance near the frequency where the driven dipole is $\frac{3}{2}\lambda$ long ($\frac{3}{4}\lambda$ for a vertical), the dipole has a fairly low impedance. The spacing of the C-R element needs to be increased to raise its impedance so that the parallel combination of the main element and C-R element equals the desired impedance (usually 50 Ω). There is also significant antenna current in the part of the main dipole extending beyond the C-R section, contributing to the total radiation pattern. As a result, this particular arrangement radiates as three $\lambda/2$ sections in phase, and has about 3 dB gain and a narrower directional pattern compared to a dipole (**Fig 3**). This might be an advantage for antennas covering bands with a frequency ratio of about three, such as 3.5 and 10.1 MHz, 7 and 21 MHz, or 144 and 430 MHz.

The other special situation is when we want to add a new frequency very close to the resonant frequency of the main dipole. An antenna for 80 and 75 meters would be an example of this. Again, the driven dipole has a fairly low impedance at the new frequency. Add the fact that coupling is very strong between these similar conductors and we find that a wide spacing is required to make the antenna work. A dipole resonant at 3.5 MHz and another wire resonant at 3.8 MHz will need to be three or four feet apart, while a 3.5 MHz and 7 MHz combination might only need to be spaced four or five inches.

Another useful characteristic of C-R antennas is that they are easily and accurately modeled by computer programs based on either *MININEC* or *NEC*, as long as you stay within each program's limitations. For example, Severns points out that *MININEC* does not handle folded dipoles very well, and *NEC* modeling is required. With ease of computer modeling, a precise answer isn't needed for the design equation given above. An approximate solution will provide a starting point that can quickly be adjusted for optimum dimensions.

The added resonators have an effect on the lengths of all conductors, due to the capacitance between the conductors. Capacitance causes antennas to look electrically shorter, so each element needs to be about 1% or 2% longer than a simple dipole at the same frequency. As a rule of thumb, use 477/f (in feet) instead of the usual 468/f when calculating dipole length, and

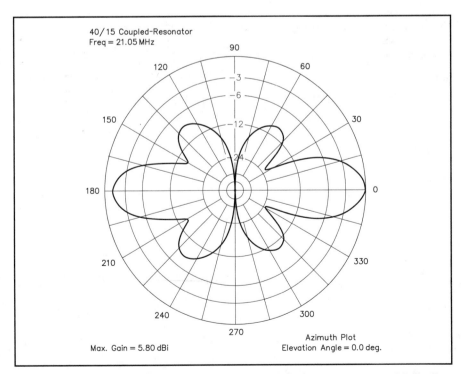

40/15 Coupled-Resonator
Freq = 21.05 MHz

Max. Gain = 5.80 dBi

Azimuth Plot
Elevation Angle = 0.0 deg.

Fig 3—Radiation pattern for the special case of a C-R antenna with the additional frequency at the third harmonic of the main dipole resonant frequency.

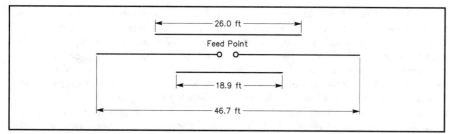

Fig 4—Dimensions of a C-R dipole for the 30, 17 and 12-meter bands.

239/f instead of 234/f for a λ/4 vertical.

A 30/17/12-Meter Dipole

To show how a C-R antenna is designed, let's build a dipole to cover all three WARC bands. We'll use #12 wire, which has a diameter of 0.08 inches, and the main dipole will be cut for the 10.1 MHz band. From the equation above (I'll skip the detailed math), the spacing between the main dipole and the 18-MHz resonator should be 2.4 inches for 72 Ω, or 1.875 inches for 50 Ω. At 24.9 MHz, the spacing to the resonator for that band should be 2.0 inches for 72 Ω, or 1.62 inches for 50 Ω. Of course, this antenna will be installed over real ground, not in free space, so these spacing distances may not be exact. Plugging these numbers into your favorite antenna modeling program will let you optimize the dimensions for installation at the height you choose. Remember, if the antenna is very close to ground (less than λ/4), *MININEC*-based programs may not give accurate results, and a *NEC*-based program should be used.

For those of you who like to work with real antennas, not computer-generated ones, the predicted spacing is accurate enough to build an antenna with minimum trial-and-error. I suggest using a nice round number just larger than the calculated spacing for 50 Ω. For this antenna, I decided that the right spacing for the height I wanted would

be 2 inches for the 18-MHz resonator and 1.8 inches for the 24.9-MHz resonator. For simplicity of construction, I just used 2 inches for both, figuring that the worst I would get is a 1.2:1 SWR if the numbers were a little bit off. Like all dipoles, the impedance varies with height above ground, but the 2-inch spacing results in an excellent match on the two additional bands, at heights of more than 25 feet.

The final dimensions of the dipole for 10.1, 18.068 and 24.89 MHz are shown in **Fig 4**. These are the final pruned lengths for a straight dipole installed at a height of about 40 feet. If you put up the antenna as an inverted V, you will need each wire to be a bit longer. Pruning this type of antenna is just like a dipole—if it's resonant too low in frequency, it's too long and the appropriate wire needs to be shortened. So, you can cut the wires just a little long to start with and easily prune them to resonance.

A final note: if you want to duplicate this antenna design, remember that the 2-inch spacing is just for #12 wire! The required spacing for a C-R antenna is related to the conductor diameter. This same antenna built with #14 wire needs under 1½-inch spacing, while a 1-inch aluminum tubing version requires about 7-inch spacing.

Summary

The coupled-resonator principle is one more weapon in the antenna designer's arsenal. It's not a well-known concept (although I'm trying to change that) and it's not the perfect method for all multiband antennas. What the C-R principle offers is an alternative to traps and tuners, in exchange for using more wire or aluminum. Although a C-R antenna requires more complicated construction, its main attraction is in making a multiband antenna that can be built with no compromise in matching or efficiency.

Notes and References

[1] Rudy Severns, N6LF, "A Wideband 80-Meter Dipole," *QST*, July 1995.

[2] Robert Wilson, AL7KK, "The Offset Multiband Trapless Antenna (OMTA)," *QST*, October 1995.

[3] Roger Cox, WB0DGF, "The Open-Sleeve Antenna: Development of the Open-Sleeve Dipole and Open-Sleeve Monopole for HF and VHF Amateur Applications," *CQ*, August 1983.

[4] Bill Orr, W6SAI, "Radio FUNdamentals," The Open-Sleeve Dipole, *CQ*, February 1995.

[5] U.S. Patent 5,489,914, "Method of Constructing Multiple-Frequency Dipole or Monopole Antenna Elements Using Closely-Coupled Resonators," Gary A. Breed, Feb 6, 1996.

[6] Gary A. Breed, "Multi-Frequency Antenna Technique Uses Closely-Coupled Resonators," *RF Design*, November 1994.

[7] *MININEC*-based programs that are readily available include *ELNEC*, from Roy Lewallen, W7EL, PO Box 6658, Beaverton, OR 97007; and *AO*, from Brian Beezley, K6STI, 3532 Linda Vista, San Marcos, CA 92069.

[8] *NEC*-based programs that are readily available include *EZNEC*, from Roy Lewallen, W7EL, PO Box 6658, Beaverton, OR 97007; *NEC/Wires*, from Brian Beezley, K6STI, 3532 Linda Vista, San Marcos, CA 92069; and *NEC-Win Basic*, from Paragon Technology, Inc, 200 Innovation Blvd, Suite 240, State College, PA 16803.

A Homebrew Seven-Band Vertical

By Paul I. Protas, WA5ABR
15709 Seattle Street
Houston, TX 77040

WA5ABR describes a versatile, efficient vertical for 40 through 10 meters. It's easy to build, costs much less than an equivalent commercial antenna, and works great!

I built this antenna to replace an aging four-band trap vertical that had been in service for about 12 years. I had looked around at various manufactured verticals but decided to build one myself. This antenna should out-perform any trap vertical and should be at least comparable to any of the "half-wave" or "elevated-feed" products on the market today. My vertical cost around $130, considerably less than most commercially available units. If you have any scrap aluminum tubing from other projects, your cost could be even less.

My design philosophy is that if a full-size λ/2 dipole is an efficient radiator, then a full-size λ/4 radiator would also be efficient, given a good ground system, of course. On 30 through 10 meters, the vertical uses six full-sized λ/4 elements. On 40 meters a loading coil is used at the 16-foot level, with a 4-foot whip above the coil. You could elect to use a full-sized 40-meter element, but this will definitely mandate the use of guy ropes.

There are no traps or matching circuits. One feed line is used, connected to all seven elements in parallel at the base. See the drawing in **Fig 1** and the photo of the assembled antenna in **Fig 2**. All elements are made using 0.058-inch-thick aluminum tubing so that they may be telescoped easily.

If you want to build a really low-cost version, you might consider making your antenna like the wire prototype I made before changing to aluminum tubing in the final design. In the prototype I used #14 wire supported by a 23-foot tall 2×2 wood support, with wooden discs at the top and bottom to separate the individual wires. I used heavy fishing line to support the shorter elements from the top disc. While it certainly was not as durable as the final aluminum version, the prototype worked fine and only cost about $20.

Construction

See **Table 1** for the Parts List to build this antenna. To start construction of the aluminum-tubing version, cut two 15-inch diameter discs out of acrylic sheet. In one disc, drill six ⁵⁄₈-inch holes at points 6³⁄₄ inches out from the center of the circle (center to center) at 60° intervals. Drill a 1¹⁄₂-inch hole in the center. This will be the bottom support. In the second disc, drill six holes as above, but make two opposing holes ⁵⁄₈ inch and the other four ¹⁄₂-inch OD. The center hole in this disc should be 1¹⁄₂ inch. This will be the top support. See the photo in **Fig 3**, showing details of the bottom support disc.

From the remainder of the acrylic sheet, cut a rectangular piece 2¹⁄₂ by 15 inches. Drill a 1³⁄₈-inch hole in the exact center of this piece

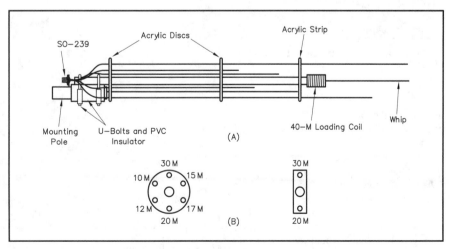

Fig 1—At A, layout of WA5ABR seven-band vertical. The acrylic discs at top and bottom support the aluminum-tubing elements. The acrylic strip at the top keeps the longer elements stable. At B, detail for the acrylic discs and strip.

113

Fig 2—Photo of assembled antenna at WA5ABR.

Table 1

Parts List

Qty	Description
3	12 foot sections, $1/2$-inch OD
4	12 foot sections, $5/8$-inch OD
1	6 foot section, $1/2$-inch OD
1	6 foot section, $1 3/8$-inch OD
1	12 foot section, $1 1/2$-inch OD
1	48 inch aluminum-wire whip
1	2 foot section, $1 1/4$-inch Schedule-40 PVC
7	solder lugs, $1/8$-inch holes
1	SO-239 connector
1	34 foot length of insulated #14 solid copper wire
1	36×18-inch sheet of $1/4$-inch thick acrylic (Plexiglas)
2	2-inch U-bolt clamps
20	$3/4$-inch hose clamps, non-rusting
9	$1 1/2$-inch hose clamps, non-rusting
7	$6-32 \times 3/8$-inch screws, lockwashers, and nuts

Note: all aluminum tubing is T-6061 material, with 0.058-inch-thick walls.

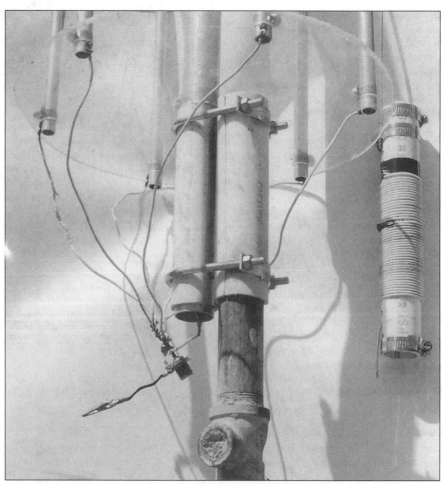

Fig 3—Photo detail of the bottom acrylic disc, with insulating "clam-shell" clamps of PVC pipe around 40-meter element and mounting pipe using U-bolts. The seven elements are connected in parallel to the SO-239 connector. In this photo an alligator clip was used to connect to the ground radials during experimentation.

and two $1/2$-inch holes, $1/2$ inch in from each end. This will support the top two longest elements.

Next, separate the aluminum tubing into three batches according to size. The $1 1/2$ and $1 3/8$-inch sections can be telescoped and set aside as one batch. The other two batches will be the $1/2$-inch and the $5/8$-inch OD pieces. Cut two 12-foot pieces of $5/8$-inch tubing in half (each 6 feet long). Cut one piece of $1/2$-inch tubing into a 4-foot and an 8-foot section. Cut another piece of $1/2$-inch tubing into two 6-foot sections. The remaining two 12-foot sections of ⁵/₈-inch tubing and the one remaining 12-foot section of ¹/₂-inch tubing will not be cut.

You should now have six sections of each size smaller tubing. Drill a ¹/₈-inch hole, ¹/₄ inch from the end, of each of these sections. Saw two opposing slots ¹/₂-inch deep into the other end of each section of ⁵/₈-inch tubing. This will be the place where the ³/₄-inch hose clamps tighten around the next telescoping section. These hose clamps can now be attached loosely. Telescope the different sections of tubing using **Table 2** as a preliminary guide. You will adjust the final length for SWR at installation. The starting lengths may seem to be short, but the length of wire attached to the SO-239 coaxial connector makes each element electrically longer.

Drill a hole, similar to the ones drilled in the ⁵/₈-inch sections, in the end of the 12-foot section of 1¹/₂-inch tubing. Also, cut two opposing slots ¹/₂-inch deep in the other end of this section. Similar slots need to be cut in the end of the 1³/₈-inch section. Cut six lengths of the #14 insulated copper wire about 11 inches long. Remove ¹/₄ inch of insulation from each end. Install solder lugs onto one end of each wire.

I made a small loop on the other end of the six wires and slid each onto a short wire soldered to the center pin of the SO-239 connector. (I crimped another solder lug on the end of this wire so the other wires didn't fall off while soldering the individual wires to the feed bus.) Also attach (either with small screws and nuts, or solder) a five-inch loop of #14 wire to the ground hole on the SO-239. This makes a bus where ground radials are attached. Use at least 8 to 12 radials, 16 feet or longer and a 4-foot or larger ground rod for lightning protection.

Now, place the bottom disc (the one with six ⁵/₈-inch holes) about eight inches from the bottom end of the 1¹/₂-inch tubing. Secure this to the tubing on either side with 1³/₄-inch hose clamps. Slide another hose clamp, the top disc, and another hose clamp after that, over the other end of the 1¹/₂-inch tubing, where the 1³/₈-inch tubing is sticking out. Secure this disc about 7¹/₂ feet up from the first disc by tightening the hose clamps.

Attach a seven-inch piece of Schedule-20 or 40 PVC pipe to the bottom end of the 1¹/₂-inch tubing. I used a piece of 1¹/₄-inch ID Schedule-40 pipe, cut in two lengthwise as the clamp-on insulator. Place the two U-bolts over the PVC at each end (so that the U-bolts don't touch the 1¹/₂-inch tubing) and temporarily tighten them. Now, attach the solder lug on the SO-239 bus wire to the hole in the end of the 1¹/₂-inch tubing with a 6-32×³/₈-inch screw, lockwasher and nut.

Next, put the two longest sections of tubing, for 20 meters and 30 meters, through

the larger holes in the uppermost disc. They should be opposite each other (at 0° and 180°). Slide a ³/₄-inch hose clamp over the bottom end and insert the element into the bottom disc. Attach one of the feed wires to the element with a screw, lockwasher and nut. Keep the disc perpendicular to the larger center element, and press the tubing against the disc and tighten the hose clamp. Also, slide another hose clamp over the smaller telescoped end of the elements and tighten at the point where the element comes out of the upper disc. Do the same with the remaining elements so that they are positioned as shown in Fig 1. Once all the elements are in place, make sure that the feed wires don't cross each other.

Next, slide a 1¹/₂-inch hose clamp, the 15-inch piece of acrylic, and another hose clamp over the top section of 1³/₈-inch tubing and over the 20-meter and 30-meter elements, stopping just below the top of the 20-meter element. This keeps these two elements from moving in the wind and affecting resonance. Place ³/₄-inch hose clamps on these two elements above the acrylic, keeping it level, and then tighten the large hose clamps also.

If you are going to use a full-size element on 40 meters, telescope the center three sections (the top section is 1¹/₄-inch OD) to an overall length of 33 feet and tighten the hose clamps. If you are using a loaded 40-meter element, as I did, you will only be using the 12-foot long 1¹/₂-inch section and a 6-foot long 1³/₈-inch section. This telescopes to a total of 16 feet long.

For the 40-meter loading coil form, cut a piece of the 1¹/₄-inch ID Schedule-40 PVC pipe 12 inches long. Cut two lengthwise slots about one inch long at right angles to each other on one end. Draw a line running the length of the pipe on one side. Drill two ¹/₄-inch holes on this line. One should be one inch from the end with the slots and the other 1¹/₂ inches from the end. Drill two similar holes on the line 2¹/₂ and 3 inches from the other end.

Starting at the end with the holes near the slots, insert five inches of #14 coil wire

through the inside hole and back out the outer hole. This end will attach to the 1³/₈-inch tubing with a sheet metal screw. Wind 48 turns of the #14 wire tightly around the PVC and put the end of the wire through the inner hole in the far end and back out the outer hole. Leave about two inches and snip off the extra. Strip one inch of insulation off this end and make about three loops with an inside diameter the same as the whip you will be attaching to the top. Press this loop close to the PVC pipe, insert the whip into the loops, and attach the whip to the PVC pipe with two 1¹/₂-inch hose clamps. Squeeze the wire loops tightly around the whip with vise grips or similar pliers.

Slip a 1¹/₂-inch hose clamp over the slotted end of the PVC and install onto the 1³/₈-inch aluminum. Drill a hole in the aluminum just below where the PVC pipe is mounted and attach a small, uninsulated loop of the lower coil wire with a solder lug and sheet metal screw.

Finally, mount the antenna on its support pipe placed in the ground and attach the ground wire loop connected to the SO-239 to the ground radial system. Using low power, check the SWR on all bands. If necessary, adjust the lengths of each element by loosening the clamps at the joints and above the acrylic discs and retighten to check the SWR again. If the 40-meter SWR is too high, you may have to remove a few turns of the coil (if changing the length of the element is not sufficient). The SWR on all bands except 40 meters will typically be 1.1:1 over most of the band and less than 2:1 at the band edges. The 40-meter bandwidth will be similar if you use a full-size element. The loaded 40-meter element I used had a 2:1 SWR bandwidth of about 90 kHz.

I have been using this antenna for about one year and have worked over 60 countries during low solar-flux conditions (including ZD7, 3B8, VR6, 5N0, V6, CY9, A3, YB, and EA9, to name a few). A vertical like this is a viable option for hams with small lots, tower deed restrictions, or limited finances. It should provide years of operating enjoyment due to its simple construction.

Table 2

Preliminary Dimensions for Telescoping T-6061 Aluminum Tubing; 0.058" Wall

Band	Overall Length	1¹/₂" OD	1³/₈" OD	⁵/₈" OD	¹/₂" OD
10 meters	7' 6"	—	—	6'	4'
12 meters	8' 6"	—	—	6'	6'
15 meters	10' 3"	—	—	6'	6'
17 meters	12' 3"	—	—	6'	8'
20 meters	15' 9"	—	—	12'	6'
30 meters	22' 10"	—	—	12'	12'
40 meters	16', plus whip	12'	6'	—	—

A Triband 75/40/30-Meter Delta Loop

By David J. Crockett, WB4DFW
820 Alpine Drive
Seneca, SC 29672
e-mail: dave.crockett@pubaff.clemson.edu

WB4DFW wanted to use his Delta Loop on other bands besides 75 meters, so he installed some coaxial traps. He also reveals an effective technique to prevent breakage when the support trees sway in the wind.

I've been using horizontally mounted, coax-fed Delta Loop antennas on 75 meters for many years. I've found them to be superior to dipoles—they are generally quieter on reception (perhaps because they are dc-grounded), exhibit broader bandwidth (typically less than 2:1 SWR over more than 200 kHz using a co-axial matching section), and are often more easily adaptable to available tree supports. They also don't fall down very often when installed correctly!

Outside of using an antenna tuner, a 265-foot-long 75-meter loop doesn't lend itself to working other bands—and I don't have a tuner. So while listening to a virtually dead 75-meter phone band one day and wishing I had a way to hit 30 meters for a long-over-due dose of CW, I had a brainstorm. Why not use traps? I'd homebrewed coaxial traps based on Sommer's article[1] several times in the past, and I knew they were easy to make and adjust.

The Traps

A quick examination of the latest incarnation of my 75-meter loop found it to be a reasonably equilateral triangle (a little over 85 feet per leg). I calculated that a 30-meter dipole should be something over 46 feet in total length (465/frequency in MHz). It appeared to be a simple enough matter to station some homebrewed coaxial traps 23 feet on each side of the feedpoint. A little more calculating found that it might even be practical to put in a pair of 40-meter traps as well.

Sommer gives a useful set of nomographs for computing trap dimensions on PVC pipe forms, using either RG-58 or RG-174 coax (the former capable of handling a kW with few problems). I opted for the larger cable. The 30-meter traps each require 46.8 inches of coax on a 1.62-inch OD section of PVC (6.75 turns). The 40-meter versions each use 61 inches of coax on a 2.25-inch OD PVC section (6.0

turns). Both PVC sections can be any convenient length. I used four-inch long pieces.

I wound the 30-meter traps in less than an hour and inserted them in the loop antenna 23 feet either side of the feedpoint. A quick SWR check found them resonant well below the bottom of the 30-meter band, but spacing the turns slightly apart and securing them with electrical tape raised the resonance easily enough, with the SWR bottoming out at 1.2:1 at 10.125 MHz.

The 40-meter traps went together with similar ease. Placement in the existing antenna was a little more challenging than with the first set of traps. The loading caused by the 30-meter traps made determining the physical location for the 40-meter traps uncertain. Grebenkemper notes that traps in a half-wave antenna tend to "pull" an antenna toward resonance,[2] even if they are not placed at exactly the optimum location. So, I simply took the formula length for 40-meter resonance, generously subtracted the physical length of the coax in each of the 30-meter traps and installed the 40-meter traps accordingly.

The approach was unscientific, but SWR

was 1.3:1 at 7.1 MHz on the first try. Dumb luck, I guess, and I opted to make no further adjustments. The final antenna layout is shown in **Fig 1**, and a close-up photo of a trap is shown in **Fig 2**.

The Matching Section

The ARRL Handbook notes that the typical impedance of a full-wave loop antenna is on the order of 100 Ω, or about a 2:1 SWR presented to a 50-Ω feed line at resonance. The original WB4DFW 75-meter Delta Loop used a quarter-wavelength coaxial matching section (also described in the *Handbook*) between the feedpoint and the coax going to the shack. The matching section is merely 40.1 feet of RG-59 75-Ω coax [(465/3.85)/2 ×.66, the velocity factor] and transforms the load of the antenna very close to the 50-Ω characteristic impedance of the RG-8X feed line I use.

I also use a five-turn loop of the matching section coax to form an RF choke near the antenna feedpoint to minimize feed-line radiation. The matching section has no perceptible effect on the new 40 and 30-meter antenna SWRs (checked empirically by

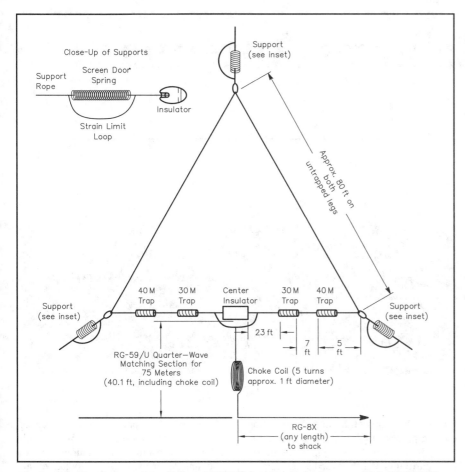

Close-Up of Supports

Support Rope — Screen Door Spring — Insulator

Strain Limit Loop

Support (see inset)

Approx. 80 ft on both untrapped legs

Support (see inset)

Support (see inset)

40 M Trap — 30 M Trap — Center Insulator — 30 M Trap — 40 M Trap

23 ft

7 ft

5 ft

RG-59/U Quarter–Wave Matching Section for 75 Meters (40.1 ft, including choke coil)

Choke Coil (5 turns approx. 1 ft diameter)

RG-8X (any length) to shack

Fig 1—Overhead view of the Triband 75/40/30-meter Delta Loop, which WB4DFW mounted 40 feet off the ground. The traps were constructed with RG-58 coax. See text for winding details.

Fig 2—Close-up photo of one of the 40-meter traps. The ends of the coax cable used in each trap have been sealed against moisture with silcone caulking. See text for construction details.

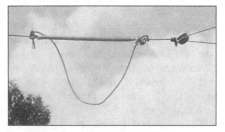

Fig 3—Photo of one of the support system springs, with rope used to limit total stretch. The rope is twice the length of the unstretched spring.

temporarily removing the section and feeding the antenna directly with the 50-Ω coax), so it has been left in service.

What effect did all this have on my original 75-meter loop? As one might expect, the resonant frequency dropped as a result of the traps' loading effects (nearly 200 kHz). But pruning about 8 feet off the far apex end of the loop brought my normal 1.2:1 resonant SWR back to my regular 3.842-MHz stomping grounds.

Keeping It Up in the Air

Here in South Carolina, pines are a great source of antenna support but their swaying during summertime thundershowers can wreak havoc on long, unprotected wire antennas. Over the years, I've found a simple shock-absorber system using a run-of-the-mill screen-door spring placed between a support rope and each antenna insulator can save a lot of grief. See **Fig 3** for a closer look at the details.

A special aspect of my system is the employment of a "strain limit rope" in parallel with each spring. This piece of rope, approximately twice the length of the unstretched spring, keeps swaying pines from overly taxing the spring and gives a nice visual reference for gauging how much tension the antenna is operating under. It also provides a fail-safe for the inevitable corrosion-caused failure of the spring after several years aloft. It makes a fine bird perch, too.

So, How Does It Play?

Performance on all three bands has been most satisfactory. The 2:1 SWR bandwidth of the 30-meter section far exceeds the band edges. Similar 40-meter bandwidth is about 100 kHz (noticeably narrower than dipoles I've used, and I suspect partially attributable to my guesswork on the 40-meter trap placement). The 75-meter bandwidth is now approximately 150 kHz (slightly narrower than before the traps were installed). Signal reports both ways on all three bands are comparable to any other non-directional HF antennas I've used.

Notes and References
[1]Robert C. Sommer, "Optimizing Coaxial-Cable Traps," *QST*, Dec 1984, pp 38-42.
[2]John Grebenkemper, "Multiband Trap and Parallel HF Dipoles—A Comparison," *QST*, May 1985, pp 29-30.

The HF Cabover Antenna

By Jim Ford, N6JF
2415 College Dr
Costa Mesa, CA 92626

N6JF gives some practical tips on designing and constructing a high-efficiency, multi-band antenna system for a mobile camper.

If you have ever had the pleasure of traveling across the country with an HF mobile in a camper, trailer or motorhome you may want this antenna. In a previous trip across the country I used two different antennas; one was a small multiband continuously loaded all-band mobile whip (the Outbacker) and the other was an 80-meter dipole fed with 450-Ω ladder line and a homebrew link-coupled antenna tuner. This antenna combination proved to be less than ideal.

The camper has limited spots on which to mount an HF mobile antenna, without making things too high. The back ladder is a convenient place to mount a small mobile whip, but there is a price to be paid in efficiency with an antenna of this type and size, particularly on 80 and 40 meters. Please note: despite the name "mobile," you still can't operate while in motion!

The efficiency of typical mobile center-loaded antennas, depending on coil Q and other assumptions, is often less than 2% for 80 meters and 10% for 40 meters. These numbers come from the excellent, easy-to-use MOBILE antenna design program by Leon Braskamp, AA6GL. MOBILE comes on the disk bundled with *The ARRL Antenna Compendium, Vol 4*.

My 80-meter dipole was very efficient and it worked great on all bands (10 meters has a high take-off angle) but it had a couple of problems. First, it often took over 40 minutes to set up and about 20 minutes to tear down, since I was dealing with sling shots and tree snags. This is too much time for a schedule or an early morning departure, although it's okay if you plan to stay for a while. Even more important, there were often too many trees or other barriers (perhaps some even social) to allow putting up the dipole at a campsite. When this happened I was stuck with the mobile antenna with poor efficiency. There had to be a better antenna for camper operation.

I briefly considered a small transmitting loop antenna, but it seemed to be more difficult to construct, especially for all-band HF operation. I decided on a simple loaded vertical. This vertical had to have good efficiency, be lightweight, and shouldn't require a power-robbing antenna tuner. Antenna tuners can have significant power loss, particularly if they feed a low-impedance load like this short vertical. See the two-part article by Frank Witt, AI1H, in the April and May 1995 issues of *QST* concerning antenna tuners and losses.

While in the planning stage for this project, I decided that a telescoping aluminum extension pole used for roller painting would make a good bottom section for the loaded vertical. These are available at many local home supply centers. Mine was 1 inch in diameter and 6 feet long, telescoping to almost 12 feet. I disassembled both sections and cross-cut a 1-inch slot in the top of the bottom section with a hacksaw to allow compression-clamping with a hose clamp. The tip of the top section was fitted with an aluminum plug that had a 3/8-24 hole tapped in it. This procedure was a simple machine-shop operation, with the help of my friend KE6OP, Dick McKee. The plug was pounded into the top section and is quite snug. I tapped some setscrews through the pole into the plug, however, just to be sure. An insulated, laydown marine antenna mount fit perfectly into the bottom of the aluminum base and was secured with a bolt that also served as the electrical connection from the capacitor matching box. See **Fig 1** showing the aluminum base plate, the laydown mount and the antenna mast itself lowered down the back of the camper. **Fig 2** shows the layout of the back of the camper, with the antenna on the right-hand side, laid down for travel.

The variable capacitor in the matching box is a surplus three-section 365-pF broadcast tuning capacitor. Two of three sections are connected in parallel and a switch parallels in the third section, along with an extra 800-pF mica capacitor for use on 80 meters. See the schematic in **Fig 3**.

The capacitor assembly was put in a cus-

Fig 1—Photo of aluminum base plate, showing details of mounting to the camper. The four banana-plug jacks on the bottom are for extra radials, if desired.

Fig 2—Photo showing the back of the camper, with the antenna in the lowered position, parallel to the ladder. The "Outbacker" antenna is shown clamped to the left side of the ladder.

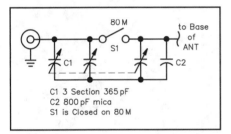

Fig 3—Schematic of tuning capacitor at base plate. C2 is a 800-pF transmitting mica capacitor. C1 is a three-section 365-pF broadcast tuning capacitor. S1 is closed for 80-meter operation.

Fig 4—Close-up photo of 80 and 40-meter coils, with top section and telescoping whip antenna with swivel bracket for tuning the top section for the higher bands. Note quick-disconnect connectors at top and bottom of both coils. The top whip is a fiberglass CB whip, used on 80 and 40 meters.

tom-glued Plexiglas box to keep out the weather and mounted to a piece of plate aluminum, along with an SO-239 connector. The aluminum plate was riveted to the camper shell using a lot of aluminum rivets. Do not use steel. I peeled back about a 4-inch wide section of the side of the camper for this direct aluminum-to-aluminum connection. People who are hesitant to modify their campers like this need to find an alternate low-resistance connection method. My camper was old enough not be to an issue.

For an extra low-resistance connection I added a 1½-inch aluminum strip from the top of the base plate to the camper as shown in Fig 1 near the 80-meter switch. The tuning knob protrudes from the side of the Plexiglas box. The four bottom holes in the plate are for banana plugs to connect ground radials if I want the extra efficiency they can provide. However, the roof of a camper is one of the better places for a mobile antenna,

so I seldom hook up the radials. When I do use them the tuning changes a little.

To keep losses down, I used coils wound with aluminum clothesline wire on old mobile coil center sections with quick disconnect fittings. See **Fig 4**, which shows both the 80 and 40-meter coils, together with the top portion of the antenna. An article by Robert Johns, W3JIP, in October 1992 *QST* described techniques for building your own loading coils. I built mine based upon calculations from AA6GL's MOBILE antenna program. The coils ended with a little more inductance than calculated and I had to remove some turns. Both coils are spaced at 4 turns per inch. The 8-inch long 80-meter coil has 18 turns. The 7-inch long 40-meter coil has 8 turns.

The matching network (which might seem to be just a capacitor) is actually an L-section step-up match, using the net inductive reactance of the antenna plus the center loading coil. The PVC coil construction technique was described in W3JIP's *QST* article and a follow-up Technical Correspondence piece in October 1992 *QST*. Basically, it consists of drilling an accurate row of slightly undersized holes along a length of ½-inch PVC pipe and then carefully sawing down the center of the row of holes with a hacksaw. Then, you trap the coil wires in the grooves between the two sawed halves and tie the two halves together with string. When you are satisfied that everything is proper, you then tighten the string and apply epoxy to make it strong and permanent.

One advantage of aluminum clothesline wire is that it is already coiled at the approximate diameter needed when you buy it and it is easy to position on the coil form. The clothes line wire I bought had a plastic coating on it and I didn't take it off except at the contact points. Once the epoxy dries, this method of construction does a good job of holding the finished coil together and it is lightweight.

The computed coil Q from the MOBILE antenna program is about 800 for the 80-meter coil and about 400 for the 40-meter coil. I accidentally made the 40-meter coil 7 inches in diameter instead of a higher-Q 6 inch diameter. Even so, the whole antenna system with a 9-foot top section calculates as being 56% efficient on 80 meters and 85% for 40 meters. AA6GL's program can change wire size, diameter, pitch, antenna size and location and a few other things to make the coil and system efficiency calculations easy. His program tells me my mistake cost me 3% in efficiency on 40 meters—not enough to do over again! The program assumes copper wire, so these coils and the whole antenna system would be slightly less efficient than indicated. Aluminum has a DC resistance about 50% more than copper but it turns out that skin effect makes a poor conductor better than you would think because the RF signal travels deeper into the wire, providing more conduction area and thereby partially compensating for the higher resistance.

The removable top section for 80 and 40 meters is a full-size fiberglass CB whip from Radio Shack. The fiberglass whip has about a #16 wire in the center of it. Be sure to sand and paint the whip for protection against UV and to protect yourself against fiberglass spurs in the hands. I tried a full-size stainless steel CB whip to get a slightly higher capacitance to ground because of larger conductor size but discovered it was far too heavy. That experience reinforced my decision that aluminum was a far more practical coil and base-section material for this project.

Quick-disconnect connectors found years ago at a hamfest were used for both coil forms and for the top section. Bands higher than 40 meters don't need any loading coils and the antenna length can be telescoped to get a ¼ wavelength. Ten meters doesn't require any top section. Be sure to use some NOALOX or similar compound to prevent corrosion and poor connections at all aluminum joints. This is especially true for the telescoping sections and the aluminum rivets. The matching capacitor is in the circuit at all times but when the 80-meter switch is off you can set the capacitor at minimum (about 14 pF) and it is effectively out of the circuit, even at 10 meters. I have not tried this antenna on power levels greater than about 100 W but the weakest link would probably be the matching capacitor. The voltage at the matching capacitor is low, so I suspect 200 W would be no problem.

A 1:1 SWR match is easy to get for 80 or 40 meters and a good SWR is obtained without retuning the base matching capacitor for approximately 100 kHz on 80 meters and most of 40 meters. The top section, however, does not have such low Q and needs to be tuned. The 2:1 SWR bandwidth on 80 meters is about 25 kHz and 150 kHz for 40 meters. Tuning is accomplished by using a telescoping FM portable radio antenna connected to the top-section whip with a stainless steel hose clamp. The maximum length of the telescoping section I used is 29 inches, and it collapses down to 7 inches. I used a whip with an elbow so I could adjust the angle of the whip as well. A telescoping whip half the length would still be long enough for the full adjustment of both bands.

The top-section adjustment is one of the penalties paid to achieve high efficiency. Substituting an automatic antenna tuner would likely lose efficiency, particularly on 80 meters since the base resistance calculates to about 10 Ω. I have not tried to make the antenna work on 160 meters. The SWR for the higher bands was good enough without any matching network.

As light as the antenna is, it still won't hold up in a moderate wind without some support guys. I used ¾-inch and ½-inch PVC support pipes in a telescoping arrangement for storage, but expanding to give an approximate 45 degree support at the top of the bottom section from both directions.

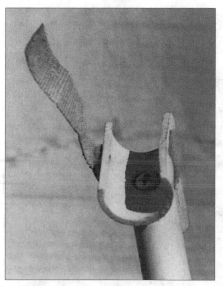

Fig 5—Close-up of one of the snap-on support brackets used to brace the antenna. Note the Velcro pieces used to ensure that the antenna doesn't pop out of the bracket in the wind.

One end of the telescoping section was connected to the camper roof with a hinge. The other end formed a snap-fit out of a PVC barrel that was cut lengthwise. See **Fig 5** and **Fig 6** for details. Even though it formed a good snap fit, I didn't trust the joint for strong winds so I glued a piece of Velcro to the joint to close up the open end. Be careful though, because Velcro deteriorates with exposure.

Another hose clamp near the top of the bottom section holds a ⅛-inch line I can take up the ladder to pull up the antenna without the need of an assistant. The disadvantage is you have to climb the roof. Use a non-slip floor mat or something similar to spread the load on the roof and to avoid slipping. Once on the roof, however, the coil is at a height for easy adjustment when the telescoping section is in the down position.

An advantage of being able to assemble the whole antenna on the roof is that you don't need a lot of swing-up room and you can clear trees easily. You can make calibration marks on the upper aluminum section for resonant lengths on the higher bands but just raise the top section up all the way for 80 or 40 meters. Marks can also be used on the small coil tuning whip for 80 or 40 meters, although different locations may require slightly different

Fig 6—Photo showing the two support bracket poles bracing the bottom section of the antenna. The top tuning whip is evident above the homemade loading coil at the top of the bottom section.

settings, due to detuning from nearby metal objects.

The overall length is about 21 feet. The top of my camper is about 10 feet high when on the truck, putting the tip at 31 feet. I can put up this antenna in less than 5 minutes and take it down in half the time. The success of this project has as much to do with knowing how and where you operate as it does paying attention to mechanical and electrical details. This antenna has been a good compromise between efficiency and convenience for me.

Two HF Discone Antennas

By Daniel A. Krupp, W8NWF
5616 Weber Rd
Whitehall, Michigan 49461

The name *discone* is a contraction of the words disc and cone. Although people often describe it by its design-center frequency (for example, a "20-meter discone"), it works very well over a wide frequency range, as much as several octaves. **Fig 1** shows a typical discone, constructed of sheet metal for UHF use. On lower frequencies, the sheet metal may be replaced with closely spaced wires and/or aluminum tubing.

The dimensions of a discone are determined by the lowest frequency of use. The antenna produces a vertically polarized signal at a low-elevation angle and it presents a good match for 50-Ω coax over its operating range. One advantage of the discone is that its maximum current area is near the top of the antenna, where it can radiate away from ground clutter. The cone-like skirt of the discone radiates the signal—radiation from the disc on top is minimal. This is because the currents flowing in the skirt wires essentially all go in the same direc-

W8NWF loves his discone, a broadband, low-angle antenna. He also believes in building his antennas really rugged! You'll enjoy seeing how he made his discones.

tion, while the currents in the disc elements oppose each other and cancel out. The discone's omnidirectional characteristics make it ideal for roundtable QSOs or for a Net Control station.

Electrical operation of this antenna is very stable, with no changes due to rain or accumulated ice. It is a self-contained antenna—unlike a traditional ground-mounted vertical radiator, the discone does not rely on a ground-radial system for efficient operation. However, just like any other vertical antenna, the quality of the ground in the Fresnel area will affect the discone's far-field pattern.

Both the disc and cone are inherently balanced for wind loading, so torque caused by the wind is minimal. The entire cone and metal mast or tower can be connected directly to ground for lightning protection.

Unlike a trap vertical or a triband beam, discone antennas are not adjusted to resonate at a particular frequency in a ham band or a group of ham bands. Instead, a discone functions as a sort of high-pass filter, coupling RF to the "ether" all the way from the low-frequency design cutoff to the high-frequency limits imposed by the physical design. The concept of not having to prune or tune an antenna was alien to my experience with ham antennas and lingered in the back of my mind even after my first discone was up and in use!

While VHF discones have been available out-of-the-box for many years, HF discones are rare indeed. I've seen some articles describing HF discones, where the number of disc elements and cone wires was mini-

mized to cut costs or to simplify construction. While the minimalist approach is fine if the sought-after results really are obtained, I believe that if you want to experience how well an antenna will work, you should build it without compromise!

In this article I will describe two no-compromise discones I built, with some on-the-air observations. I will also include vertical radiation patterns produced by the *NEC/Wires* antenna program by Brian Beezley, K6STI.[1]

History of the Discone

A couple of months after building my first discone antenna, I obtained the July 1949 and July 1950 issues of *CQ* magazine. Both contained excellent articles on discones. The first article, by Joseph M. Boyer, W6UYH, said that the discone was developed and used by the military during World War II.[2] The exact configuration of the top disc and cone was the brainchild of Armig G. Kandonian. Boyer described three VHF models, plus information on how to build them, radiation patterns, and most importantly, a detailed description of how they work. He referred to the discone as a type of "coaxial taper transformer." I recommend reading Boyer's article, since it really helped me understand how the discone functions.

The July 1950 article was by Mack Seybold, W2RYI.[3] He described an 11-MHz version he built on his garage roof. The mast actually fit through the roof to allow lowering the antenna for service. (I wonder how many hams would do that today?) Seybold's construction techniques could

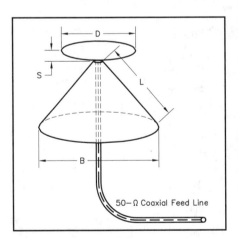

Fig 1—Diagram of VHF/UHF discone, using a sheet-metal disc and cone. It is fed directly with 50-Ω coax line. The dimensions L and D, together with the spacing S between the disc and cone, determine the frequency characteristics of the antenna. L = 246 / f_{MHz} for the lowest frequency to be used. Diameter D should be from 0.67 to 0.70 of dimension L. The diameter at the bottom of the cone B is equal to L. The space S between disc and cone can be 2 to 12 inches, with the wider spacing appropriate for larger antennas.

(In figure: D, S, L, B, 50—Ω Coaxial Feed Line)

still be used today and all materials are readily available. However, I think a more modern approach is to use PVC pipe for the hub, with tubular aluminum radial spreader elements, making a more streamlined and lighter-weight disc assembly.

Seybold stated that his 11-MHz discone would load up on 2 meters but that performance was down 10 dB compared to his 100-MHz Birdcage discone. He commented that this was caused by the relatively large spacing between the disc and cone. I believe the performance degradation he found was caused by the wave angle lifting upward at high frequencies. The cone wires were electrically long, causing them to act like long wire antennas. See **Fig 2**.

In the chapter on Multiband Antennas, *The ARRL Antenna Book* contains a construction article for a 7 to 29.7-MHz discone using a 36-foot high support mast.[4] This design had a 26 foot, 7 1/4 inch diameter top disc. Eight 1-inch OD aluminum tubing spreaders were used as spokes for the top disc, with wires connected between them. This substantial disc assembly was mounted to the vertical mast using two circular metal plates sandwiching a phenolic insulator. The 24 cone wires also served as guys for the mast.

The WARC Bands Made Me Do It!

Like many other ham radio operators, I went through a period of time when I put raising youngsters before my hobby. At the end of 1991, I asked my wife, Sandy, if she would give me a hand stringing up one end of a simple "V" wire antenna (remnants of a 1970-vintage rhombic). I was back on the

air again and really enjoying it!

The upper HF bands were hot at that time and I was eager to try out the new WARC bands. My enthusiasm for hamming continued to build as I started using 17 and 12 meters. However, I still had to adjust the antenna-matching network every time I changed bands or tried to QSY a good amount. I also knew my "V" was doomed to come down someday, because its apex was rather precariously attached to the mast holding up our TV antenna. I started thinking about a broadband (and preferably multiband) antenna.

While reviewing all my radio handbooks, I came across "The Low-Frequency Discone" in *The Radio Handbook*, 14th edition, by William I Orr, W6SAI.[5] This article contains a drawing showing three different HF discones, a construction sketch for a 20-meter discone, and a graph showing the wonderful broadband SWR from 13.2 to 58.0 MHz. That convinced me to build one cut for just below 14 MHz. I was convinced that it would work well for the 20, 17, 15, 12 and 10-meter bands.

I decided not to stretch the antenna size to include 30 meters because I hadn't proved my particular style of construction. I felt it was more important to be successful on a smaller scale first.

My First Discone: the A-Frame Discone

A word of warning: building an HF dis-

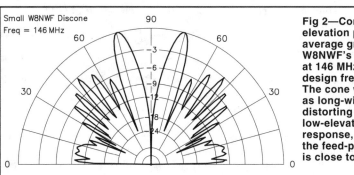

Fig 2—Computed elevation plot over average ground for W8NWF's small discone at 146 MHz, ten times its design frequency range. The cone wires are acting as long-wire antennas, distorting severely the low-elevation angle response, even though the feed-point impedance is close to 50 Ω.

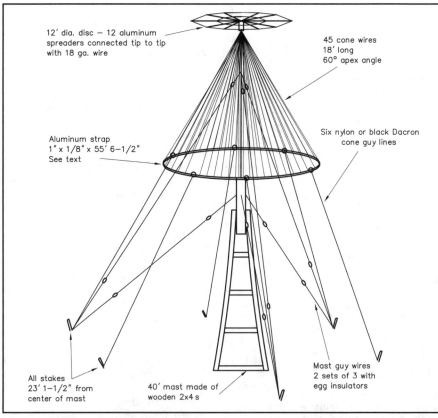

Fig 4—Detailed drawing of the A-frame discone for 14 to 30 MHz. The disc assembly at the top of the A-frame is 12 feet in diameter. There are 45 cone wires, each 18 feet long, making a 60° included angle of the cone. This antenna works very well over the design frequency range.

Fig 3—Photo of W8NWF's original A-frame mounted HF discone.

cone will take more than an afternoon! The good news is that you won't spend any time adjusting and tuning it. I recommend making a project out of it and savoring every minute of the planning and construction, as I did.

My first discone was one that would fulfill my need to cover 20 through 10 meters without using an antenna-matching unit. The cone assembly used 18-foot long wires, with a 60° included apex angle and a 12-foot diameter disc assembly. See **Fig 3** and **Fig 4**. I assembled the whole thing on the ground, with the feed coax and all guys attached. Then with the aid of a lot of friends, it was pulled up into position.

I used a 40-foot tall wooden "A-frame" mast, made of three 22-foot-long 2×4s. I primed the mast with sealer and then gave it two coats of red barn paint to make it look nice and last a long time. The disc hub was a 12-inch length of 3-inch schedule-40 PVC plumbing pipe. The PVC is very tough, slightly ductile, and easy to drill and cut. PVC is well suited for RF power at the feedpoint of the antenna.

For the 12-foot diameter top disc, I purchased three 12-foot by 0.375-inch OD pieces of 6061 aluminum, with 0.058-inch wall thickness. These were cut in half to make the center portions of the six telescoping spreaders. I also bought four twelve foot by 0.250-inch OD (0.035-inch wall thickness) tubes and cut these into 12 pieces, each 40 inches long. This gave extension tips for each end of the six spreaders.

See **Fig 5** for details on the disc hub assembly. I started by drilling six holes straight through the PVC for the six spreaders, accurately and squarely, starting about two inches down from the top and spaced radially every 30°. Each hole is 0.375 inches below the plane of the previous one. Take great care in drilling—a poor job now will look bad from the ground for a long time! It's a good idea to make up a paper template beforehand. Tape this to the PVC hub and then drill the holes, which should make for a close fit with the elements. If you goof, I would recommend starting over with a new piece of PVC—it's cheap, after all.

I drilled each six-foot spreader tube exactly in the center to clear a 6-32 threaded brass rod that secured the elements mechanically and electrically. A two-foot long by 1/4-inch OD wooden dowel was inserted into the middle of each six-foot length of tubing. The dowel added strength and also prevented crushing the element when the nuts on the threaded rod were tightened.

The 40-inch long extensions were inserted four inches into each end of the six-foot spreaders. I marked and drilled holes to pin the telescoping tips, plus holes big enough to clear #18 soft-drawn copper wire. This was for the inner circumferential wire for the disc. I also drilled a single hole for

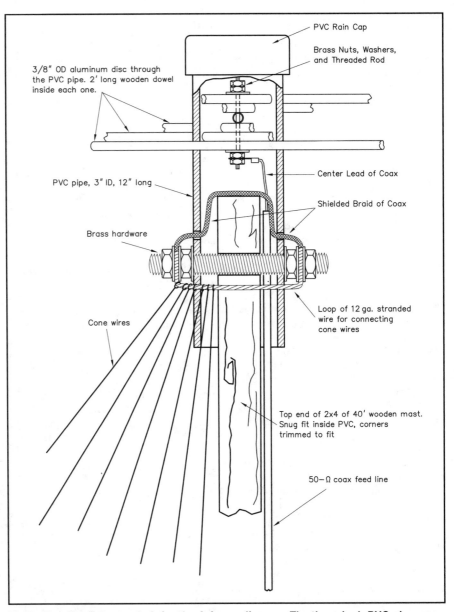

Fig 5—Details of the top hub for the A-frame discone. The three-inch PVC pipe was drilled to hold the six spreaders making up the top disc. Connections for the center conductor of the feed coax were made to the disc. The coax shield was connected to the cone-wire assembly by means of a loop of #12 stranded wire encircling the outside of the PVC hub.

#18 wire about 1/4 inch from each extension element tip, through which passed the outer circumferential wire. Finally, I inserted all six-foot elements into the PVC hub and lined up the holes in the center so the brass rod could be inserted through the middle to secure the elements.

My next step was to "chisel to fit" the top of my wooden mast to allow the PVC to slide down on it about six or seven inches. For convenience, I placed my whole mast assembly in a horizontal position on top of two clothesline poles and one stepladder.

The disc head assembly was placed over the top of the mast, but wasn't secured yet. This allowed for rotation while adding the

disc spreader extensions. A tip for safety: tie white pieces of cloth to the ends of elements near eye level. Just remember to remove them before raising the antenna!

For a long-lasting installation, I used an anti-corrosion compound, such as Penetrox, when assembling the aluminum antenna elements. As the extensions were added, I secured them in the innermost of the two holes with a short piece of #18 wire. Then I ran a wire through the remaining holes looping each element as I went. This gave added support laterally to the elements. Next I added a #18 wire to the tips of the extensions in the same fashion. This provided even more physical stability as well as making electrical connections.

The PVC disc hub was then pinned to the wooden mast with a $^3/_8$-inch threaded rod. This was also the point where the cone wires were attached, using a loop of #12 stranded copper wire around the PVC. Each cone wire was soldered to this loop, together with the coax shield braid. I made sure the loop of #12 wire was large enough to make soldering possible without burning the PVC with the soldering iron.

The coax center conductor was connected to the disc assembly by securing it with the same 6-32 threaded rod that tied all the disc elements together. I made sure to use coax-seal compound to keep moisture out of the coax. The coax was run down the mast and secured in a few places to provide strain relief and to keep it out of the way of the cone wires.

Two sets of three guy wires were used. These were broken up with egg insulators, just to be sure there wouldn't be any interaction with the antenna. I used 45 wires of #18 soft-drawn copper wire for the cone, 18 feet long each. These were cut a little long so they could be soldered to the connecting loop.

A difficult task was now at hand—keeping all the cone wires from getting tangled! I soldered each of the 45 cone wires to the loop of #12 wire, spacing each wire about $^1/_4$ inch from the last one for an even distribution all the way around.

The cone base was 18 feet in diameter to provide a 60° included angle. At the base of the cone, I used five 12-foot long aluminum straps, 1 inch wide by $^1/_8$ inch thick, overlapping $8^1/_4$ inches and fastened together with aluminum rivets. Holes were then drilled along the strap every 15 inches to secure the cone wires.

I made sure to handle the aluminum strap carefully while fastening the cone wire ends; too sharp of a bend could possibly break it. I fastened six small-diameter nylon lines to the cone-base aluminum strap to stabilize the cone. These cone-guys shared the same guy stakes as the mast guy lines. After cutting the nylon lines, I heated the frayed ends of each with a small flame to prevent unraveling. I also applied several coats of clear protective spray to the disc head assembly, after checking that all hardware was tight. A rain cap at the top of the PVC disc hub completed construction.

Putting It All Up

The reward for all the hard work was close at hand! Speaking of hands, if you built the antenna I just described, you are going to need a lot of them now to raise the antenna. Have the whole process fully thought out before trying to raise it.

You should have the spot selected for the base of the mast and some pipes driven into the ground to prevent the mast from slipping sideways as it is being pulled up. The three guy stakes should be in place, 23 feet, $1^1/_2$ inches from mast center. The guys should have been cut to the correct length, with some extra, of course. Be sure the coax transmission line will come off the mast where it should. I added a long length of rope to an upper and lower guy line to pull up the whole works.

I used the old trick of standing an extension ladder vertically near the antenna base with the pull lines looped over the top rung to get a good lift angle. The weight added to the mast from the antenna disc assembly and cone wires is about 26 pounds, most of it from the cone assembly. Use two strong people to pull up the antenna slowly so that the other helpers on the guy wires and cone guy lines have time to move about as required. As the antenna rises to the vertical position, if there are no snafus, the guy lines can be secured. Then tie the six cone lines to stakes.

That's it! Solder on a coax connector while someone else is passing out the pizza and beer or other refreshments. My antenna crew consisted of Rick Dial, AA8JZ; Mike Chapel, AA8KA; Chris Hanslits, KA8UNO and my son Steve Krupp and his friend Wyatt Miel. Believe me, all helping hands were very much appreciated.

First QSOs

Even before I could get my transceiver set up on the picnic table beside the antenna site, my son had parked his car nearby and had connected the coax to his CB. (I hope that someday he will get his ham ticket.) My own first QSO using the discone was with Marty, WB9DCM, in Orlando, Florida, on 17-meter SSB. We talked for about 10 minutes and exchanged 57 reports. Then I said to AA8JZ, "Why don't you give it a try?" He got on CW and started a QSO with someone in England. That's when I noticed he was operating on 30 meters. I said, "That antenna wasn't designed for 30 meters. What's the SWR?" Rick replied that "It looks a little higher than you'd like to see, but it's working, isn't it?"

The next day I checked the SWR on 10 through 20 meters. I found a maximum SWR of only 1.6:1. Not bad, so I checked 30 meters and found it to be about 5:1 there. I later found it was possible to load up even on 75 meters using an antenna matching unit. I talked to stations on both East and West coasts. It was really a pretty small antenna for 75 meters!

During the year that I had it up, I used the antenna mainly on 17 meters, my choice of

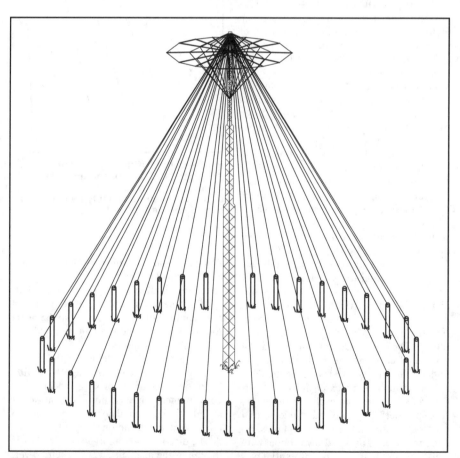

Fig 6—The large W8NWF discone, designed for operation from 7 to 28 MHz, but useable with a tuning network in the shack for 3.8 MHz.

Fig 7—Photo showing details of the hub assembly for the large discone, including the threaded brass rod that connects the radial spreaders together. The 10-inch PVC pipe is drilled to accommodate the radial spreaders. Each spreader is reinforced with a three-foot long wooden dowel inside for crush resistance. Note the row of holes drilled below the lowest spreader. Each of the 36 cone wire passes through one of these holes.

Fig 8—Details of the copper pipe slipped over the feed coax. The coax shield has been folded back over the copper pipe and secured with two stainless-steel hose clamps. The cone wires are also laid against the copper pipe and secured with additional hose clamps.

bands while the sunspot cycle was still pretty good. I was very pleased with the convenience of operation and with the good signal reports on all five bands.

Now, A Really Big Discone

I wondered how far the idea of covering all bands with one antenna could be carried. The thought of one discone antenna built for the 160-meter band and useful all the way up to 10 meters was intriguing. I'm afraid it's not very practical because of the pattern deterioration discussed earlier.

There was still the article in *The ARRL Antenna Book*—the discone covering 40 through 10 meters. I kept wondering about building a larger one. An opportunity finally came my way—a woman living nearby wanted to sell her 64-foot self-supporting TV tower.

My new tower had eight sections, each eight feet long. Counting the overlap between sections, the cone wires would come off the tower at about the 61.5-foot mark. The tower was installed on a cement base, which I made larger than normal since we have rather sandy ground.

I took some liberties with the design of this larger discone compared to the first one, which I had done strictly "by the book." There were two major changes. The first change was to make the cone wires 70 feet long, even though the formula said they should be 38 feet long. Further, the cone wires would not be connected together at the bottom. With the longer cone wires, I felt that 75 and 80-meter operation might be a possibility.

The second major change was to widen the apex angle out from 60° to about 78°.

This should produce a flatter SWR over the frequency spectrum and would also give a better guy system for the tower.

The topside disc assembly would be 27 feet in diameter and have 16 radial spreaders, using telescoping aluminum tubing tapering from $5/8$ to $1/2$ to $3/8$ inches OD. All spreaders were made from 0.058-inch wall thickness 6063-T832 aluminum tubing, available from Texas Towers. A section of 10-inch PVC plumbing pipe would be used as the hub for construction of the disc assembly.

Construction Details for the Large Discone

While installing the tower, I had left the top section on the ground. This allowed me to fit the disc head assembly precisely to it. Detailed preparation always makes final assembly a simpler and more pleasant job! **Fig 6** shows the overall plan for the large discone.

The 10-inch diameter PVC hub was designed to slip over the tower top section, but was a little too large. So a set of shims was installed on the three legs at the top of the tower for a just-right fit. Drilling the PVC pipe for the eight $5/8$-inch OD elements was started about an inch down from the top. I purposely staggered the drilled holes in the same fashion as the hub for the smaller antenna. See **Fig 7**.

Again, three-foot sections of $1/2$-inch wooden dowel were used to strengthen the $5/8$-inch center portion of each spreader. Instead of using a loop of #12 wire for connecting the cone wires, as had been done on the smaller discone, I drilled 36 holes in the PVC hub. These holes are small enough so

that the PVC hub would not be weakened appreciably. The circles of holes for the cone wires were drilled about six inches below the disc spreaders.

I prepared a three-foot long piece of RG-213 coax, permanently fastened on one end to the antenna, with a female type-N connector at the other end. Type-N fittings were used because of their superior waterproofing abilities. The coax center lead was connected with a terminal lug under a nut on the brass threaded rod securing the disc spreaders. The coax shield braid was folded back over a six-inch long copper pipe and clamped to it with a stainless-steel hose clamp. See **Fig 8** for details.

The plan was that after the top disc assembly had been hoisted up and attached at the top of the tower, individual cone wires would be fed, one at a time, through the small holes drilled in the PVC. They were to be laid against the copper pipe and secured with stainless-steel hose clamps.

The $1/2$ and $3/8$-inch OD spreader extension tips were secured in place with two aluminum pop-rivets at each joint. Again, antioxidant compound was used on all spreader junctions. A hole was drilled horizontally near the tip of each $3/8$-inch tip all around the perimeter to allow a #8 aluminum wire to circle the entire disc. A small stainless-steel sheet-metal screw was threaded into the end of each element to secure the wire.

In parallel with the aluminum wire, I ran a length of small-diameter black Dacron line, securing it in a couple places between each set of spreaders with UV-resistant plastic tie-wraps. The reason for doing this was to hold the aluminum wire in position and to prevent it from dangling, in case it should break some years in the future. Two coats of clear protective spray were applied for protection.

A truss system helps prevent the disc from sagging due to its own weight. See **Fig 9** for details. This shows the completed disc assembly mounted on the top of the tower. I used a three-foot length of two-inch PVC pipe for a truss mast above the disc assembly, notching the bottom of the pipe so that it would form a saddle over the top couple of spreaders. This gave a good foothold. I cut a circle of thin sheet aluminum to fit over the 10-inch PVC to serve as a rain cap. The cap had a hole in the center for the two-inch PVC truss mast to pass through, thereby holding it down tight. A few light coats of paint were sprayed over the PVC for protection from ultraviolet radiation from the sun.

Sixteen small-diameter black Dacron ropes were connected at the top of the truss support mast, with the other ends fastened to the disc spreaders, halfway out. Another rain cap was added to the top of the two-inch PVC truss mast. Eight lengths of the same small diameter Dacron rope were added halfway out the length of every other spreader. These ropes were meant to be tied back to the tower, to prevent updrafts from blowing the disc assembly upward. I used small egg insulators near the spot where the eight bottom trusses were tied to the disc spreaders, just to be sure there was no RF leakage in rainy weather.

Hoisting the completed disc assembly to the top of the tower was done easily, with the assistance of my son Steve, his friend Wyatt, and my wife Sandy. The trickiest part was to get the disc assembly from its position sitting flat on the ground to the vertical position needed for hoisting it up the tower without damaging it. The disc assembly weighs about 35 pounds. My son Steve volunteered to go to the top of the tower to receive the disc as it was hoisted up by gin pole, and to mount it on the tower top. All went very well, and all the holes lined up too!

I had prepared three six-foot long metal braces going over the outside of the PVC to fasten to the tower legs. They really beefed things up. Once that was accomplished, we had pizza and refreshments, as the last glimmer of evening sunset glowed on that beautiful tower and disc assembly!

In plastic irrigation pipe buried between the house and tower base, I ran 150 feet of 9086 low-loss coax to the shack. For cone wires, I was able to obtain some #18 copperclad steel wire, with heavy black insulation that looked a lot like neoprene. The cone system takes a lot of wire: 36×70 feet = 2520 feet, plus some extra at each end for termination. You'd be well advised to look around at hamfests to save money.

As I connected each cone wire at the top of the tower, a helper would place the other end at its proper spot below. The lower end of each cone wire is secured to an insulator screwed into a fencepost. See **Fig 10**. There are 36 treated-pine fenceposts, each standing about 5$^{1}/_{2}$ feet tall, 45 feet from the tower base to hold the lower end of the cone wires. This makes mowing the grass easier and the cone wires are less likely to be tripped over too.

On the final trip down the tower, the eight Dacron downward-truss lines were tied back to the tower about six feet below the disc assembly. The tower has three ground rods driven near the base, connected with heavy copper wire to the three tower legs.

Performance Tests

On the air tests proved to be very satisfying. Loading up on 40 meters is easy—the SWR was 1:1 across the entire band. I can work all directions very well and I receive excellent signal reports from DX stations. I often switch to my long (333 foot) center-fed dipole for comparison while listening. I find

Fig 9—Photo of the spreader hub assembly, showing the truss ropes above and below the radial spreaders. This is a very rugged assembly!

Fig 10—Photo showing some of the fence posts used to hold individual cone wires to keep them off the ground and out of harm's way. The truck in the background is carting away the A-frame discone for installation at KA8UNO's QTH.

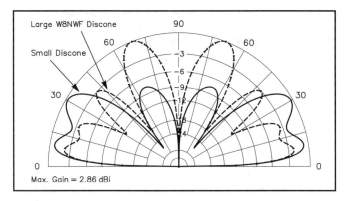

Fig 11—Computed patterns showing elevation response of small discone at 28.5 MHz compared to that of the larger discone at 28.5 MHz. The cone wires are clearly too long for efficient operation on 10 meters, producing unwanted high-angle lobes that rob power from the desirable low-elevation angles.

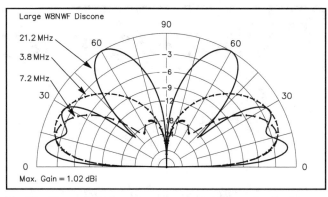

Fig 12—Computed elevation-response patterns for the larger W8NWF discone for 3.8, 7.2 and 21.2-MHz operation. Again, as in Fig 11, the pattern degrades at 21.2 MHz, although it is still reasonably efficient, if not optimal.

the dipole is much noisier and that received signals are weaker. During the daytime, nearby stations (less than about 300 to 500 miles) can be louder with the dipole, but the discone can work them just fine also.

I am happy to report that this antenna even works well on 75 meters. As you might expect, it doesn't present a 1:1 match. However, the SWR is between 3.5:1 and 5.5:1 across the band. I use a Ten-Tec 229 matching unit to operate the discone on 75. It seems to get out as well on 75 as it does on 40 meters.

The SWR on 30 meters is about 1.1:1, but since I haven't operated much CW lately, I have nothing to report for that band. On 20 meters the SWR runs from 1.05:1 at 14.0 MHz to 1.4:1 at 14.3 MHz. The SWR on the 17, 15, 12 and 10-meter bands varies, going up to a high of 3.5:1 on 12 meters. My on-the-air results are not as good as with the smaller discone. Of course, part of the reason is the decline of the sunspot cycle in 1994 and 1995.

Naturally, I had to try it on 160 meters, just for fun. I used the antenna-matching unit and could manage to load it. Reception for nearby stations was down about 4 or 5 S units compared to my dipole, and transmitting gave similar results. Listening to DX was a different story—I could hear DX stations Q5 that couldn't be heard at all on the dipole, because of the noise. I did receive a 53 report from CT1ESV on 160 meters using the discone.

My original intention for this antenna certainly did not include 160 meters, and it clearly does not transmit as well as an antenna designed specifically for that band. Sometimes, however, I use the discone to

receive on 160 meters, while using the dipole for transmitting.

Radiation Patterns

From modeling using *NEC/Wires* by K6STI, I verified that the low-angle performance for the bigger antenna is worse than that for the smaller discone on the upper frequencies. See **Fig 11** for an elevation-pattern comparison on 10 meters for both antennas, with average ground constants. The azimuth patterns are simply circles. Radiation patterns produced by antenna modeling programs are very helpful to determine what to expect from an antenna.

The smaller discone, which was built by the book, displays good, low-angle lobes on 20 through 10 meters. The frequency range of 14 through 28 MHz is an octave's worth of coverage. It met my expectations in every way by covering this frequency span with low SWR and a low angle of radiation.

The bigger discone, with a modified cone suitable for use on 75 meters, presents a little different story. The low-angle lobe on 40 meters works well, and 75 meter performance also is good, although an antenna matching unit is necessary on this band. The 30-meter band has a good low-angle lobe but secondary high-angle lobes are starting to hurt performance. Note that 30 meters is roughly three times the design frequency of the cone. On 20 and 17 meters there still are good low-angle lobes but more and more power is wasted in high-angle lobes.

The operation on 15, 12, and 10 meters continues to worsen for the larger discone. The message here is that although a discone may have a decent SWR as high as 10 times the design frequency, its radiation pattern is

not necessarily good for low-angle communications. See **Fig 12** for a comparison of elevation patterns for 3.8, 7.2 and 21.2 MHz on the larger discone.

What Have I Learned?

A discone antenna built according to formula will work predictably and without any adjustments. One can modify the antenna's cone length and apex angle without fear of rendering it useless. I believe the broadband feature of the discone makes it attractive to use on the HF bands. The low angle of radiation makes DX a real possibility. I am also pleased to report that my discone is much less noisy than my dipole on receive.

Probably the biggest drawback to an HF discone is its bulky size. There is no disguising this antenna! However, if you live in the countryside you should be able to put up a nice one. One last reminder: this antenna will transmit harmonics or other undesirable signals from your transmitter, so keep your signal clean.

Notes and References

[1] *NEC/Wires*, Version 1.54, by Brian Beezley, K6STI, 3532 Linda Vista, San Marcos, CA 92069.
[2] Joseph M. Boyer, W6UYH, "Discone—40 to 500 Mc Skywire," *CQ*, Jul 1949, pp 11-15, 69-71.
[3] Mack Seybold, W2RYI, "The Low-Frequency Discone," *CQ*, Jul 1950, pp 13-16, 60.
[4] "An HF Discone Antenna," *The ARRL Antenna Book*, 16th Edition (Newington: ARRL, 1991), pp 7-17 to 7-20.
[5] William I. Orr, W6SAI, editor, "The Low-Frequency Discone," *Radio Handbook*, 14th Edition (Editors and Engineers, 1956), p 369.

Propagation and Ground Effects

The NVIS Propagation Mode and the Ham

By Jacques d'Avignon, VE3VIA
965 Lincoln Dr
Kingston, ON
K7M 4Z3 Canada

We amateurs implicitly assume that the HF bands are only suitable for long-range communications. We tend to regard the spectrum between 1.5 and 30 MHz as useful only for communicating with stations we cannot reach by VHF/UHF. Most textbooks discuss the use of HF for medium to long-range circuits. Very seldom do we consider what sort of system we might use to communicate in the geographical area beyond reliable VHF/UHF coverage, yet inside the skip zone that encircles every HF transmitter.

A skip zone always occurs between the limit of HF ground-wave radiation and the first return to Earth of the skywave component of a transmission. The HF skip zone also covers an area where, under normal propagation conditions, VHF/UHF systems don't work, since the two terminals are not within direct line-of-sight of each other.

There is, however, an HF propagation mode called *NVIS*, which stands for "Near Vertical Incidence Skywave." Commercial and military operations use NVIS to fill the wide gap between the end of VHF/UHF coverage and the end of the HF skip zone. See **Fig 1**, which shows typical coverage for VHF/UHF, NVIS and regular HF communication.

While NVIS is not mentioned in many textbooks dealing with communications, it can be very useful in times of emergency, since it can cover areas that are not reached by a regular VHF/UHF network. Even if a repeater network is well established, it may be overloaded with emergency traffic or non-operational because of power outages.

NVIS also has other names. It is sometimes referred to as the "Showerhead Propagation Mode." The Australians call it "District Propagation Mode" in their propagation forecasting software, *ASAPS*. I have also seen NVIS called the "Jungle Broadcasting Mode."

VE3VIA describes an underutilized mode of HF propagation with the intimidating name NVIS, meaning "Near Vertical Incidence Skywave."

If we look at the analogy of a showerhead, we would see a point source of radiation expanding on the way down, after refraction by the ionosphere, to cover a wide area on the ground. If the ionosphere were perfectly homogeneous and parallel to the Earth's surface, the radiated energy directed straight up from an antenna would come back directly to the source with very little dispersion. However, the ionosphere is nei-

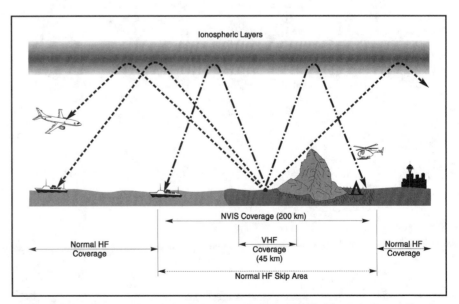

Fig 1—Near Vertical Incidence Skywave propagation. NVIS bridges the gap between 45 km, the nominal maximum reliable coverage for VHF/UHF, and 200 km, the "skip zone" for typical HF propagation. The term "showerhead propagation" is sometimes applied to NVIS, easy to visualize for signals launched directly overhead.

ther homogeneous nor parallel to the Earth's surface, and thus we have a large area irradiated around the RF source.

If you want to experiment and see for yourself how NVIS works, you can try this very simple, and wet, experiment. When you water your garden, point your hose straight up. (It's best to do this while standing under an umbrella!) Look at the water coming back down and how much it disperses in a large area compared to the size of the hose nozzle launching the water.

The name "District Propagation" is also used to describe the area illuminated around the NVIS transmitter site. In Australia, a "district" is an administrative entity smaller than a province or a state and NVIS is used to communicate within that particular area. The size of an Australian district could be easily related to an emergency area in a state or province in North America.

USES OF NVIS

Broadcasting

In the tropical areas of the world, the extreme attenuation of the normal broadcasting ground-wave mode by the dense, and in some cases jungle-like vegetation, makes ground-wave broadcasting impossible to use.

The lateral attenuation has been demonstrated by Hagn and Barker[1] to be an increasing function of frequency:

$$A_1 (dB/m) \approx 0.009 \times f(MHz) + 0.1 \qquad (Eq\ 1)$$

Using Eq 1, we find that at 3.0 MHz the lateral attenuation due to vegetation would be 127 dB/km. At 10 MHz, along the same path, the attenuation is now up to 190 dB/km! In the same reference, Hagn and Barker cite the total 3.0-MHz losses using NVIS as being only 97 dB for 600 km. This amounts to 0.16 dB/km along the ionospheric virtual path, a very substantial difference compared to ground wave!

If the normal HF ionospheric mode were used for local broadcasting, the "local" audience, widely scattered in small pockets around the transmitter site, would usually be located inside the skip zone of the HF transmitter. From Eq 1, it is obvious that it would not be commercially viable to increase the HF transmitter power to use the ground-wave mode. Increasing the power might not ensure that you would reach your audience, but it would warm up the vegetation surrounding the transmitter site! So for tropical broadcasting, we are left with one option: NVIS.

"District" Propagation

Some stations in the interior of Australia have been broadcasting using NVIS for years. Their audiences are widely scattered and "District" propagation was the only way to reach them. These Australian stations are broadcasting in their "tropical domestic" band, extending from 2.300 to 2.495 MHz. The transmissions are designed to be local in nature and fulfill this role very well. They can even be heard in Eastern North America in the early morning hours at certain times of the year.

The NVIS broadcasting technique is also common in Africa and South America for daily domestic broadcasting use. Check your shortwave receiver in the "Tropical Bands." Listen to stations from those regions in the late afternoon if you live on the East Coast of North America.

Other specific tropical bands besides those

```
ASAPS DISTRICT/NVIS FORECAST
=============================================================================
ASAPS V2.4 Best Usable Frequency Prediction ------ 12 May 1995
=============================================================================
Circuit 1: VE3VIA-district      Distance: 0-300 km  Date: 15 Sep 1995
Tx: VE3VIA      44.22 283.42 Bearings: 0 180    T-index: 11
Rx: VE3VIA      44.22 283.42 Path: District/NVIS  TxAntenna: HHWD
FSet: amateur          Noise: -150  dBW/Hz   RxAntenna: HHWD
Required S/N: 0  dB   %Days: 90  Power: 1000W    Min.Angle: 3 deg.
Modes: 1F 1E               BandWidth: 3.0kHz
=============================================================================
UT     MODE   PROB  ANGLE  NOISE   S/N   SN@MUF  SN@OWF   BUF     MUF    OWF
0000   1F     99    90     -25     48    47      48       3.775   5.8    4.5
0100   1F     99    90     -25     48    47      48       3.775   5.1    4.0
0200   1F     99    90     -19     44    44      46       1.900   4.2    3.3
0300   1F     99    90     -16     40    41      42       1.900   3.5    2.7
0400   1F     99    90     -16     40    41      41       1.900   3.2    2.4
0500   1F     99    90     -16     40    41      41       1.900   3.1    2.4
0600   1F     99    90     -15     39    41      40       1.900   3.0    2.3
0700   1F     99    90     -15     39    40      40       1.900   2.9    2.3
0800   1F     99    90     -15     39    40      40       1.900   2.8    2.3
0900   1F     99    90     -15     39    39      40       1.900   2.6    2.1
1000   1F     99    90     -18     42    43      43       1.900   2.8    2.2
1100   1F     99    90     -21     46    45      48       1.900   3.8    3.0
1200   1E     99    90     -21     42    41      41       1.900   2.1    2.0
1300   1F     99    90     -23     39    42      39       3.775   5.1    4.1
1400   1F     99    90     -33     46    50      47       3.775   5.4    4.3
1500   1F     99    90     -34     45    51      48       3.775   5.6    4.4
1600   1F     99    90     -34     43    51      47       3.775   5.8    4.4
1700   1F     99    90     -34     42    50      46       3.775   6.0    4.5
1800   1F     99    90     -34     42    50      47       3.775   6.0    4.6
1900   1F     99    90     -34     44    51      48       3.775   6.0    4.7
2000   1F     99    90     -34     46    52      50       3.775   5.9    4.8
2100   1F     99    90     -34     49    53      52       3.775   6.0    4.9
2200   1E     99    90     -30     51    49      49       1.900   2.2    2.1
2300   1F     99    90     -25     46    47      47       3.775   6.0    4.7
=============================================================================
```

between 2.300 and 2.495 MHz are found from 3.200 to 3.400 MHz and from 4.750 to 5.060 MHz. In addition, there are the "domestic bands" found between 3.900 to 3.950 MHz in Asia and 3.950 to 4.000 MHz in Europe. Yes, that is the top of the 75-meter ham band! In Europe this is a very popular broadcast band used by many international broadcasters. While NVIS is used commercially for local broadcasting, there's no assurance that a signal will not be heard half way around the world—an NVIS circuit is not a secure circuit!

Marine Use

Let's look at another use of NVIS. Worldwide, maritime coastal stations normally have VHF capability to communicate with ships in their area, but there is a limit to the possible VHF coverage of such stations. Even with a coastal station's antenna and a ship's antenna each located 30 meters over a perfectly flat ground/sea surface the maximum communication range would be:

$$D\ (km) = 4.124\left(\sqrt{h_1} + \sqrt{h_2}\right) \qquad (Eq\ 2)$$

where h_1 and h_2 are in meters.

$$D\ (km) = 4.124\left(\sqrt{30\ m} + \sqrt{30\ m}\right) = 45\ km$$

This is an ideal theoretical maximum distance with no physical or meteorological obstructions between the two antennas. The range may be increased by meteorological ducting but this is not a reliable means of communication.

Coastal stations also have many HF frequencies that can be used. However, their regular HF antennas are designed for multihop ionospheric propagation. Further, a limited number of operating frequencies are assigned to them, with a high-frequency set between 8 to 26 MHz for daytime use and a low-frequency set between 2 to 8 MHz for nighttime use. A skip zone will extend, around the clock, from the typical 45 km limit of VHF coverage to about 4 to 500 km, where normal HF coverage skywave would take over. The exact size of this skip zone depends on the frequency in service, the time of day, the 10-cm flux value, geomagnetic storms, the exact antenna radiation pattern and other factors, but there will always be an HF skip zone where normal HF communication is unavailable.

However, by using the proper frequency in the maritime assignment and the correct NVIS antenna, coast stations can communicate easily and reliably with ships within this skip zone. Success is highly dependent on using the proper antenna. The regular maritime frequency assignments in the 2 to 6-MHz marine bands are ideal for NVIS operations.

One US Coast Guard station on the Atlantic coast, NIK/NMF in Boston, uses NVIS for facsimile transmissions to ships in the immediate vicinity, using the 6 and 12-MHz marine bands. On the 6-MHz band the take-off angle is very close to vertical and at 12 MHz the angle is still fairly steep. These NVIS transmissions are effective as far west as the Great Lakes.

Other Uses

Another use of NVIS is for two-way communication with low-flying helicopters or aircraft to establish circuits between remote campsites in very mountainous terrain, where it is impossible or uneconomical to install VHF/UHF repeaters for line-of-sight communication. This particular use of NVIS is for both commercial and military purposes. For commercial use, think of the surveying and/or geological survey camps scattered across rugged terrain. NVIS has also been used for military tactical communications in similar terrains.

Regular Ionospheric Modes

Before we look at the mechanics and specific requirement of the NVIS mode, let's summarize how "normal" multihop ionospheric propagation modes are used and the antenna/frequency requirements imposed by these modes.

If you read a textbook explaining HF propagation modes, you will learn that there is a maximum frequency, which will be reflected by the ionosphere directly over a station at any particular time. If we transmit on a frequency much above this critical frequency, our signal will get lost in space, as the ionosphere will become in effect a transparent window and will no longer act as a mirror for this frequency and any frequencies above it.

When using "normal" HF propagation modes for long distance, the optimum radiation angle from an antenna will be in the range from about 3° to about 20° above the horizon. This is for a variety of antennas, ranging from $1/4$-λ vertical radiators mounted over excellent ground planes to huge Sterba-curtains, log-periodic or Yagi arrays. On 3,000 to 10,000 km long-distance circuits, the shallower the take-off angle, the less the number of hops will be needed between the ionosphere and the ground on the journey between the terminals. Less hops means less power will be lost between the transmitter and the receiver. So, for long-distance circuits, we aim for low to very-low radiation angles from our antennas. Refer to any antenna handbook and you will note that they recommend low radiation angles for DX work.

So, in summary for "normal" HF propagation modes we want:

a) A frequency that is as high as possible to minimize losses and yet is still refracted by the ionosphere along the circuit path. It must definitely be lower than the HPF (Highest Possible Frequency). If you are operating a commercial circuit, the frequency should be as close as possible to the OWF (Optimum Working Frequency) for best reliability.

b) We will also want an antenna with a low take-off angle to reduce the number of hops along the circuit path.

But note that a very low take-off angle will also create a very wide skip zone, starting just at the outer edge of the ground-wave region and extending for many hundreds of kilometers in all direction from your antenna.

NVIS and the Ham

For the ham, two conditions have to be met to use NVIS: the right antenna has to be available and the right frequency should be used.

Antennas

For NVIS we need an antenna that will have its major radiation lobe as close to 90° above the horizon as possible. Refer to **Fig 2**. What we need is an antenna that is often called a "cloud warmer." That is definitely not a vertical or a mobile whip! A simple half-wavelength dipole or an inverted V will suffice. The dipole should be installed 0.15 λ above ground. For a discussion on the proper antenna to use for this mode, refer to Farmer.[2]

While researching the antenna requirements for the efficient use of NVIS, a colleague in the UK came across two antennas that have been specifically designed for jungle use. The description of one of them reads: "... high angle skywave antenna for use in close country such as jungle, giving good communications in the range 0-320 km..." That is an extremely good description of what NVIS is all about.[3]

These antenna designs are 1950 vintage, but I have very recently seen references to them in the literature. These specialized antennae are called the "Shirley Antenna" and the "Jamaica Antenna." Details and construction diagram

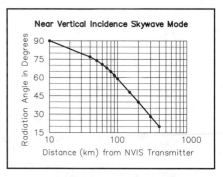

Fig 2—Range, in km, from NVIS antenna versus elevation angle of radiated signals. For signal launched at 40°, the NVIS range is 200 km, while a signal launched at 80° will cover out to 60 km from the transmitter.

can be found at the end of this article for anyone inclined to do some homebrewing.

If you have access to a catalog from any large industrial communication antenna manufacturer, you will find antennas specially designed for NVIS operation. One manufacturer has many antennas in his latest catalog that carry the following wording in the description: "Broadband Dipole" and one of their listed uses is "Short Range." If you look closely at the radiation diagrams, they all have superb main radiation lobes located from 75° to 90° above the horizon — these antennas have obviously been designed with NVIS mode in mind. I find it strange that the description of their obvious use "for NVIS operations" is not mentioned anywhere in the catalog.

The catalog that I consulted recently did not have a price list—we all know what that means. However, a well-designed and installed dipole will prove to be all that is needed for good use of NVIS. If you want to experiment, you could try a Shirley or a Jamaica antenna.

Frequency Choices

The choice of frequency for NVIS operations is simplified for the ham, since we only have small segments in the 1.8, 3.5 and 7.0- MHz parts of the spectrum that can be utilized. If you are working with the Armed Forces, chances are that you would have access to an ALE (Automatic Link Establishment) system and you could have a better choice of frequencies, but I do not believe that many hams have access to an ALE yet!

What you are looking for to operate properly in NVIS, is the Highest Possible Frequency (HPF) that can be radiated straight up and that will neither be absorbed nor pass through the ionosphere. This frequency is the overhead value of f_oF_2.

The most important question is now: how and/or where do we find the optimum frequency to use in NVIS mode? The easiest way to obtain these frequencies is to run a propagation prediction, using software in the "nowcasting" mode. This will give you a good and reliable MUF (Maximum Usable Frequency) when the two terminals of the circuit are 50 km or less apart. It is regrettable that some propagation forecasting software cannot accept frequencies below 3.0 MHz. For amateur NVIS operations, the 160-meter band is important.

This forecast shown in **Fig 3** was produced by *ASAPS*. The two terminals are located at the same geographical position, but for computation purposes there is a one kilometer difference automatically entered in the equation. The output gives you the BUF (Best Usable Frequency), in addition to other information that may be of some help in establishing your NVIS network. This forecast not only gives you the best frequency to use but also many other useful parameters.

If your propagation-prediction software gives you only an OWF or FOT (Optimum Working Frequency or *Frequence Optimale de Travail*, in French), you would be operating at approximately 85% of the desired MUF. Do not operate above the calculated MUF, as your transmissions will likely end up in outer space. In this forecast it is interesting to note that the software predicts F-layer refraction most of the day, except for the hour around sunrise and sunset where E-layer refraction is forecasted to be the dominant mode.

In this forecast, it was assumed that both antennae, transmitting and receiving, were half-wave dipoles and the transmitting mode was SSB with a bandwidth of 3.0 kHz. The radiation angle is computed as 90°.

Some propagation-prediction software will not allow you to put the two terminals of a circuit so close together, while other software will automatically assign a≈distance of 1 km between the transmitter and the receiver if you ask for a NVIS/District forecast. If you have a choice of software to calculate your NVIS frequency, use one that will allow you to put the two ends of the circuit as close as possible.

Probably the most practical method of finding the proper frequency for NVIS operations is to make a scheduled contact on the 3.5 MHz and then on the 1.8-MHz bands in succession. The lower frequency will work better at dawn and the higher one will be of some use during the day and evening. There might be a window where it will be difficult or sometimes nearly impossible, to use NVIS, and that is from local sunrise until about 3 to 4 hours after.

Why Would Hams Use NVIS?

In this era of highly sophisticated communication methods, such as VHF/UHF repeaters, satellite links and packet radio, the HF NVIS mode may not seem very sophisticated or even necessary. But in an emergency like the aftermath of a natural disaster like a hurricane or earthquake, if your main power lines feeding your repeater are down, VHF/UHF links are of very little or no value. If your repeater system cannot carry all the required traffic, what do you do to establish an area network able to carry your emergency traffic?

Most hams or clubs have or can rapidly have access to a portable power supply and thus can run a state/province-wide network

```
ASAPS DISTRICT/NVIS FORECAST
================================================================
ASAPS V2.4 Best Usable Frequency Prediction ---------- 12 May 1995
================================================================
Circuit 1: VE3VIA-district    Distance: 0-300 km  Date: 15 Sep 1995
Tx: VE3VIA      44.22 283.42 Bearings: 0 180    T-index: 11
Rx: VE3VIA      44.22 283.42 Path: District/NVIS TxAntenna: HHWD
FSet: amateur         Noise: -150 dBW/Hz  RxAntenna: HHWD
Required S/N: 0  dB  %Days: 90  Power: 1000W    Min.Angle: 3 deg.
Modes: 1F 1E                   BandWidth: 3.0kHz
================================================================
UT    MODE  PROB  ANGLE  NOISE  S/N  SN@MUF  SN@OWF  BUF    MUF   OWF
0000  1F    99    90     -25    48   47      48      3.775  5.8   4.5
0100  1F    99    90     -25    48   47      48      3.775  5.1   4.0
0200  1F    99    90     -19    44   44      46      1.900  4.2   3.3
0300  1F    99    90     -16    40   41      42      1.900  3.5   2.7
0400  1F    99    90     -16    40   41      41      1.900  3.2   2.4
0500  1F    99    90     -16    40   41      41      1.900  3.1   2.4
0600  1F    99    90     -15    39   41      40      1.900  3.0   2.3
0700  1F    99    90     -15    39   40      40      1.900  2.9   2.3
0800  1F    99    90     -15    39   40      40      1.900  2.8   2.3
0900  1F    99    90     -15    39   39      40      1.900  2.6   2.1
1000  1F    99    90     -18    42   43      43      1.900  2.8   2.2
1100  1F    99    90     -21    46   45      48      1.900  3.8   3.0
1200  1E    99    90     -21    42   41      41      1.900  2.1   2.0
1300  1F    99    90     -23    39   42      39      3.775  5.1   4.1
1400  1F    99    90     -33    46   50      47      3.775  5.4   4.3
1500  1F    99    90     -34    45   51      48      3.775  5.6   4.4
1600  1F    99    90     -34    43   51      47      3.775  5.8   4.4
1700  1F    99    90     -34    42   50      46      3.775  6.0   4.5
1800  1F    99    90     -34    42   50      47      3.775  6.0   4.6
1900  1F    99    90     -34    44   51      48      3.775  6.0   4.7
2000  1F    99    90     -34    46   52      50      3.775  5.9   4.8
2100  1F    99    90     -34    49   53      52      3.775  6.0   4.9
2200  1E    99    90     -30    51   49      49      1.900  2.2   2.1
2300  1F    99    90     -25    46   47      47      3.775  6.0   4.7
================================================================
```

Fig 3—Printout from propagation-prediction program *ASAPS*, showing all pertinent information for local NVIS propagation for a signal launched straight overhead. Note that the "Best Usable Frequency," BUF, is typically found on either 80 or 160 meters, with 160 meters dominating during the nighttime. Note also that the propagation mode changes from a single F-layer hop to a single E-layer hop at local sunrise and sunset.

of NVIS stations for the duration of an emergency. VHF/UHF repeaters have limited coverage, even if they are linked in a network. On the other hand, one well-positioned NVIS station can cover a whole state (well, maybe not Alaska or Texas—those states are always different!). This station can be part of a nation-wide emergency network operating in NVIS mode.

Summary

Let me relate a short story about NVIS. In the 1950s there was a full network of low-power stations serving the fire-protection services in an area along the St Lawrence river in Quebec. The frequency used was in the vicinity of 1.7 MHz and the forest fire towers had small battery-operated transceivers, with output power of about 5 W. The main station was using 5 kW and all stations were using horizontal dipoles.

It was necessary to maintain an extremely reliable communication network. If a fire tower failed to answer the roll call twice in succession, it was standard practice to send an aircraft to check that the fire wardens were safe. The roll calls were twice a day and aircraft did not go out very often to check out the fire towers.

One time I remember that fire wardens could not get to their radio on top of the fire tower because a bear family had decided to camp out on the front porch of the house. In that instance, the failure to communicate was not because NVIS was not working properly!

Certainly NVIS is not new and neither is it a panacea for difficult communication circuits. Try it now; don't wait until you have a critical need for it! Establish a state/province-wide network now and see how well NVIS can work for you.

Appendix

While working on this project, a colleague in the UK sent me some interesting information on two antennas that are often referred to as excellent for NVIS operations: the Shirley and the Jamaica antennas. Both of these have been designed to work well for distances up to 300 km in mountainous or jungle surroundings.

It is necessary to have a fairly large area to build these antennas and the ground under the array should be as flat as possible. In both cases, the height above ground should be more than $1/8$ but less than $1/4$ λ high. Obviously, at low frequencies, the amount of land necessary is important, but remember NVIS is a mode that will work well only in the low-frequency end of the HF spectrum.

Both dipoles in each antenna have to be fed in phase in order to work properly. When the antenna is properly phased any signal arriving from low angles will be canceled. The gain is 4 dB over a single dipole and this gain is concentrated straight up, where you want your maximum radiation for NVIS.

The Shirley Antenna and the Open-Wire Shirley

The Shirley antenna is an array of two $1/2$-λ dipoles, spaced 0.65 λ apart and fed in phase with equal-length feeders. It is supported by four supports, 0.125 to 0.25 λ high and produces a very high-angle radiation pattern. As you can see from **Fig 4A**, this antenna can be constructed using folded dipoles or using regular dipoles, as shown in **Fig 4B**.

It was interesting to read in the construction notes that the folded-dipole version of a Shirley antenna could be built using "lamp cord." The old twisted lamp cord used in the home had an approximate impedance of 75 to 200 Ω. (The impedance varied according to the manufacturer and the country of manufacture.) It would appear that this antenna can be built fairly easily with almost any wire you have on hand. As long as the antennas and feeders are balanced and are the same length, it will work well.

The Open-Wire Shirley in **Fig 5** is similar to the basic Shirley, but the dipoles are not folded. The feeders are made using open-line construction. The computed elevation response for a 0.125-λ high Open-Wire Shirley on 1.83 MHz is shown in **Fig 5B**. The reference consulted did not specify the impedance of the feeders, it just specified the spacing, although I imagine that the impedance is not critical, since an antenna tuner must be used at the bottom of the feeders.

The Jamaica and the Half-Jamaica Antenna

Now, if you have some real space you can experiment with the Jamaica antenna! The concept is simple: two full-λ dipoles, $1/2$ λ apart and fed in phase with open-wire feeders. Again, the height above ground is $1/8$ to $1/4$ λ. The gain of this antenna is better than the Shirley if it is erected full size. If space is at a premium, the Half Jamaica can be erected, but according to the reference this is not as efficient as the Shirley version. In this case the smaller Shirley seems easier to erect.

The feed line for the Jamaica is impressive. The wire is one inch in diameter, the spacing calls for two feet between the wires and a spacer every three feet along the line! Now this is serious construction.

At 1.8 and 3.5 MHz, the dimensions of these antennas are pretty big. Unless you want to dedicate a special antenna for this

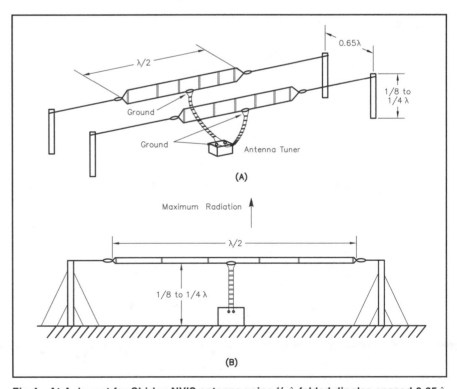

Fig 4—At A, layout for Shirley NVIS antenna using $1/2$-λ folded dipoles spaced 0.65 λ apart and 0.125 to 0.25 λ high. At B is same antenna viewed from the side. The transmission line is a balanced line, and can even be constructed with "lamp cord" according to the literature describing this antenna. Regular open-wire "ladder" line is likely to be a better choice for the amateur, especially in climates where ice or snow might be a problem.

service, the low dipole found in a typical ham's backyard should be sufficient to keep a circuit viable.

Notes and References

[1]Hagn, G. H. and G. E. Barker, 1970, "Research-Engineering and Support for Tropical Communications," AD-889-169, Final Report, Contract DA-36-039 AMC 00040(E), SRI Project 4240, Stanford Research Institute, Menlo Park, CA.

[2]Farmer, Ed, AA6ZM, "A Look at NVIS Techniques," *QST*, Jan 1995, pp 39-42.

[3]Goodman, John M. *HF Communication, Science and Technonlogy* (New York: Van Nostrand Reinhold, 1992). An excellent reference book for HF communications, extensive bibliography on various subjects including NVIS.

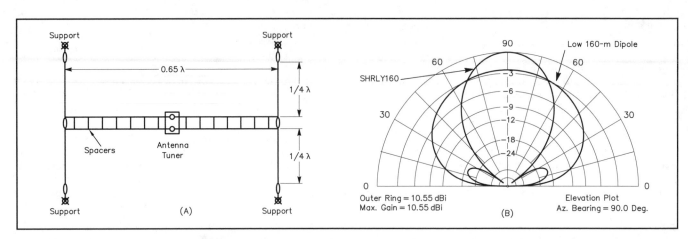

Fig 5—At A, top view of Open-Wire Shirley NVIS antenna. At B, computed elevation response of this antenna, 70 feet high on 1.83 MHz over average ground, compared to single dipole at the same height. The radiation is directed straight overhead, making this an excellent "cloud warmer," just what is needed for effective NVIS communication!

The Equatorial Ionosphere

By Carl Luetzelschwab, K9LA
1227 Pion Road
Fort Wayne, IN 46845

N o doubt you've heard of trans-equatorial propagation (TE). Reports of it are quite regular in W3EP's monthly *QST* column, "The World Above 50 MHz." One of the more common transequatorial paths is from the Caribbean to Argentina. Although most of us in the continental US are too far north to directly use this mode (as we'll see later), it can help us. For example, if you've ever worked an LU on 10 meters in the late afternoon or early evening, chances are good that transequatorial propagation occurred.

What is transequatorial propagation? What makes it go? What advantages does it offer? Let's take a look at the ionosphere in the equatorial region.

Characterizing the Ionosphere

Although the ionosphere is comprised of distinct *layers* (E, F_1, and F_2), in reality these layers are maximums (peaks) in a plot of electron density versus height. See **Fig 1** for a typical summer's day profile.

Instead of trying to characterize the ionosphere in terms of the electron density profile, it is easier to use the concept of *critical frequency*. The critical frequency of each

> ## K9LA explains the how and why of transequatorial propagation.

layer is the frequency where a vertically incident pulse of RF energy penetrates the layer instead of being returned to Earth. The critical frequency is directly related to the maximum electron density of that layer. Critical frequencies are abbreviated f_oE for the E layer, f_oF_1 for the F_1 layer, and f_oF_2 for the F_2 layer.

The time it takes a pulse just below the critical frequency of a given layer to travel up and be returned is indicative of the height of the peak electron density for that layer. The height information, along with the critical frequencies, allows us to characterize the ionosphere for propagation analysis.

For the analysis of transequatorial propagation in this article I will be presenting only data for the F_2 layer—the critical frequency f_oF_2 and its corresponding height. The effects of the E and F_1 layer on transequatorial propagation are minimal. They will not be presented (but certainly will be taken into account in later ray traces, even if they are minimal).

A Mid-Latitude Path

Let's first look at a path in the mid-latitudes far removed from the equatorial region. **Fig 2** shows how f_oF_2 and its height vary along a path from W9 (Indiana) to KH6 (Hawaii) for mid-October at 20 UTC (3 PM local time in Indiana) for a Smoothed Sunspot Number (SSN) of 20. This level of solar activity is typical for the minimum of a solar cycle. This f_oF_2 and height data, and all subsequent data, comes from a book entitled *Ionospheric Predictions* by the Office of Telecommunications.

Note that critical frequency f_oF_2 is well behaved along the entire path. It steadily increases going toward KH6 because of the influence of the sun on the F_2 layer. Note

also that the F_2-layer height is fairly constant along the entire path.

The *MUF* (Maximum Usable Frequency) can be computed by borrowing a concept from optics and using a simple curved-earth/curved-ionosphere geometry. For low elevation angles the MUF is approximately three times the value of f_oF_2 and the maximum single-hop length approaches 4000 km. Relating this to the W9-to-KH6 path indicates we could expect 20 meters (14 MHz) and possibly 17 meters (18 MHz) to be open for this path. Although this 7000-km path could be covered in two hops, it is more likely that three hops will be the dominant mode for average installations. This is due to the difficulty radiating sufficient energy at the extremely low elevation angle needed for two hops at these frequencies: this will be evident in a later ray trace.

A Path Crossing the Equator

Now let's take a look at f_oF_2 and its height for a path from KP4 (Puerto Rico) to LU (Argentina) for the same conditions (mid-October at 20 UTC with an SSN of 20). **Fig 3** does this, and the differences compared to Fig 2 are dramatic.

The graph of f_oF_2 shows two peaks along the path centered over the *geomagnetic equator*. At this distance (3400 km south of KP4) the F_2-layer height increases significantly. For reference, the geographic equator is approximately 2000 km south of KP4.

Remember that the Earth's magnetic field can be approximated by a large bar magnet, tilted at an angle to the rotational axis of the Earth. The tilt is about 11° and results in a geomagnetic equator that is different from the geographic equator. In the Western Hemisphere the geomagnetic equator is about 12° south of the geographic equator.

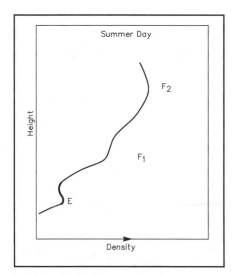

Fig 1—A typical profile of the ionosphere's electron density for a summer's day.

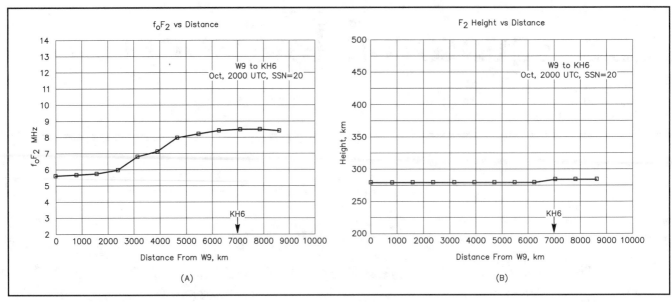

Fig 2—At A, the critical frequency f_oF_2 for a mid-latitude path from Indiana to Hawaii at 3 PM Indiana time (20 UTC) in October, for low solar activity. At B, the height of the F_2 layer for the same conditions.

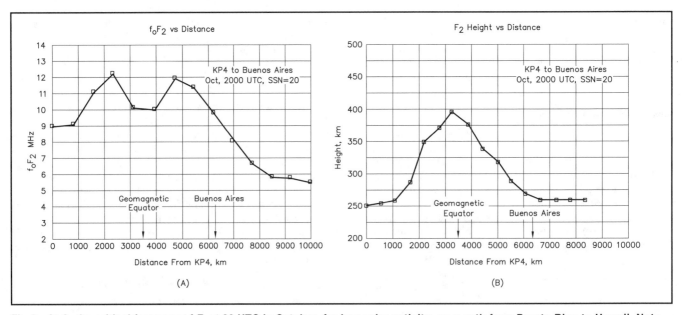

Fig 3—At A, the critical frequency f_oF_2 at 20 UTC in October, for low solar activity, on a path from Puerto Rico to Hawaii. Note the dip in the MUF over the geomagnetic equator. At B, the height of the F_2 layer for the same conditions. Note that the F_2-layer height peaks over the geomagnetic equator.

This 12° of latitude is the difference between 3400 km and 2000 km cited in the previous paragraph.

The Fountain Effect

Why does f_oF_2 and its height take on the characteristics shown in Fig 3? The process that results in these characteristics is due to the interaction of direct solar radiation and the magnetic field, and can be likened to that of a fountain. Electrons produced near the geomagnetic equator by the sun's ultraviolet radiation drift upwards under the com-

bined influence of electric and magnetic fields. As they rise, they encounter the horizontal lines of force of the Earth's magnetic field. They then move down these field lines and reenter the main body of the ionosphere, where the field lines cut through the F region. This results in large "clumps" of electrons at latitudes 10° to 20° from the geomagnetic equator. These clumps are the f_oF_2 peaks seen in Fig 3.

Ray Tracing

Using ray-tracing techniques[2] and a de-

tailed model of the ionosphere, we can visually see the impact that the equatorial ionosphere has on propagation. **Fig 4** is the result of a ray trace on the mid-latitude W9-to-KH6 path. The MUF for this path, determined by ray tracing, is about 16.5 MHz. This agrees quite well with our earlier prediction. Any frequency much higher than 16.5 MHz penetrates the ionosphere. Note that a two-hop mode (2° elevation angle out of W9) and a three-hop mode (8.5° elevation angle out of W9) are possible for this 7000-km path at 16.5 MHz. Radiating sufficient energy at a 2° elevation angle at

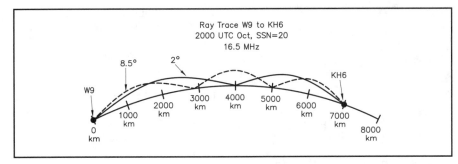

Fig 4—A ray-tracing of the 7000-km path from Indiana to Hawaii resulting in a MUF of 16.5 MHz.

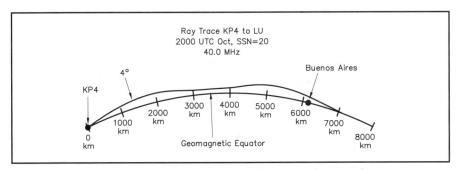

Fig 5—A ray-tracing of a 7000-km path from Puerto Rico to Argentina resulting in an MUF of 40 MHz. This 4° single hop is characteristic of transequatorial propagation.

16.5 MHz requires a very high horizontal antenna. This is why I said earlier that the three-hop mode is most likely the dominant mode for average installations.

Fig 5 is the result of a ray trace on the KP4-to-LU path. The MUF for this path, again determined by ray tracing, is about 40 MHz. Any frequency much higher than 40 MHz penetrates the ionosphere. The 7000-km distance is covered in only one very long hop requiring a 4° elevation angle out of KP4. This is the classical picture of transequatorial propagation. The 7000-km path will propagate our signal past Buenos Aires. In order to communicate with Buenos Aires (6200 km) a slightly different elevation angle is required. I have not included this ray trace to Buenos Aires on Fig 5 in order to keep the graph less cluttered.

Benefits of the Equatorial Ionosphere

From the ray-tracing analysis, the benefits of transequatorial propagation are three-fold:
• Fewer lossy ground reflections
• Fewer transitions through the absorbing D region down at 50 to 90 km
• Propagation of much higher frequencies.
This all comes about because of the clumps of electrons (f_oF_2 peaks) on either side of the geomagnetic equator and the corresponding increase in the F_2 layer height.

There are limits to transequatorial propagation. First, it is confined to the equatorial regions—about 3000 to 4000 km either side of the geomagnetic equator. This means that those of us in the continental United States are too far north to directly take advantage of this propagation mode. We have to rely on an initial hop by some other means to get our signals into the equatorial regions, where we can take advantage of TE. This can be done by a normal F-layer hop, a normal E-layer hop, or a sporadic-E hop. Second, a path nearly perpendicular to the geomagnetic equator is required. Regardless of these two limitations, it still can give those of us in the continental US some great openings deep into South America.

Daily, Seasonal, and Solar Cycle Variation

All of the analysis so far were computed at 20 UTC for the month of October, during a period of minimum sunspot activity. For other times, Figs 6-8 show the daily, seasonal and solar-activity related changes.

Fig 6 shows the daily variation for October. Prior to 16 UTC, f_oF_2 does not exhibit the double-humped characteristic. From 16 UTC through 0000 UTC, f_oF_2 has taken on this shape. By 02 UTC, the double-humped characteristic has collapsed. What this means is that transequatorial propagation is a possibility from mid-afternoon until early

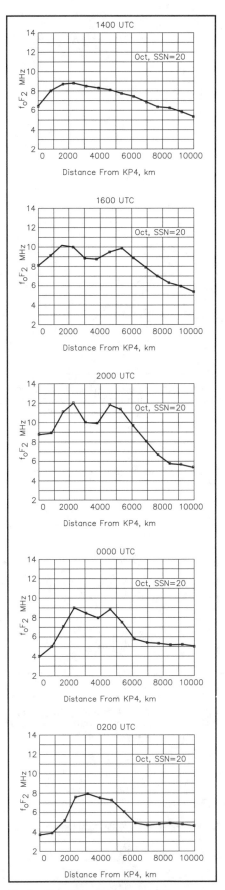

Fig 6—The daily variations of the critical frequency, f_oF_2, on a path from Puerto Rico to Argentina in October for a low level of solar activity.

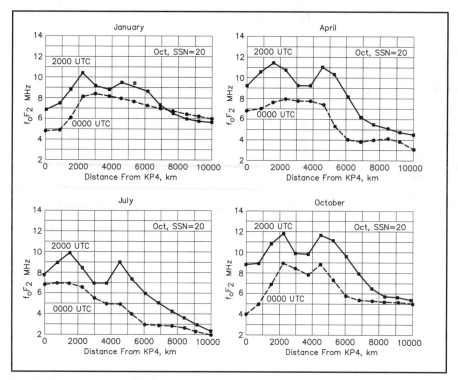

Fig 7—The seasonal variations of the critical frequency f_oF, on a path from Puerto Rico to Argentina for a low level of solar activity.

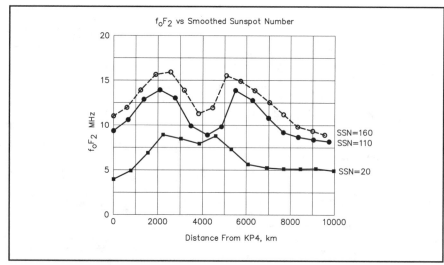

Fig 8—The critical frequency, f_oF_2, varying as a function of the solar cycle. The higher the Smoothed Sunspot Number, the better the chance of transequatorial propagation, especially at VHF.

evening local time for this path.

Fig 7 shows the variation for a winter month (January), a spring month (April), a summer month (July), and a fall month (October). Spring and fall are the best months, but don't rule out winter and summer months—the frequencies propagated may not be as high, but TE propagation still exists. In fact, TE propagation is what allows the East Coast and Midwest to work the early morning 10-meter long path to JA. This is particularly true in the summer months during higher levels of solar activity.[3] October appears to offer the longest window of opportunity throughout the day.

Fig 8 shows the variation for three levels of solar activity—minimum (SSN = 20), the peak of an average cycle (SSN = 110), and the peak of an above-average cycle (SSN = 160). With propagation supported at 40 MHz for an SSN of 20, scaling the f_oF_2 values indicates propagation is possible well above 50 MHz (6 meters) near the peak of a sunspot cycle.

Final Comments

The graphs of f_oF_2 presented are *median* values. In other words, on any given day in a given month, f_oF_2 could be several MHz above or below the stated value. In addition, the activity of the Earth's magnetic field greatly impacts the ionosphere. Thus transequatorial propagation could occur when you don't think it should, or it may not occur when you think it should!

Even at a very low level of solar activity, don't give up on transequatorial propagation. It should be there somewhat regularly to South America on 10 meters, with occasional surprises giving us even higher frequency openings.

References

[1] *Ionospheric Predictions*, Telecommunications Research and Engineering Report 13, Office of Telecommunications, Boulder, CO, 1971
[2] See *HF Ray Tracing* article elsewhere in this book.
[3] C. Luetzelschwab, K9LA, "10-Meter Long Path During Solar Cycles 21 and 22," *The ARRL Antenna Compendium* Vol 4 (Newington: 1995)

HF Ray Tracing of the Ionosphere

By Carl Luetzelschwab, K9LA
1227 Pion Road
Fort Wayne, IN 46845

Computer aficionados take notice! K9LA takes you through his ray-tracing computer programs, where you can find some valuable insights into the mechanisms involved in HF propagation.

This article describes an HF ray-tracing program developed in conjunction with my article about 10-meter long path in the last volume of *The ARRL Antenna Compendium*.[1] The genesis of my program was from NM7M's *PRTYBIRD* program, which traced signals from satellites to the ground. K2ARO modified *PRTYBIRD* for earth-ionosphere-earth applications, and I modified it further for general use.

The ray-trace program included on the disk is written in BASIC, and is called *RAYTRACE.BAS*. It uses Snell's Law in differential form to trace a ray through the ionosphere. This will give you a visual display of the various ionospheric modes conveying a signal from your transmitter to a distant station. This program does not tell you which mode is the most likely one (best signal)—it only tells you which modes are possible, based on the condition of the ionosphere.

Necessary ionospheric parameters are also on the disk, imbedded in another program, called *IONPARA.BAS*. This stands for "IONospheric PARAmeters." The theory portion of the article will be followed by an example of a path from my location in Ft Wayne, Indiana, to Antigua (V2A).

Procedure

This program is not a "one keystroke and out pops a ray-trace" program! It requires some math in the form of curve fitting, and a beginner's knowledge of BASIC. The steps involved are:

1. Run *IONPARA.BAS* for the desired path, date, time, and solar flux level. This gives you the critical E-layer frequency f_oE, the extraordinary wave critical F_2-layer frequency $F_2MUF(Zero)$, and the height of the maximum F_2-layer ionization density h_mF_2 all along the path.
2. Curve fit the above data to define equa-

tions for f_oE, $F_2MUF(Zero)$, and h_mF_2 versus distance along the path. There are many curve-fitting software packages available—I use EasyPlot[2] on an IBM PC in the subsequent example.
3. Insert the equations into *RAYTRACE.BAS* and run a ray trace.

Note that any change in path, date, time, or solar flux level will give different values of f_oE, $F_2MUF(Zero)$, and h_mF_2. This means that new equations are needed.

This program is obviously not intended to replace one of the fine, user-friendly propagation programs that give relatively quick predictions of MUF, LUF, modes, or signal levels. My program is designed for instructional purposes and for a deeper look at how your signal actually gets from point A to point B.

Ionosphere Model

The model of the ionosphere contained in *RAYTRACE.BAS* is defined as an electron density profile. The maximum electron density of the E-layer is directly related to the critical frequency f_oE. The height of this peak is taken as constant at 110 km.

The electron density below the peak is modeled as a parabola with a semi-thickness of 20 km. This transitions to a decaying exponential.

The maximum electron density of the F_2-layer is directly related to the $F_2MUF(Zero)$. The height of this peak is h_mF_2. The electron density above the peak is modeled as a parabola with a semi-thickness of 100 km. It too transitions to a decaying exponential.

Between the E-layer peak and the F_2-layer peak, the electron density is modeled as a function of $\sin^2$. This provides an accurate and smooth transition from the E-layer peak to the F_2-layer peak, resulting in a no-kinks ray trace.

No F_1-layer characteristics are used, as the ionization in the F_1 height region is accounted for in the E- and F_2-layer parameters.[3] Sporadic-E effects are not included either.

The ray-trace algorithm in *RAYTRACE.BAS* uses spherical coordinates, with variables R for height and TH (short for theta) for angle. It uses Snell's Law in differential form. The geometry is diagrammed in **Fig 1**, showing the incremental change in R (line 500 of the program) and TH (line 505) for an incremental

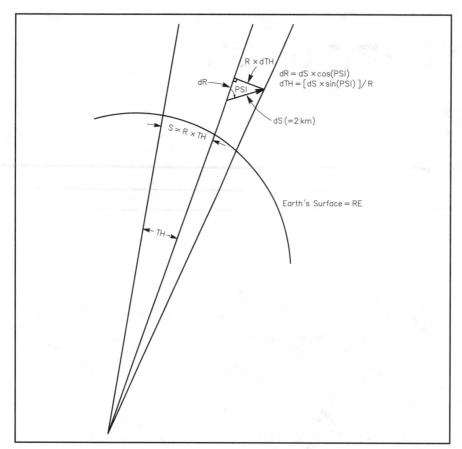

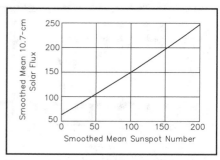

Fig 2—Plot showing relationship between Smoothed Mean Sunspot Number (SSN) to Smoothed Mean 10.7 cm Solar Flux (SF).

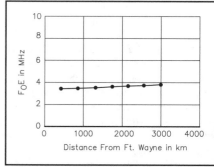

Fig 3—Plot of raw data from Table 1 of f_oE as a function of distance from Ft Wayne, IN, on the path to Antigua. The plot looks like a straight line, which can be represented by a first-order polynomial fit.

Fig 1—Geometry of an ionospheric path, related to earth-center coordinates. The important relations are the incremental changes: in height R and range S, subtended angle TH and angle PSI.

change in PSI (line 510) due to a change in refractive index MU.

Snell's Law in differential form requires not only the index of refraction (calculated from the electron density NE) but also its spatial variation (dMU/dR). In this program, spherical symmetry is assumed. This imposes a limitation on the use of the program, which will be discussed later.

The program advances the ray in 20-km steps. At every 20-km step, the electron density profile (NE versus height R) is computed, given the critical frequencies (f_oE and $F_2MUF(Zero)$), the layer heights (110 km for the E-layer and h_mF_2 for the F_2-layer), and the assumed variation (parabolic, exponential, $\sin^2$). The refractive index and its derivative are then calculated every 20 km. With this information, the program computes how much the ray is refracted (bent) in each 20-km segment.

The ray trace continues until either the ray goes off the graph or hits the ground. If it hits the ground, it is turned around and sent back up at the same angle at which it hit— this allows multiple hops.

The *IONPARA.BAS* Program

IONPARA.BAS gives f_oE, $F_2MUF(Zero)$, and h_mF_2 along the chosen short path for the

desired month, time, and solar flux level. All three parameters come from algorithms developed by R. Fricker of the BBC External Services.

The critical frequency f_oE is calculated as a function of sunspot number and solar zenith angle.[4] Two equations are used: one for daytime and one for nighttime. The critical frequency $F_2MUF(Zero)$ is the most complicated of the three parameters, and is derived from one main function plus 25 modifying functions.[5]

The F_2-layer height h_mF_2 also comes from Fricker's work in Reference 5. It is based simply on the sunspot number, with some modifiers for the geomagnetic equatorial region and seasonal variations. It is accurate to the nearest 25 km.

Installing the Programs

In addition to *IONPARA.BAS* and *RAYTRACE.BAS*, the Power BASIC program *PB.EXE* is included. I recommend creating a subdirectory (C:\RAYTRACE, for example), and copying RAYTRACE.BAS and IONPARA.BAS there. Run these programs using the QBasic Program supplied with DOS.

Detailed Ray Trace Example

The best way to explain how to do a ray

trace is to work through an example. For this I will do a ray trace from my location in Ft Wayne, Indiana (41° N, 86° W) to Antigua V2A (18° N, 62° W) for February 15 at 1600 UTC (11:00 AM, Ft Wayne local time). I assume a smoothed sunspot number of 110 (considered to be the peak of an average 11-year solar cycle).

Once in your subdirectory, execute the QBasic program, by typing QBasic. Then open *IONPARA.BAS*. Finally, run *IONPARA.BAS*. Follow the prompts, making sure your printer is turned ON. **Fig 2** is a plot of smoothed mean sunspot number versus smoothed mean solar flux, in case you need to convert from sunspot number to solar flux.[6]

The result is a printout of the three ionospheric parameters versus distance along the path from Ft Wayne to Antigua. **Table 1** shows a printout for this example. The latitude and longitude of the various points along the path are also printed for reference. After the data is printed, you will be told to hit any key for a graphics display. This display is a Mercator-projection map showing the path selected, the terminator, the geomagnetic equator, and the sub-solar point. The center horizontal line is 0° latitude, and the center vertical line is 0° longitude. This

Table 1

Ft Wayne, Indiana to Antigua, Feb 15 at 1600 UTC and SSN = 110

Distance km	Lat Deg N	Long Deg W	$F_2MUF(Zero)$ MHz	f_0E MHz	h_mF_2 km
429	38.4	82.3	9.9	3.4	312
857	35.7	78.8	10.3	3.4	312
1286	32.9	75.6	10.6	3.5	312
1714	30.0	72.6	10.8	3.6	312
2143	27.1	69.8	11.0	3.7	312
2572	24.1	67.0	11.1	3.7	312
3000	21.1	64.5	11.2	3.8	312

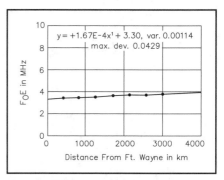

Fig 4—Plot of f_0E versus distance for the straight-line curve fit described by the equation shown at the top of the graph. The fit looks good.

simple picture gives you insight into where your selected path is with respect to daylight/darkness and the geomagnetic equator.

The second step is to define equations for the data of Table 1. We'll start with f_0E. **Fig 3** is a plot of the data using the program *EasyPlot*. The data points appear to fall on a straight line. So let's try a first-order polynomial. This is a straight line of the form $y = ax + b$. **Fig 4** shows the resulting curve fit—not too bad. The resulting equation, using HDIST as the distance variable, is:

$$f_0E = 3.30 + (1.67E{-}4) \times HDIST.$$

Now let's move on to $F_2MUF(Zero)$. **Fig 5** is a plot of the data. This plot has a gradual curve to it. Let's try a third-order polynomial for this one. **Fig 6** shows the resulting curve fit—very good indeed. The resulting equation is:

$$F_2MUF(Zero) = 9.40 + (.00131) \times HDIST - (3.45E{-}7) \cdot HDIST^2 + (3.55E{-}11) \times HDIST^3$$

Finally, the data for h_mF_2 shows it is constant versus distance. Thus no curve fitting is required. It will simply be entered as 312 km.

The last step is to insert the equations (actually two equations and one constant in this example) into the ray-trace program. To do this, load *RAYTRACE.BAS*. Go to line 815. This line is used for defining h_mF_2 versus distance. Since h_mF_2 in this example is a constant, simply enter 312 after the equal sign.

Now go to line 1046. This line is for defining $F_2MUF(Zero)$ versus distance. Note that $F_2MUF(Zero)$ is called FXF2 in the program. Enter the equation after the equal sign.

Finally, go to line 1430. This line is for defining f_0E versus distance. Enter the equation after the equal sign.

Fig 7 shows these three lines (815, 1046, and 1430) with the curve-fitted equations. Note the use of parenthesis to better group the various portions of the equations. Make sure the number of left and right parenthesis

is correct—if not, the program won't run.

Preliminaries

Before running a ray trace, there are several additional inserts to enter into *RAYTRACE.BAS*.

Line 260 and 330 are used to label the ray trace plot for documentation purposes. For this example, "February 15," "1600," and "155" are inserted in line 260. "Ft Wayne," "Antigua" and "28.3" are inserted in line 330.

Line 405 sets up labels along the horizontal axis. The first one is "Ft Wayne" at 0 km. The second one is the target location Antigua placed at 3400 km (distance taken from Fig 3). Column "23" puts the label "Antigua" at the 3400-km point.

Line 550 sets where the ray trace ends. Since the horizontal axis goes to 10,000 km, you could run into a situation on a shorter path where the curve fits are good for the path but go wild past your target location. This can result in the program crashing. Line 552 gives you a guide as to what the number in line 550 should be. For this example, we don't want the ray trace to go much farther than 3400 km, so "250" (= 4400 km) is chosen.

The last insert is the desired frequency in line 80. We'll start on 10 meters, so "28.3" is entered.

The Actual Ray Trace

To run the ray trace, simply type F9, once you've entered all the proper data. You will be prompted for an elevation angle. Let's

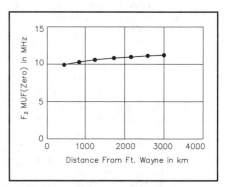

Fig 5—Plot of $F_2MUF(Zero)$ from Table 1 versus distance from Ft Wayne, IN. This looks like a curve that could be represented with a third-order polynomial fit.

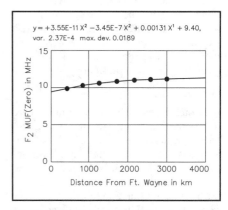

Fig 6—Result of third-order polynomial fit of the data in Fig 5. The equation is shown at the top of the graph.

```
815 HMF2=312

1046 FXF2=9.4+(.00131)*HDIST-(3.45E-7)*HDIST^2+(3.55E-11)*HDIST^3

1430 FOE=3.30+(1.67E-4)*HDIST
```

Fig 7—Changes made to BASIC statements in RAYTRACE.BAS to enter the curve-fitting equations derived for the example from Ft Wayne, IN, to Antigua.

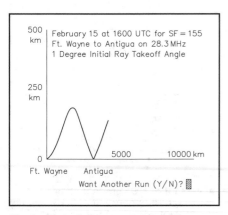

Fig 8—Plot of ray trace created by RAYTRACE.BAS program for a 1° takeoff angle at Ft Wayne. The higher the sunspot level, the better the chance of transequatorial propagation, especially at VHF.

start with 1.0°. **Fig 8** is the resulting ray trace, showing that a 1° takeoff angle gives a one-F_2 hop from Ft Wayne to Antigua. Answer the "Want Another Run (Y/N)" prompt with a "Y," and select 12.42°. This will give you a higher angle, one-F_2 hop called the *Pedersen Ray*. Now try 12.5°, where the ray will penetrate the ionosphere and go off into space.

Change the frequency to 20 meters in line 80 ("14.2") and you'll see E-layer hops and F-layer hops, depending on the elevation angle selected. A comment is in order about the ray trace screen. The ray trace is plotted on a rectangular coordinate system. This is why the 1° elevation angle in Fig 8 doesn't look like 1°. If the results were plotted in a spherical coordinate system, as it really is, everything would look right.

Limitations

As mentioned earlier in the article, there is a limitation to this ray-trace program. This comes about because spherical symmetry is assumed. This means that the refractive index MU, which is derived from the critical frequencies f_oE and $F_2MUF(Zero)$, varies only as a function of height. In other words, f_oE, $F_2MUF(Zero)$, and h_mF_2 must be somewhat constant all along the desired path.

This generally won't be a problem except for some special cases, like crossing the geomagnetic equator in the late afternoon or early evening hours. Under these conditions, $F_2MUF(Zero)$ varies substantially along the path, as does h_mF_2. In order to do a proper ray trace, a more complicated version of *RAYTRACE.BAS* must be used. This would include the spatial variation of MU with respect to distance (dMU/dTH).

There is a self-limiting aspect of this simpler *RAYTRACE.BAS* program: if $F_2MUF(Zero)$ and h_mF_2 vary too wildly, you won't be able to curve fit them with a single polynomial (even with a higher-order one).

Accuracy

I compared the angles predicted by *RAYTRACE.BAS* against those predicted by IONCAP[7] for several paths and conditions (month, time, solar flux). The shorter paths (out to several thousand kilometers) were in very good agreement with IONCAP. As the path got longer and longer (up to 10,000 kilometers), the angles predicted by *RAYTRACE.BAS* diverged more and more from those of *IONCAP*.

This is to be expected for two reasons: the ionospheric models are not exactly the same in the two programs, and the method of determining elevation angles (modes) is quite different.

IONCAP uses a large data base of ionospheric parameters to describe the ionosphere in terms of hourly, seasonal, and solar cycle variations. *RAYTRACE.BAS* essentially uses equations that were devised by Fricker to fit this large data base. Several propagation programs now on the market use these equations.

IONCAP methodology for paths beyond 10,000 km identifies sample areas along the path based on path distance—either one, three, or five sampled areas are used for all predictions, including elevation angle (mode) predictions. *RAYTRACE.BAS* determines the elevation angle based on a ray going continuously through the ionosphere along the entire path, and not at discrete sampled areas.

So which is more accurate, *IONCAP* or *RAYTRACE.BAS*? For short paths, I believe *RAYTRACE.BAS* is as good as *IONCAP*. For

the longer paths, I'd be inclined to believe in *IONCAP* a little more because of the compromises in fitting equations to enormous amounts of data, but this is somewhat tempered, since *RAYTRACE.BAS* takes the entire path into account. Regardless of these issues, remember that both programs give statistical results. They use *median* parameters of the ionosphere; that is, results with probabilities tied to them.

Enhancements

For those of you with BASIC programming experience, it would be easy to add some enhancements to the *RAYTRACE.BAS* program. For example, at a given frequency the elevation angle could be stepped from 1° to 20° to see the effect of angle on skip distance. Or at a given angle, the frequency could be stepped through several amateur bands to see the effect of frequency on skip distance.

Closing Comment

Although this is a simple ray trace program, it can provide valuable insight into the mechanics of propagation. Have fun with it, and always subject the results to the test of common sense!

Notes and References

[1] C. Luetzelschwab, K9LA, "10-Meter Long Path During Solar Cycles 21 and 22," *The ARRL Antenna Compendium, Vol 4* (Newington: ARRL, 1995), pp 138-145.

[2] *EasyPlot*, Version II, Spiral Software, Brookline, MA.

[3] P. A. Bradley and J. R. Dudeney, "A Simple Model of the Vertical Distribution of Electron Concentration in the Ionosphere," *Journal of Atmospheric and Terrestrial Physics*, 1973, Vol 35, pp 2131-2146.

[4] R. Fricker, *IONPRED*, Ver 1.4.

[5] R. Fricker, "A Microcomputer Program for the Critical Frequency and Height of the F Layer of the Ionosphere," *Fourth International Conference on Antennas and Propagation*, IEE (UK), 1985, pp 546-550.

[6] M. Leftin, "Ionospheric Predictions, Vol 1, The Estimation of Maximum Usable Frequencies from World Maps of $F_2MUF(Zero)$, $F_2MUF(4000)$, and $EMUF(2000)$," Institute for Telecommunications Science, Boulder, CO, 1971.

[7] *IONCAP*, Ionospheric Communications Analysis and Prediction Program, Institute for Telecommunications Science, Boulder, CO, 1983.

Measurements and Computations

More On the 3-Meter Impedance Measuring Bridge

By Peter Dodd, G3LDO, Technical Editor, *Radcom Magazine*
37 The Ridings
East Preston
West Sussex BN16 2TW
United Kingdom

G3LDO revisits the "3-Meter Method"—to improve graphics and to improve measurement accuracy.

A method of measuring impedance, using fixed reference components in an RF bridge, was originally described by W8CGD and called the "3-Meter Method."[1] I used the 3-Meter Method for measuring impedance for many years, long before the noise bridge made an appearance. I have instinctively felt that it was giving good results, mainly because they correlated well with other observations, such as SWR measurements.

However, there is a demand now for improved accuracy in all areas of test equipment and measurement of antenna feed impedance is no exception. This is exemplified in the article by Wilfred Caron describing an admittance bridge in *The ARRL Antenna Compendium, Vol 3*. Not only does it give details of the construction of an accurate instrument but it also provides a method to check the accuracy.

In *The ARRL Antenna Compendium, Vol 4* I showed how a computer program could be used to extract impedance data from scalar voltage measurements made using the 3-Meter Method. This article assesses the accuracy of the method. I also improved the graphics display in the software, which is included on the disk accompanying this volume of *The ARRL Antenna Compendium*.

The traditional way of checking an impedance measuring instrument is to use a range of pre-calibrated plug-in impedances. This is fine, provided you have a method for the initial calibration. The Caron approach uses a length of coaxial cable with a resistive load twice or half the characteristic impedance of the coaxial cable. Even without knowing the electrical length of the test cable some idea of the accuracy of the measurements will be apparent by any deviation from the 2:1 SWR contour when the results are plotted on an impedance diagram. The ideal results are shown in **Fig 1** and the results of a set of measurements

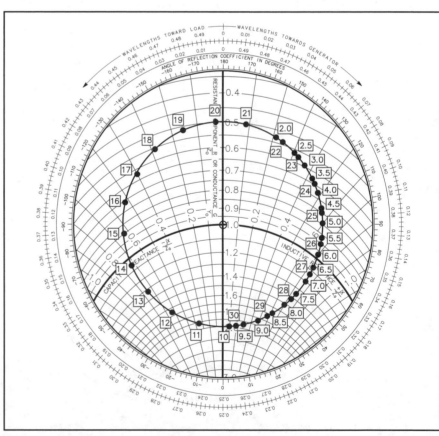

Fig 1—Idealized impedance plot, measured at different frequencies over a lossless length of terminated coaxial cable. The termination is arranged to give an SWR of 2:1. This is from *The ARRL Antenna Compendium, Vol 3*.

made on the Caron admittance bridge are plotted in **Fig 2.**

I decided to use the Caron method to test the accuracy of the 3-Meter Method.[2] But before doing this, I wanted to improve the program (called GRAPH, in the G3LDO subdirectory on the disk included with *The ARRL Compendium, Vol. 4*) used to graphically display the impedance readings.[3] Originally, I had scaled the graph display so that the vertical axis ranged from 0 to 150 Ω resistive and the horizontal scale ranged from ± 150 Ω reactive, as shown in **Fig 3**. I found that the reactance range was too large for most of my antenna measurements, so I reduced the reactance range to ± 100 Ω. This has the advantage of making the chart more accurate and it also made the SWR contour circular and improved the appearance of the display. (Note that the 1:1 SWR point is not at the center of the circle as in the Smith Chart, but is at the 50 + *j* 0 Ω point.)

A further improvement in chart resolution was to change the screen from CGA to EGA. This is as far as I could go while still maintaining GW-Basic/QuickBasic compatibility. All the programs will run in QuickBasic in spite of the line numbers and they can be compiled into *.EXE files, as has already been done for you on the disk.

You may have to change the source code for the SCREEN number in line 90 to match your screen and then recompile the program. You can also add color, at the beginning of the appropriate subroutine if you like a more fancy display. The on-disk file is called GRAPHEGA.BAS, with the executable program called GRAPHEGA.EXE.

I have included a calibration file on disk called GRAPHCAL.DAT. The actual impedance values calculated from this file are plotted using GRAPHEGA in **Fig 4**. I put the calibration values in the FREQUENCY column. Some improvement in the accuracy of the plot may be possible by experimenting with the values of factors in lines 980 and 990 of GRAPHEGA.BAS.

So, how did the 3-Meter impedance box shape up in the Caron test? The data from the first test on a 12 feet, 5 inches length of RG-213 terminated with a 100-Ω resistor is plotted in **Fig 5**. The results don't fall on the 2:1 SWR circle as well as I thought they should have. Note that there is nothing particularly significant about the coax cable length; I just happened to have a piece that long. I then did some tests to find the reason for these errors, finding two main causes:

1. harmonic content in the excitation source
2. a value of reference component inappropriate for the value being measured.

I made further measurements at the frequencies where the errors were greatest, using an ATU to reduce the harmonic content that may be present. I then altered the value of the fixed reference capacitor for each band. Previously, I had thought replacing the capacitor for every frequency band was unnecessary. In my new version I used 2000 pF for 160 meters, 560 pF for 80 meters to 30 meters and 200 pF for bands from 20 meters to 10 meters, rather than the values shown in the earlier article.

Both these measures improved the accuracy, as shown in **Fig 6**. However, replacing the reference capacitor at each measured frequency is, to say the least, inconvenient. A method recommended by G3HCT over-

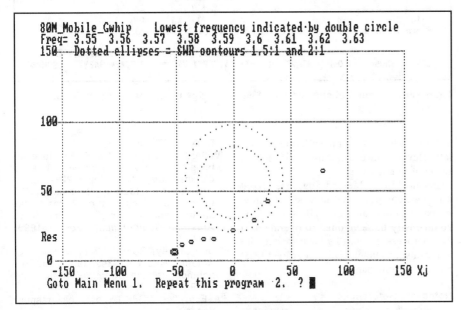

Fig 2—Set of measurements made using the Caron admittance bridge. This is from *The ARRL Antenna Compendium, Vol 3.*

Fig 3—Original GRAPH program scaled the graph display with the vertical axis ranging from 0 to 150 Ω resistive and the horizontal scale ranging from ± 150 Ω reactive.

comes this problem.[4] He uses a plug-in reference capacitor using a PCB connector strip. It might also be useful to make the reference resistor plug-in when measuring high or low impedances. In other words, the reference component must be roughly equal to the impedances being measured if greater accuracy is required.

Impedance Measurements at VHF

Some time ago I tried to measure impedance at VHF (145 MHz) using the 3-Meter box. The results were approximate and I was unsure of their reliability. I tried the Caron test using the same coaxial cable/resistor termination setup. See **Fig 7**. The data from the impedance measurements made on the two highest frequencies were automatically modified by the program because a non-intersection error occurred. (For an explanation, see Reference 3.) Errors on the two highest frequencies are apparent. These errors appear to be caused by the capacitance of the diodes, which become more significant as frequency, and the value of the impedance being measured, is increased. I think that microstrip layout and microwave diodes may help; as would appropriate valued chip capacitors for dc isolation and decoupling.

Excitation Level

It is not necessary to use 5 V across the reference resistor for the 3-Meter Method to work. To use some other, lesser voltage might overcome the difficulty of adjusting the transmitter level to exactly 5 V before measurements could commence. However, the graphic method and the computer programs derived from the graphic method are based on a reading of 5 (that is, an input scaled by 10 to be 50).

The program ZCALCB on the accompanying disk was modified to accomplish this. New versions of TODISKB and TABLEB are also included on the disk, each with an additional matrix box to accommodate the $E_R(B)$ variable. I have checked the system by reducing excitation voltage. It appeared to give good results, but errors increased sharply if any one of the measured voltages fell below 1 V. This level may depend on the type of diode being used but ZCALCB will give a good indication of the errors.

Conclusion

The method of displaying the measured impedances relative to an SWR contour indicates that the 3-Meters Method could, with care, have the same degree of accuracy as the admittance bridge. WD8KBW suggested that, for greater accuracy, the attenuators should be external to the bridge.[5] I found that a switched variable attenuator is

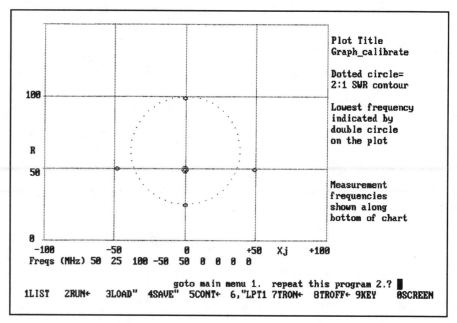

Fig 4—Calibration data plotted using GRAPHEGA, showing new horizontal axis ranging from ± 100 Ω reactive.

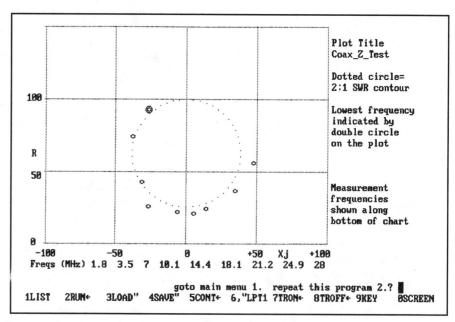

Fig 5—First attempt at checking the 3-Meter bridge using the Caron method.

also useful because of the variation in excitation level required when measuring a wide range of impedances.

Because the 3-Meter bridge uses fixed reference components it may be capable of measuring impedances at VHF. The method could easily be developed to operate as a microprocessor-controlled instrument with a five or six channel ADC and a direct LCD readout of impedance.

Notes and References

[1]D. Strandlund, W8CGT, "Measurement of R+jX," *QST*, Jun 1965, p 24.

[2]Wilfred N. Caron, "The Hybrid Junction Impedance Bridge," *The ARRL Antenna Compendium, Vol 3*, 1992, p 223-230.

[3]Peter Dodd, G3LDO, "Measuring RF Impedance Using the Three-Meter Method and a Computer," *The ARRL Antenna Compendium, Vol 4*, 1995, p 175-9

[4]John Bazley, G3HCT, "Measurement of Antenna Radiation Resistance and Reactance," *Radio Communication*, June 1979.

[5]A. E. Weller, WD8KBW, private correspondence.

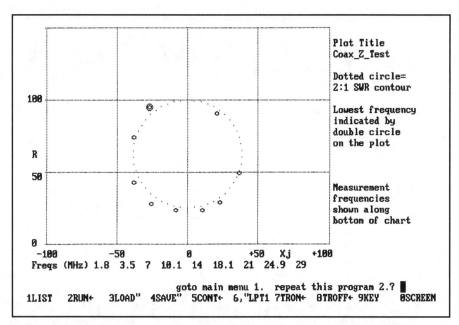

Fig 6—Second attempt at checking the 3-Meter bridge, after identifying the causes of some of the errors: transmitter output harmonics and less-than-appropriate values for reference components.

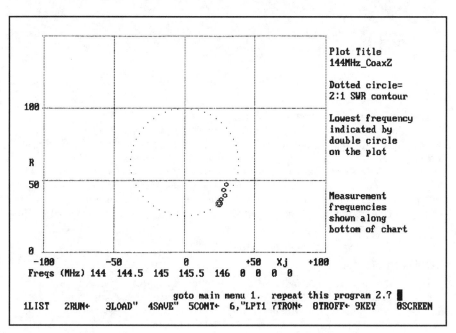

Fig 7—Errors on the two highest VHF frequencies become apparent, probably caused by diode capacitance.

Measuring Antenna SWR and Impedance

By Francis J. Merceret, PhD, WB4BBH
1265 Emma St
Merritt Island, FL 32952

WB4BBH demonstrates that what you see on the SWR meter isn't always what's really on the transmission line.

For thirty-five years as a licensed amateur, I have measured standing wave ratio (SWR) with commercial SWR bridges. For many of those years I used the kind that requires setting a *sensitivity* control for a full-scale reading on forward power, and then switching to reverse power to obtain an SWR reading. For the last decade, I have been using *cross-needle* meters that simultaneously measure forward and reverse power, with separate needles that move in opposite directions on the meter face. These meters indicate SWR at the intersection of the two needles without requiring any operator intervention—like switching or sensitivity setting.

Recently, I began to question the reliability of such SWR measurements. This article presents the results of a study to determine the accuracy and limitations of various methods available to amateurs for measuring the SWR and impedance of their antennas.

The inquiry began with a new commercial multi-band antenna that had an SWR greater than 2.5:1 on 20 meters. The manufacturer suggested that I change the length of transmission line to the antenna. My understanding is that the SWR on a transmission line is determined by the ratio of the antenna impedance to the characteristic impedance of the line, and that line length should have little to do with SWR. When insertion of a short piece of RG-8 lowered the SWR to 2:1, I got curious enough to purchase a noise bridge to explore this some more. I also wanted to measure the actual complex impedance. When the SWR computed from the noise-bridge measurements (about 1.7:1, independent of line length) didn't agree with that measured with the cross-needle bridge, I decided to do a serious study!

I constructed a set of resistive and reactive loads from discrete components so I could measure their impedance and SWR, using a variety of meters typically used by amateurs. *The ARRL Antenna Book* describes the use of these SWR-measuring devices in detail.[1] My results indicate that SWR bridges often do not measure capacitive loads or high SWR values accurately. I found that a noise bridge is more accurate, but that it is also much more difficult to use. The most accurate and easiest instrument to use is a modern, microprocessor-based "RF analyzer."

Experimental Setup

The instruments evaluated in this test are listed in **Table 1**. I selected them strictly on the basis of availability—I used every meter I could get my hands on.

I built the loads from 51-Ω metal oxide (non-reactive) 3-W 5% resistors, disc ceramic capacitors, and small inductors. For each load, a resistance made of series and/or parallel combinations of the resistors was placed either in series or in parallel with a single reactive component. These were mounted on dual banana plugs and plugged into an adapter, which connected the load either to coaxial cable or directly to an instrument. The composite resistors were painted with a color code to indicate their value.

Before conducting the experiment, I measured the individual components. The resistances were measured on two separate digital multimeters, as well as on the Autek and Palomar instruments. The DMM readings

Table 1

Instruments Evaluated

Manufacturer	Model	Type
Autek	RF-1 (1.7 - 30 MHz)	RF analyzer
Bird	43 (2 - 30 MHz)	RF Wattmeter
Daiwa	CN-101L (1.8 - 30 MHz)	Cross-needle SWR
Daiwa	CN-410M (3.5 - 150 MHz)	Cross-needle SWR
Daiwa	CN-620A (1.8 - 150 MHz)	Cross-needle SWR
MFJ	949	Set full scale SWR
Palomar	RX-100 (1 - 100 MHz)	Noise Bridge

Table 2

Resistor-Reactor Pairs Used in the Experiment

Resistor (Ω) (color code)	Reactor	Spreadsheet File
12.7 (Brown)	1 µH	R013L001.WK1
37.8 (Red)	1000 pF	R038C1K.WK1
50.3 (Orange)	470 pF	R050C470.WK1
75.5 (Yellow)	1 µH	R076L001.WK1
100.8 (Green)	220 pF	R101C220.WK1
151.7 (Blue)	470 pF	R152C470.WK1
201.3 (White)	100 pF	R201C100.WK1

Fig 1—Photo of test setup used by WB4BBH to make his measurements.

Table 3

Results for Purely Resistive Loads

R, Ω	12.7			100.8			201.3		
F, MHz	1.8	3.5	7	1.8	3.5	7	1.8	3.5	7
Design SWR	4.0	4.0	4.0	2.0	2.0	2.0	4.0	4.0	4.0
Autek SWR	4.2	4.6	4.7	2.0	2.1	2.1	4.1	4.1	4.1
Bird SWR	3.9	3.9	3.4	2.5	2.5	2.6	4.3	4.4	4.7
CN101 SWR	4.2	4.0	3.5	1.5	1.7	2.0	2.8	3.1	4.2
CN410 SWR	4.0	4.0	4.0	1.8	1.9	1.9	3.0	3.1	3.2
CN620 SWR	3.6	3.8	3.5	1.6	1.9	1.9	2.8	3.1	3.4
MFJ949 SWR	1.7	2.5	2.8	3.0	1.1	1.0	2.6	2.0	1.2
Palomar SWR	—	—	—	2.0	2.0	2.0	4.4	4.4	4.5
Design Z, Ω	12.7	12.7	12.7	100.8	100.8	100.8	201.3	201.3	201.3
Autek Z, Ω	17.5	13.0	14.0	101.0	100.5	100.5	201.0	199.0	196.5
Palomar Z, Ω	—	—	—	100	100	100	220	220	221.3
Design R, Ω	12.7	12.7	12.7	100.8	100.8	100.8	201.3	201.3	201.3
Autek R, Ω	12.0	11.1	11.0	101.6	97.8	97.8	199.7	193.8	191.4
Palomar R, Ω	—	—	—	100	100	100	220	220	220
Design X, Ω	0	0	0	0	0	0	0	0	0
Autek X, Ω	3.6	6.8	8.7	—	23.1	23.1	23.0	45.0	44.7
Palomar X, Ω	–0.2	–0.1	0	–0.2	–0.2	0	–0.2	–0.1	24.4

and RF measurements below 10 MHz all agreed within 5%. I measured the capacitors and inductors with the Autek and the Palomar instruments. Readings agreed within 20% below 10 MHz. Above 10 MHz, both the resistors and the reactive components began to deviate from ideal behavior, probably due to stray capacitance and lead inductance. As a result, I conducted all measurements in this study below 10 MHz.

I used a spreadsheet to compute the impedance and SWR relative to 50 Ω for series and parallel combinations of the components at 1.8, 3.5 and 7 MHz. Eight resistor-reactor pairs were selected having series and parallel impedances covering a broad range at these frequencies. For each pair, I created a spreadsheet to record the calculated and measured data for comparison. **Table 2** shows the combinations selected. In addition, I used pure resistances of 12.7, 100.8 and 201.3 Ω.

For the impedance measurements with the Autek and the Palomar, the load was connected directly to the instrument. Neither of these measurements required use of a transmitter, but the Palomar required a receiver. The remaining instruments all required transmitting an unmodulated signal for the duration of the measurement. To reduce the effort required, all of the SWR meters and the Bird Wattmeter were connected in series with short pieces of RG-58 coaxial cable.

I determined by trial and error that altering the order of the meters made no difference in the readings, as long as the MFJ-949 was closest to the transmitter. It raised the SWR of instruments placed closer to the source, but did not affect those farther from the source at any frequency. I usually used 10 W of forward power for the SWR measurements. **Fig 1** is a photograph of the experimental setup.

For each pure resistance, and for each series and parallel combination of each re-

sistor-reactor pair, I made measurements at 1.8, 3.5 and 7 MHz. These were entered into the spreadsheet for the components used. Given the characteristic impedance (here, 50 Ω) and any two of the set SWR, Z, R or X, you can calculate the remaining two, using the equations in Chapter 24 of *The ARRL Antenna Book*. After collecting the data for all impedances, I compiled the results in a summary spreadsheet for analysis. I created least-squares fits and graphic plots of design-versus-measured values of SWR, Z, R and X.

Results—Resistive Loads

Table 3 shows the results for three purely resistive loads. The imprecision in the Bird measurements is due in part to the difficulty reading low reflected power levels accurately. Note that 1.8 MHz is below the rated frequency range of several of the instruments. Also note that the 12.7-Ω load is below the effective resistance range of the Palomar and at the lower limit of the Autek. I did not correct the Autek readings to compensate for the unit's input capacitance. The results suggest that the SWR bridges become inaccurate even for resistive loads as the SWR rises towards 4:1, although the results for 201.3 Ω at 7 MHz also suggest that the load may have been somewhat reactive at that frequency.

Results—All Loads

Fig 2 shows the measured versus design SWR for the SWR bridges. **Fig 3** shows the SWR results for the Autek, Bird and Palomar. Clearly, the bridges fare poorly as the SWR increases—they are essentially incapable of quantitative measurement above 10:1. They are inaccurate above 4:1.

Figs 4 presents the measured versus design resistance and reactance for the Autek and Palomar instruments. Both seem to do a competent job. The zero values for the Autek at 50, 100 and 200 Ω are artifacts of the spreadsheet. These artifacts occur for reactive loads where X >> R. The Autek manual cautions against calculating R from measured Z and SWR under such circumstances.

You may derive several quantitative mea-

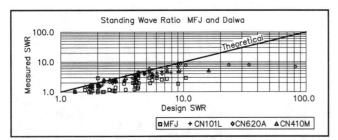

Fig 2—Measured SWR vs SWR for known loads for MFJ 949, Daiwa CN-101L, Daiwa CN-620A and Daiwa CN-401M SWR meters.

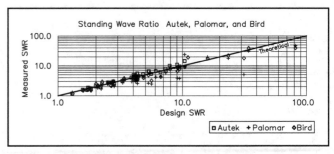

Fig 3—Measured SWR vs SWR for known loads for Autek RF-1 RF Analyst, Palomar RX-100 noise bridge and Bird Model 43 Wattmeter, with 2-30 MHz slug.

sures of agreement between the measured and design values using the *technique of least squares*.[2] This technique determines a constant and a slope that best fit the measured data in an equation of the form:

$$Y_{fit} = \text{constant} + \text{slope} \times Y_{design}$$

This minimizes the mean-square difference between the measured and fitted values. In addition to the constant and slope, the technique provides a quantity called the *correlation coefficient*, r, which measures the scatter between the data and the fit. The square of the correlation coefficient is proportional to the *variance* of the data about the fit. For a perfect fit (all observations fall on the line specified by the equation), $r^2 = 1$. If the data are totally random then $r^2 = 0$. For a perfect instrument, constant = 0, slope = 1 and, of course, $r^2 = 1$.

Table 4 presents the fits for the entire data set, and also for a subset containing just those loads for which the SWR did not exceed 4:1. The values of r^2, constant, and slope are consistent with our visual impression from the figures. The SWR bridges, especially at high values of SWR, do not provide highly accurate measurements. The offset (c) in the Autek is probably due to its input capacitance, a matter mentioned in the manual. As mentioned previously, I did not apply the correction suggested in the manual to the data here.

Using the Instruments

Both the cross-needle meters and the Autek are extremely easy to use. The Autek's digital display is especially convenient. The Autek display became weak and the frequency unstable when the battery was low. The first time I noticed this was during the 1995 WPX CW 'test. At the time, I thought there must be "RF interference" from the nearby kWs of our multi-multi station (WC4E, operating from the outstanding station of W1CW/W1YL/K1ZX). When I got home after the contest, I measured about 7 V on what was supposed to be a 9-V battery. A new battery cured the problem.

I had forgotten how painful setting the sensitivity is for older types of SWR

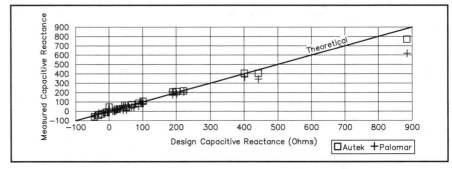

Fig 4—Measured capacitive reactance vs known capacitive loads for Autek RF-1 RF Analyst and Palomar RX-100 noise bridge.

Table 4
Linear Regression Parameters

Unit	N	r^2	Slope	Constant
Autek X	51	0.99	0.92	7.48
Palomar X	51	0.97	0.78	2.58
Autek R	46	0.96	0.89	4.24
Palomar R	43	0.98	1.01	−0.08
Autek Z	51	0.99	0.88	11.85
Palomar Z	51	0.96	0.79	15.76
Autek SWR	46	0.96	1.22	−0.54
Palomar SWR	43	0.65	0.68	0.67
Bird SWR	51	0.83	0.55	2.09
CN101 SWR	38	0.78	0.62	0.77
CN620 SWR	37	0.87	0.67	0.17
CN410 SWR	37	0.89	0.67	0.58
MFJ Set SWR	39	0.44	0.24	0.85
MFJ Cmp SWR	51	0.78	0.31	1.70
SWR Below 4.0				
Autek X	20	0.91	1.06	1.16
Palomar X	20	0.84	0.75	−3.89
Autek R	20	0.98	0.92	5.52
Palomar R	20	0.98	1.04	−4.81
Autek Z	20	0.99	0.95	4.91
Palomar Z	20	0.98	1.10	−14.00
Autek SWR	20	0.94	1.03	−0.04
Palomar SWR	20	0.68	0.68	0.53
Bird SWR	20	0.79	0.97	0.20
CN101 SWR	20	0.81	0.84	0.10
CN620 SWR	20	0.86	0.71	0.35
CN410 SWR	20	0.92	0.81	0.22
MFJ Set SWR	20	0.10	0.23	0.76
MFJ Cmp SWR	20	0.35	0.49	0.35

bridges. I had to adjust the pot for every change in frequency or power level.

The noise bridge is the most difficult to use properly. The null is often extremely narrow and not always as deep as I'd like. In many cases, I would never have found a null if I hadn't known in advance about where it was supposed to be. Such advance knowledge is a luxury we don't have while troubleshooting a misbehaving antenna.

Conclusions

The results suggest that we should use our SWR bridges with caution. Although the cross-needle types provide a quick, hands-off continuous measurement valuable for real-time monitoring, they may seriously mislead the amateur if used quantitatively. SWR bridges do not accurately measure high-SWR conditions, nor do they accurately measure even moderate SWR when the load is capacitive.

They also require that you transmit a signal into the system under test—not always easy or advisable.

Noise bridges give better accuracy than SWR bridges and also give more complete information, since you may determine both the resistive and reactive components of the impedance. These require an external receiver, and thus are not stand-alone instruments. There is also some risk—if you forget to remove them from the line and accidentally transmit into them, the consequences may be expensive! Their worst feature is the difficulty determining the null, which may be both narrow and relatively shallow. Noise bridges can also be slow and tedious to use.

The RF analyzer used in this experiment proved to be an accurate and easy-to-use stand-alone instrument. It can be carried up the tower to the antenna or used in the shack with equal ease. Its digital readout is unam-

biguous. It provides most of the information provided by a noise bridge, but without the hassle. [Note: The Autek cannot determine the sign of an unknown impedance directly. It measures two scalar quantities: SWR and the magnitude of the impedance. From these you may compute the magnitude of resistance and reactance, but not the sign of the unknown reactance.—*Ed.*]

Based on the results presented here, my recommendation is use a cross-needle SWR meter for routine real-time monitoring of your transmissions, but when you need to troubleshoot or tune an antenna system, use an RF analyzer.

Notes and References

[1] *The ARRL Antenna Book*, 17th Edition (Newington: ARRL, 1994), pp 27-2 to 27-30.
[2] Allen Edwards, *An Introduction to Linear Regression and Correlation* (San Francisco: W.H. Freeman & Co., 1976), p 213.

A Programmable Calculator Azimuth-Elevation-DX Routine

By Brian B. Turner, K2SJM
3210 Wildmere Place
Herndon, VA 22071

> *Ever been somewhere in the field, away from your big computer and needed to know exactly where to point your satellite antenna? Here's K2SJM's tiny tracking program, perfect for a hand-held calculator.*

A few years ago I figured out an algorithm to determine the true azimuth and great circle distance from one location to another for a hand calculator. It is based on Dr. Martin (K2UBC) Davidoff's math and logic in *The ARRL Satellite Experimenter's Handbook*.[1] Having dutifully reinvented the wheel, I was mildly surprised to find that the same algorithm may be found in the "Antenna Orientation" chapter of *The ARRL Operating Manual*, written in BASIC.[2]

I like to rationalize that good minds flow in the same channels, rather than admit that I have not read *The Operating Manual*. However, this particular article goes a step further than the others and permits you to compute the elevation angle to a satellite of known subsatellite point and height. It also frees up a desktop computer from having to make simple calculations, if computers are in short supply, and gives the user shirt-pocket portability and freedom from power mains.

The algorithm was written for my programmable Casio calculator because I wanted something that was portable and that could be carried into the field in support of some military communications exercises, where computers were not always available. Besides, as a ham, I wanted to "roll my own."

It really worked! I could tell the military guys which way to point their HF antennas simply by poking in the latitude and longitude of the distant end. We also used geostationary satellites for communication. I relied on graphs similar to Figure 5-10 in *The Satellite Experimenter's Handbook* to tell the guys what azimuth and elevation angle was needed for a given satellite.

Since our antennas had broad beamwidths, a few degrees of error from the graphs was insignificant.

Then one day it occurred to me—the plane of the great circle between two earth stations is the same as one connecting an earth station and a satellite whose subsatellite point corresponds to the other earth station. I could calculate the azimuth toward a satellite simply by poking in the latitude and longitude of the subsatellite point. Since the great circle arc to the subsatellite point is also generated by the algorithm, knowledge of the height of the bird plus some simple geometry and trigonometry enable us to calculate the elevation angle. All it takes is one extra line of code in the algorithm.

Applications

This tiny and inexpensive package has already proved useful in many ways, and many other applications come to mind.

Some of these are:

1. Pointing an HF antenna toward a new DX station or a maritime mobile when you don't have a computer or the right software.
2. In a Field Day (or other contest) situation, pointing HF antennas when the computer(s) is/are busy logging contacts.
3. Pointing VHF-UHF-microwave antennas at cooperating stations while mountain-topping or trying to set distance records. I've done something like this with research vessels in the Gulf of Mexico who report their latitude/longitude over VHF communications channels, allowing the shore-based station to optimize orientation of directional antennas.
4. Determining azimuths and elevations for pointing antennas at satellites, whether in geosynchronous orbit or not

(the latter requires knowledge of the ground track and of height changes along that path). I've used the algorithm for both military field units and private companies.

5. Calculating exact line-of-sight azimuth-elevation bearings for airplanes of known position, where highly directional antennas or systems are being used. I've used this algorithm as a cross-check in precise radiolocation studies conducted by a well-known university.
6. Calculating Fresnel diffraction (knife edge) distances between terrestrial microwave stations and various objects or terrain from map information alone.
7. Calculating frequent flyer miles between airports.
8. Blue water sailboat navigation, when electricity is scarce.
9. VFR aircraft navigation.
10. Astounding your friends with trivial geographic knowledge of how many kilometers it is from your house to the Kremlin, or wherever.

Constraints

I use a Casio fx-4000p programmable hand calculator that cost about $30 five years ago. Any other programmable calculator will do, so long as you understand its language and can program the algorithm that follows. If you prefer, you can also write your own program for a computer using this algorithm, but you lose the shirt pocket portability that makes this "system" so appealing.

All distances (input, output, and constants) in the algorithm are in kilometers. If you use heights in feet (for mountains or airplanes) or in miles (for satellites), they must be converted to kilometers. The algorithm is not a true geodetic creature. It assumes a perfectly spherical Earth with a radius of 6371 kilometers, whereas the Earth is somewhat flattened at the poles.

All latitudes and longitudes must be entered in degree, decimal format. That is, a latitude of 40° 30' North would have to be converted to 40.5° N. For HF and most VHF-UHF work, it is adequate to estimate the number of degrees by visual interpolation of a map grid. If you must use it for very accurate work, then you may have to carefully measure and mathematically convert all values.

There is another convention that must be followed. North latitude is a (+) number, and South latitude is a (−). East longitude is a (+) number, and West longitude is a (−). Since all of North America is in West longitude, it is necessary to enter the values for North America as negative numbers. I have to admit, this fouls me up more than anything because I forget to use the minus sign about 10% of the time.

Programming Symbols

The Casio programmable calculators use certain symbols in their programming language. To understand the algorithms, you should become familiar with these:

":" is the end of an instruction or line of code.

"Δ" (darkened triangle) is to display the results of a calculation in a line of code. On Casio calculators, one must push the "EXE" key to end the current display and go on to the next.

" = " means "equals" but Casio uses an arrow for it.

"Lbl n" is a label, with "n" being some whole number, such as "Lbl 4". It is used in conjunction with a "Go to n" command, which is used to jump either forward or backward to that label within the routine.

"?" is a prompt for the operator to enter a value.

"Abs" means the absolute value of a number.

"a>b" means "*if* the value of a is *greater than* the value of b."

"a<b" means "*if* the value of a is *less than* the value of b."

" => Go to 2:Go to 3" means "*then*, Go to 2; *else*, Go to 3." The first symbol is a double-tailed arrow, which does not come out well with my word-processing software.

The symbol "*" (asterisk) is chosen to denote multiplication in the following algorithm so that it will not be confused with the letter "x." This convention is commonly followed in BASIC, Fortran and other programming languages.

U and V are your latitude and longitude.
X and Y are the distant latitude and longitude.

H is the height of the object in kilometers (0 for a ground station).

B is the great circle arc in degrees.

D is the great circle distance in kilometers.

T is the true azimuth toward the other point in degrees.

L is the elevation angle to any object above the horizon (a positive number).

If the value of L is negative, it is the *plunge angle* into the earth of a vector (as defined in structural geology) from your station to the other object. A negative value signifies that the object is below the horizon.

Go into the "WRITE" mode of your calculator, select a number from 0 to 9 as the name of your program, and enter the following code. Notes on the right side of the page in parentheses are solely to clarify and should not be entered.

The Az-El-DX Algorithm

___.___ = U: (make a permanent entry of your latitude as part of the program)

___.___ = V: (make a permanent entry of your longitude as part of the program)

Lbl 1:

? = X: (enter the other station's latitude on prompt when running the program; use a negative value for the Southern Hemisphere)

? = Y: (enter the other station's longitude on prompt when running the program; use a negative value for the Western Hemisphere)

? = H: (enter height of the object in km; if a surface station, enter "0")

$\cos^{-1}((\sin U * \sin X) + (\cos U * \cos X * \cos(V-Y))) = B$:
  (degrees of arc along the great circle between the two points)

$B * 111.2 = D$ Δ (darkened triangle forces D to be displayed; numerical answer is the great circle distance in km)

$\cos^{-1}((\sin X - (\sin U * \cos B))/(\cos U * \sin B)) = A$:
  (interim azimuth value)

$V - Y = E$:

Abs E>180=>Go to 2:Go to 3:
  (if the absolute value of E is greater than 180, then go to label 2, else go to label 3)

Lbl 2:

E>180=>E−360 = E:E+360 = E:
  (if the value of E is greater than 180, then E−360 is the new value of E, else E+360 is the new value)

Lbl 3:

E>0=>Go to 4:A = T Δ
  (if the value or new value of E is greater than 0, then go to label 4, else the interim value of the azimuth is the true value, which is displayed)

Go to 5:

Lbl 4:

360−A = T Δ (if the value or new value of E is greater than 0, the interim value of the azimuth is subtracted from 360 to give the true value, which is displayed)

Lbl 5:

$\tan^{-1}((H + 6371 - (6371 * \cos B)) / (6371 * \sin B)) − B = L$ Δ
  (the elevation angle in de-

grees to the satellite or other object is displayed; a negative value is the plunge angle into the earth of a vector from your station to the other point which is below the horizon)

Go to 1   (go back to the start of the program and enter new values of latitude/longitude/height; note that there is no ":" after this line)

Operation

1. After returning to the normal (calculate) mode of the calculator, press "Prg." The display will show which programs (named 0 to 9) contain routines you have written. Select the number for this routine that you chose earlier.
2. Push the "EXE" key and "?" appears. Enter the latitude of the distant station, remembering that south is negative.
3. Push the "EXE" key again, and another "?" appears. Enter the longitude of the distant station, remembering that west is negative.
4. Push the "EXE" key again, and a third "?" appears. Enter the height of the sta-

tion, satellite, airplane, etc in kilometers. Enter "0" if it is at ground level.
5. Push "EXE" a fourth time, and a number appears. This is the distance in kilometers between the two ground points.
6. Push "EXE" a fifth time, and a number which is the true azimuth from your station to the other is displayed.
7. Push "EXE" a sixth time, and the number of degrees of elevation to the satellite or airplane above the horizon appears as a positive number. A negative value is the plunge angle into the earth of a vector from your station to the other point, which is below the horizon. The term plunge is strictly defined in structural geology for vectors going into the earth. Unless you are a geologist or geophysicist using this program to calculate plunges, you may wish to ignore negative values.
8. If you push "EXE" again, you return to step 2 (above) and go through another cycle for another station. Push "AC" (clear all) to exit.

Caveat

If you are going out on a Field Day, operating maritime mobile, or traveling away

from the home QTH, remember to get your new latitude and longitude from a map or a GPS receiver and enter them in the first two steps of the algorithm.

Conclusion

I think you will be pleased with this little package. It fits in your pocket or your briefcase. It can be used under all sorts of field conditions. It frees you from a desktop computer, and it frees a computer from a trivial computation. It is remarkably robust and accurate, and it depends on small batteries or light for power. It will give you the aura of being a powerful, knowledgeable person to say to your mates, "Point 'er over there, boys." Of course, the computer geeks in the local radio club will scoff at your puny machine, but you'll have them cold. Your $30 machine, not a $3000 Pentium, is all it takes to do Az-El-DX calculations, and they can't put theirs in a pocket.

Notes and References

[1]Martin R. Davidoff, *The Satellite Experimenter's Handbook* (Newington: ARRL, 1984), pp 9-5, 9-6.
[2]Steve Ford, and Paul Danzer, editors, *The ARRL Operating Manual*, 5th Ed. (Newington: ARRL, 1995), p 4-5.

Special Antennas

Plastic Antennas —a Contradiction?

By Patrick E. Hamel, W5THT
1157 E. Old Pass Road
Long Beach, MS 39560

The usual materials used in an amateur beam antenna are aluminum tubing and castings, fiberglass, stainless bolts, and special machined fittings. There are numerous designs published that can be built with the aid of a machine shop and a reputable aluminum supplier. What I hope to show in this article is that by using PVC pipe and hand tools, a beam antenna can be constructed affordably. Necessary materials can be found in almost any community. There are no expensive machined, mail-order aluminum, fiberglass or phenolic items required!

There are two parts to an antenna design: electrical and physical. I have made the electrical design easy with my program *WIDEBAND*. This program computes the input impedance of different antenna element shapes for different frequency ranges. It uses the math from *The ARRL Antenna Book* to design LPDA (Log Periodic Dipole Array) antennas. *WIDEBAND* quickly recalculates and changes the input impedance and size/shape of the antenna until you are satisfied with the design. Once a basic design has been generated, any number of parameters can be changed and everything is quickly recalculated. The user's manual file WIDEBAND.DOC is on the disk accompanying this book.

I will use my latest HF LPDA as an illustrative example. My LPDA is designed to cover 20 through 10 meters. **Fig 1** is a photo of the completed LPDA at ground level, showing the "trestle-bridge" style of trusses supporting the long PVC boom. You can build almost any kind of beam (Quagi, Yagi, or Quad) using the PVC construction techniques that follow.

Most of the elements of my LPDA are conventional, made of aluminum tubing. However, the growth of the trees in my backyard has made it necessary for me to develop a unique form of loaded 20-meter element— what I call the "Capacitive-U-PVC-and-

So you thought beam antennas had to be made of aluminum tubing or perhaps wire? W5THT gives practical details on how PVC pipe and aluminum tape can be used, even at HF!

Fig 1—The W5THT 20 to 10-meter LPDA prototype at ground level. Note the "trestle-bridge" type of truss support system used to support the 20-foot long PVC boom.

Aluminum-Tape" element. You can see how the long element in **Fig 2** is bent back on itself at each end. If you like, you can construct an entire antenna from PVC, aluminum tape, rope, turnbuckles, some fiber-

glass epoxy (such as used on boats or car repairs), PVC glue, and paint!

Physical Forces on the Boom

There are four physical forces that must be

Fig 2—Photo of W5THT installation. At top of mast are two horizontally stacked Quagis for 2 meters. The wideband PVC and aluminum tape LPDA is at the top of the tower. Note the rear element, which employs "Capacitive-U" end loading to allow the antenna to be rotated without hitting nearby trees. W5THT reports that he wishes he had used turnbuckles on the truss ropes for the end element—although the element has sagged since installation due to rope stretching, it still works fine!

Fig 3—Details of the PVC LPDA, showing the boom-to-mast bracket, a tee-saddle used to hold the vertical truss support posts to the boom and the through-boom element mounting scheme.

Table 1
Overhead Truss Support Spacing for PVC Pipe Holding Water

Nominal Pipe Size (ID, Inches)	Support Spacing (Feet)
< 1	2.5
1	3
1.25 to 2	4.5
3	5
4	6
6	6.5
8	7.5

This spacing information is very conservative, since the PVC pipe used for amateur antennas will not normally be filled with water!

considered when constructing something; *strain* (pulling-apart force), *compression* (pushing-together force), *torsion* (twisting force), and *shear* (cutting-off force). An antenna boom or element is subject to all these forces, coming from gravity and wind.

A boom is usually supported on the support mast with a plate and U-bolts. The top of the boom is in strain, resulting from the pull of gravity trying to bend it downward. The bottom of the boom at the mounting point is in shear, from the weight on the boom trying to push it through the U-bolts.

The same conditions apply where each element is attached to the boom. Failures due to the concentrated forces at the junction of the boom and element was my biggest problem until I developed the "strong center insulator."

There are ways of strengthening a boom or element from inside by increasing the effective wall thickness. Dave Leeson, W6QHS, wrote a book on the subject, entitled *Physical Design of Yagi Antennas*. I decided to use the easily constructed method of supporting the weight of the an-

tenna using trusses. Leeson mentions this in Chapter 6.6.

I was unable to get the necessary strength-of-materials information to do Leeson's math for PVC pipe, but was able to find information for supporting PVC pipe when it is full of water from the book *Uni-Bell Plastic Pipe Association Handbook of PVC Pipe Design and Construction*. A summary of part of the original table is included in **Table 1**. The book's stated 2:1 safety factor is probably 10:1 because the pipe used for the boom is empty!

PVC pipe is strong and supple—the "supple" part is both a blessing and a curse. The fittings used for PVC pipe are much more brittle than the pipe itself. Fittings fail much more easily than the pipe, especially from shock. You should not depend on a fitting under shear, torsion, or strain to support the weight of the antenna, although fittings under compression seem to fare well. If a joint is needed, use the "bell" at the end of a pipe rather than a molded coupling. Attaching vertical truss supports to a boom should be done with a "T" or saddle fitting, without cutting the boom.

Since pipes are designed to hold water, there is a lip inside the fittings to butt up against the cut end of the pipe to ensure smooth water flow. The lip must be filed off to get a good glue contact between the fit-

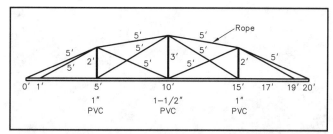

Fig 4—The "trestle-bridge" truss support system for the 20-foot long boom. Individual truss ropes go left and right from each vertical post, plus there is a long truss rope from one end of the boom to the other, over all the support posts. Dimensions for ropes are approximate; turnbuckles are used to adjust rope tension.

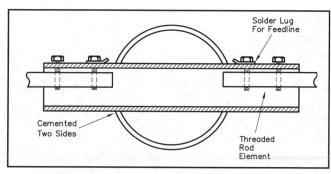

Fig 5—Details of the through-boom element-to-boom mount/insulator, with small-diameter aluminum elements bolted through the mount at each end.

ting and the uncut boom going through it. An alternative is to cut off half a straight tee and make a saddle tee; then you don't have so much plastic to file off. I used saddle-tee fittings for the truss vertical supports shown in Fig 1 and in more detail in **Fig 3**.

PVC cement works by dissolving the two pieces of plastic. When the solvent evaporates, the two pieces become one chemically. "Multipurpose cement" works for water, but not as well as the solvent weld produced by using purple PVC cleaner, followed by true PVC cement. PVC cement, pipe and fittings can be bought at a plumbing supply store. Caution: the vapors from the cleaner and cement are harmful—read the labels, please.

The entire weight of any antenna is concentrated at the connection of the boom to the mast. The vertical posts used to support the boom trusses must be spaced at the proper distance; see Table 1. A post height between $^1/_3$ and $^1/_2$ the distance between the posts results in a stable design.

Look at the diagram of the 20-foot long boom shown in **Fig 4**. The weight of the end on the left is supported by a truss running over the left post to the bottom of the center post. The bottom of the left post is supported by a truss rope over the center post to the bottom of the right post. The entire weight is transferred thus to the center post and then the mast plate. To keep the posts from bending and losing strength, there is a truss rope run from one end of the boom to the other, over all the vertical posts. All truss ropes are wrapped around the boom to keep the joints in compression to spread the force. They are whipped together with net-making twine and are painted black, making them almost rigid and preventing mildew and rot.

PVC is subject to slow UV deterioration. Painting the boom with a nonconductive black paint to shield it from UV is probably worth the time, cost and effort. It also looks less massive from the ground.

The Elements

While the boom supports itself and the elements, the elements support only their own weight. The aluminum elements I bought (at up to $4 a foot) seem to do it well if you buy real 6061-T6 material. To make a supple PVC-pipe element rigid enough to support itself without sagging, the same truss scheme is used. A single top truss may be the only one needed, with lateral support from elements on each side. A lucky accident (PVC cement sets up quickly...) showed me how to laterally guy the back (or front) element. By moving the truss support posts slightly backward from vertical, the element is bent away from the other elements. This allows a rope from the midpoint of each side of the back or front element to the boom to pull it back straight and keep tension on it.

To make PVC radiate, I cover it with aluminum tape. Go to a local air conditioning and heating shop and make sure that you get real aluminum tape. I use Polyken brand $2^1/_2$-inch wide tape, 60 feet to the roll. It costs about $8 per roll.

The truss support system can also be made of wire, attached electrically to the element. This will "fatten" the element and broadband it in the manner of a *cage* antenna. To get the broadbanded element to electrically look like the original element, use the "cylinder" portion of the *WIDEBAND* program. Run the original element with a larger effective diameter and change the length to reestablish resonance.

I have tried several methods of attaching elements to PVC booms. The simplest is one I used to construct the 2-meter Quagi from *The ARRL Antenna Book*, which uses $^1/_8$-inch OD welding-rod parasitic elements. I used a $^1/_{16}$-inch drill to drill through the PVC pipe at the proper point. I then cut the welding rod to the correct length, chuck it in a drill, and let the end drill its own final-size hole through the PVC. The rod slides in easily this way. Once drilling stops, the PVC will cool and "grab" the rod so that I can't pull it out by hand. Unfortunately, HF antennas need elements bigger in diameter than what can be chucked into a hand drill, so another approach is needed!

Mounting Larger-Diameter Elements Through a PVC Boom

Drilling a large hole in a PVC boom will weaken it, unless a solvent weld is used to close the hole. For larger elements, I create combination center-insulator/boom-to-element brackets by running a PVC pipe through the boom hole, followed by a solvent weld. Success depends on making the hole in the boom a tight fit around the smaller PVC pipe.

For VHF and 10-meter antennas using aluminum-tubing elements with outside diameters less than $^1/_2$ inch, the small weights and windloads mean that these elements may be run through empty $^1/_2$-inch ID PVC pipe sections as element-to-boom mounts, with only two screws to hold them in place. These mounts serve to spread out the weight of the tubing element and to prevent cracks from forming in the hole in the boom. A split-dipole element can also be mounted the same way on a boom. See Fig 5. In my LPDA the first two elements are aluminum rods, tapped for mounting bolts going through holes in the PVC pipe. These lock the element halves in place and make the electrical connections.

When we get to 15-meter or bigger elements constructed out of PVC pipe and aluminum tape, the weight and windload increase. These elements will break if unreinforced, hollow schedule-40 PVC is used as an element-to-boom mount/insulator, especially if a screw is used to secure the elements to the boom. I discovered that epoxy was available in quarts at a marine supply store. My first successful element-to-boom insulator for 15 meters was a piece of schedule-40 $^1/_2$-inch PVC pipe filled with epoxy and solvent-welded through the boom. The PVC elements were then solvent-welded to each end of the element-to-boom insulator. See **Fig 6**.

I say "successful" because this mount bent but did not crack. For strength, I found

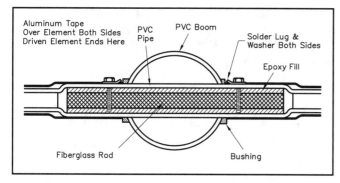

Fig 6—Details of the through-boom element-to-boom mount/insulator, with the end-bells of PVC element solvent-welded to each side. Solid fiberglass rods from old CB whip was used to fill and reinforce the element-to-boom mount.

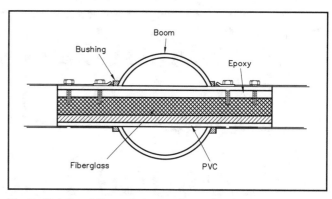

Fig 7—Details of through-boom element-to-boom mount, with aluminum-tubing element fitted over the mount at each end, with securing self-tapping screws threaded into the mount.

a rigid filler: fiberglass rod. In my case it was from an old "Avanti" CB antenna, but smaller fishing rod blanks will do. The resulting element-to-boom insulator is light and strong and will not crack or bend.

For strong element-to-boom mounts using $1/_2$- or $3/_4$-inch PVC pipe, I wrap aluminum tape at the bottom of a 2-foot piece of schedule-40 pipe to seal it. I then make a "funnel" at the top with aluminum tape and then chuck the pipe vertically in a vise. I fill it with water and check it for leaks. Then I put in the fiberglass rod and remove the pipe from the vise to measure the remaining water. This gives the volume of epoxy to mix. Then I dry the pipe and rods. Following safety rules (gloves, ventilation, etc), I mix the epoxy and put it into the pipe. I slowly add the fiberglass rods, allowing the epoxy to flow up around them, being careful not to push too hard on the tape holding the bottom. The epoxy should be up in the funnel a small amount when the rods are fully inserted (the epoxy will shrink during cure). I then put the assembly aside for at least two days. Remember: epoxy gets HOT when it cures.

For 1-inch element-to-boom mounts, I first make a $1/_2$-inch pipe insulator as above, and then use it as the filler inside the 1-inch PVC. To attach a PVC element, simply clean all epoxy off the PVC insulator and solvent-weld the Bell end of the element to it. Remember, the entire weight of the element will be on this joint, use the bell end of the element pipe rather than a molded coupling.

You can also hold an aluminum tubing element to this kind of element-to-boom mount, providing that the tubing fits over the outside of the mount. See **Fig 7**. First, you solvent-weld the mount through a hand-filed close fitting hole in the boom. Drill and tap the mount for a $1/_4$-20 thread about 4 inches from the boom and simply bolt through the aluminum. The bolt threads will fill up with the epoxy and the assembly will be stronger than a loose-fitting hole.

Loaded elements must be "guesstimated long" and measured electronically, because the formulas don't cover bent elements. Elements with their ends bent downward like mine will introduce additional torque from the wind at the attachment point on the boom. I used the "Capacitive-U" loading arrangement shown in Fig 2 to reduce torque.

Lessons Learned

Water loves holes. Make sure that the tape is as evenly stuck to the PVC pipe as possible, or water will get between the tape and pipe and increase the weight of the element. Be sure the boom and elements can drain any water that does manage to get in. A 20-foot long 3-inch pipe can theoretically hold 790 pounds of water! That much weight can make a real mess coming down in a storm.

Use strong rope. An attempt to use multiple small strands of twine failed, one strand at a time. And don't forget the turnbuckles—the rope stretches in the first week or so.

Notes

[1]One source of these mail order specialty products is United States Plastics, 1390 Neubrecht Rd, Lima, OH 45801, tel 419-228-2242.

[2]David B. Leeson, W6QHS, *Physical Design of Yagi Antennas* (ARRL, 1992). See Chapter 6.6.

[3]*Uni-Bell Plastic Pipe Association Handbook of PVC Pipe Design and Construction*, Uni-Bell Plastic Pipe Association, 2655 Villa Creek Dr, Suite 150, Dallas, TX 75234. Table 36.

[4]Polyken Technologies, Westwood, MA 02090.

[5]Dr James L. Lawson, W2PV, *Yagi Antenna Design* (ARRL, 1982).

[6]Leeson tells how to drill precise holes in his Figure 6-23, p 6-42. It works!

[7]Clear Cote, Clear Cote Corporation, 4242 31st St, North St. Petersburg, FL 33714, tel 813-822-4677.

[8]Daniel DiFonzo, "Reduced Size Log-Periodic Antennas," *The Microwave Journal*, Dec 1964.

My Pasturized Antenna–An All-Band, Multi-Wavelength Horizontal Loop

By Stephen M. McCoy, AA0SH
3310 Plateau Road,
Longmont, CO 80503

> *A down-low, easy-to-make, super-quiet antenna!*

The official technical name for this antenna should be something like "A Pasturized Horizontal Loop." Aside from its mundane day-to-day function, which is to contain horses in my pasture, it also radiates my signal! See **Fig 1**, which is a photograph of the fence.

The fence around the pasture is wire-grid horse-fencing, supported each 10 feet by steel posts. On top of the fencing is a single #17-gauge galvanized-iron electric-fence wire, with plastic insulators near the top of each post. This top wire is the center of our high-tech discussion here. Most of the time, the wire is charged with a current-limited 18,000 V. This makes touching it very unpleasant, for both horses and humans.

Recently, my son Marty, WB0TCZ, was visiting me from Cheyenne, Wyoming. I remarked to him that I'd often wondered how I could turn this fence into some sort of "super-antenna," since all that wire seemed to be going to waste. Without batting an eye, he said, "Easy," and began to talk about turning the top electrified wire into a horizontal loop. It would be near-perfect, he said, since the top wire was well insulated from ground by stand-off insulators, and it formed a continuous rectangular loop. See **Fig 2**.

Marty sketched a feed line that he had heard about from Dale Putnam, WC7S. This feed line, a pair of RG-58 coaxes, is connected to the balanced-line output of an antenna tuner. The shields of the coaxes are grounded, and the center conductors are connected to the tuner output. A DPDT knife-switch at the barn disconnects the fence charger from the top fence-wire, and connects the center conductors of the coaxes to the top fence-wire. The wire becomes a

loop, open only at the knife switch. The coax is easily obtained from Radio Shack, as is the knife switch.

If you have a fence without a "hot wire," the wire and fence-post insulators can be obtained from nearly any feed store. Cost for a quarter-mile of 17-gauge galvanized-iron fence wire is about $10, and insulators are about 15 cents apiece.

Practically Invisible

Perhaps you live somewhere where putting up a 1500 foot long loop antenna is a bit of a problem...either due to aesthetics or neighbor preferences. This is an unobtrusive, or at least not a very obvious, antenna, even if you don't own livestock! And the hot-wire along the top keeps undesirable animals from leaning on the fence.

Fleet-footed animals, like deer, are not a nuisance to the fence, since they usually just go *over* it. However, since the 17-gauge wire is nearly invisible to deer, they sometimes do manage to break the wire as they leap over it, especially at dawn or dusk.

If you use the wire to keep livestock back, and if it does get broken, it will touch the ground or one of the posts, thus shorting out the high voltage. This becomes immediately obvious, because the red light on the fence charger will go out, telling you that there is

Fig 1—Photo of AA0SH's fence. It doesn't much look like an antenna, does it?

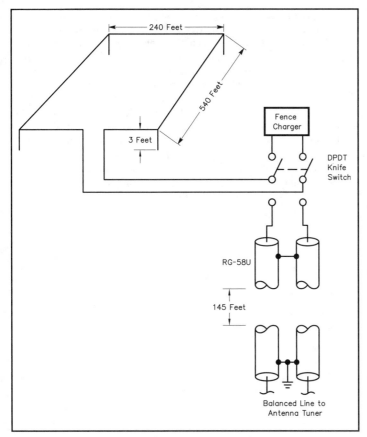

Fig 2—Electrical layout of AAØSH's "Pasturized Horizontal Loop." He walks to the barn to throw the DPDT knife switch, disconnecting the electric fence charger, when he wants to operate the radio.

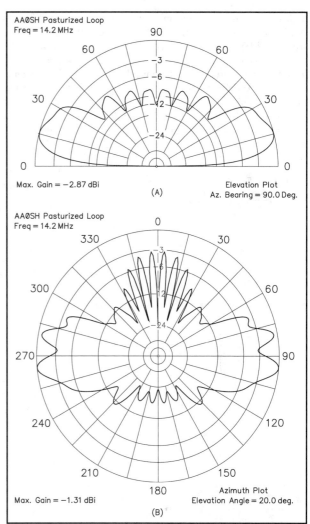

Fig 3—At A, elevation plot for the Pasturized Horizontal Loop on 20 meters. At B, azimuth plot on 20 meters. Not too bad for an antenna only three feet off the ground!

a break somewhere. If you're not using a fence charger, then testing the loop with an ohmmeter before each radio use will tell you whether or not there are any breaks in it.

My loop resistance varies between 200 and 1200 Ω, depending upon temperature and humidity. The other nice part about this antenna's unobtrusiveness is its height from the ground: approximately one meter! The down-low profile escapes a lot of background noise and QRN. In fact, the noise level is about seven S units below that of my 400-foot long skywire.

It's easy to achieve a low SWR using a simple antenna tuner to feed this loop. I have been able to use it from 160 through 10 meters. This loop is exceptionally quiet for receiving. I have found that it doesn't always bring in the same signals that my 400-foot skywire does, and vice versa. I often switch back and forth between the skywire and the pasturized loop. Some signals are stronger on the skywire; some are stronger on the pasturized loop. I have reasoned that this is because of the "diversity capture effect" of the two antennas. They both cover such different and such large physical areas that a signal may strike one, while missing the other entirely.

Safety From High Voltage

I use a double-pole, double-throw knife switch to connect the fence loop either to the feed line coaxes or to the electric-fence high voltage. It is impossible to connect the high voltage to my receiver or transmitter (see Fig 2), since the loop itself is connected to the center poles of the DPDT switch.

Loop Resistance

As I mentioned before, my loop resistance seems to vary between 200 and 1200 Ω. There are several splices in my loop because of the deer problem. They manage to break the wire as neatly as if someone had cut it with wire cutters. Since I have the top wire pulled fairly snug to begin with, so that sag is at a minimum, I must often splice in a piece three or four inches long. The wires are joined using "telegraph splices." My son Marty experimented in his shop by shining the ends of two pieces of this wire with emery paper, and then

soldering the telegraph splice together. He met with moderate success, but this was in an ideal shop environment: no wind, rain or snow to help him out.

My outdoor splices are solderless, at least so far. Our electric fence has three gates. Each gate in the electrified portion of the fence uses a spring-loaded hook. This is unhooked and rehooked when opening and closing the gate. Of course, even though the hooks have a galvanized finish, there is still a small amount of corrosion on them. The corrosion seems to vary in resistance, depending upon the temperature and humidity on any given day. I have not, however, been able to notice any difference in the ability to achieve a 1.1:1 SWR, nor is there any correlation to resistance and the effectiveness of my transmission.

Therefore, in the interests of practicality, I ignore the varying resistance. If one wanted to be a purist about it, it certainly would be easier if I weren't trying to contain livestock with my antenna! I would merely make a continuous run of #14 copper-clad steel wire around the fence, with no splices or "hook" gates. (The copper-clad wire can be obtained from Cable Experts, Inc, telephone 1-800-828-3340. Ask for #14-gauge copper-weld steel wire. This sells for 9 cents per foot in quantities of 100 feet or more.) Even if deer or other animals do manage to break the wire, it is much easier to make a good splice with a portable torch.

Easy Test For Integrity

Before each use, I go down to the barn to throw the knife switch and turn off the electric-fence voltage. I use this antenna about once a week. If the red light on the fence-charger is on, I know that the top wire is continuous and not down somewhere. Just out of curiosity, I sometimes put an ohmmeter across the two coaxes back up at the house to check on the resistance.

I am really quite pleased with this antenna. I have been able to work coast-to-coast in the US, and have snagged Canada and Mexico. In my library I have books on antennas and antenna theory. Many contain complex analyses, full of heavy-duty mathematics and antenna patterns, not to mention gnat's-eyebrow measurements to make them tune up just right. Now, I wouldn't dare to argue with any of those learned discussions, but sometimes it seems to me that my Pasturized Horizontal Loop flies in the face of those theories! [Around ARRL HQ, we have a saying, attributed to a certain Mr Gooch: "RF's gotta go somewhere." **Fig 3** shows an elevation and azimuth pattern for a 3-foot high, 1500-foot long loop over average ground on 20 meters. Not too bad!—*Ed.*]

The Reflected, Stacked Double Zepp

By John Stanley, K4ERO
ARRL Technical Advisor
8495 Hwy 157
Rising Fawn, GA 30738

This article was written in response to requests from a number of hams who have worked packet or SSB with KC6WH in the last few months. It describes the curtain antenna in use in Palau, and discusses the various design choices when building curtain antennas. This antenna was designed by Joe Fay, KØOO, and myself and is used by Joe from Palau.

Who says you've got to have rotatable directional arrays for HF? K4ERO goes back to broadcast-radio basics, describing several broadband varieties of high-gain wire curtain arrays.

Curtain Antennas

Curtains have been around for a long time, but in recent years have seen relatively little use in amateur circles. Yet they have a definite place in ham operations. Most of the various types of ham curtains that have been used in the past work on one or perhaps two harmonically related bands. However, it is not difficult to make a curtain that will work well over at least a 2:1 frequency range. A curtain like this is competitive with a rhombic in performance, while requiring only two supports and being much smaller in size.

Wide bandwidth is more important now than it was in years past when amateur bands were generally an octave apart. Now we have as many as five bands in little more than one octave, for example between 14 and 29 MHz. One antenna working over that frequency range is often more practical than a separate antenna for each band.

One trade-off, or course, with any fixed-direction antenna is that you must decide ahead of time which direction is important and realize that other azimuths will not be available. While a slewable curtain is quite practical, it is beyond the scope of this article. Some curtains are reversible, giving two azimuths to choose from. If you are fortunate, both of these directions may be of interest to you. One example of that would be the long and short paths to a given area of the world.

There are occasions in which a single direction of propagation is of major inter-est. One example would be for an East Coast station intent on relaying traffic to the West Coast. Another would be the situation that gave rise to the present work—designing an antenna suitable for a packet link from KC6 (Palau) to the US mainland. Palau is located in the Western Caroline Islands, about half-way between Guam and the southern Philippine Islands. From there, the States looks like a target about 30° wide and generally Northeast.

Asia and other areas to which the packet links were to be established were easily within range with a simple trap vertical. Stateside signals, however, were just too weak for reliable packet operation.

Since 14 MHz was expected to bear the brunt of the operation, the antenna is optimized for that band. However, acceptable operation on both 10 and 7 MHz was desirable as well. The final result was very gratifying. Successful operation on 14 MHz was immediately established and the other bands are usable when propagation denies the use of 20 meters. Once the sunspots get back to high levels, an 18 to 29-MHz version of the same antenna can be easily added.

Our plan was to begin with a simple antenna, and gradually add elements. We would thereby add gain, while maintaining a broad bandwidth. Curtains consist of a number of dipoles fed in phase and stacked either vertically, horizontally, or both. Various types of reflectors may be added, to provide unidirectional gain.

To illustrate the evolution of a curtain, we begin with consideration of the operation of the so-called end-fed Zepp antenna. This is a dipole, end fed using open line. See **Fig 1**. This antenna is inherently broadband. It presents some real impedance, roughly 30 to 3000 Ω, for any frequency where the dipole represents $\lambda/4$ or more in length. If fed with 300-Ω open line, the SWR will be less than 10:1, which is acceptable because of the low loss of the line. (An antenna tuner will be needed at the transmitter end to reduce the SWR to an acceptable level, as well as to transform from balanced to unbalanced line. A tuner is assumed for all the antennas discussed in this article.)

If we examine the pattern of the above antenna, we find that it has a single lobe broadside to the wire for frequencies up to

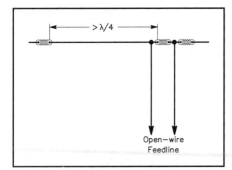

Fig 1—End-fed "Zepp" antenna, with open-wire transmission line to the antenna tuner at the transmitter. This antenna is broadband in that for any frequency where the length is greater than λ/4 the feed-point impedance has a resistive part greater than 30 Ω.

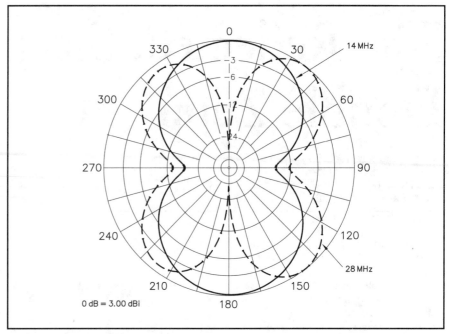

Fig 2—Azimuth patterns for end-fed Zepp, fed at 14 and 28 MHz in free space. The pattern at the second-harmonic frequency breaks up into a cloverleaf, an undesirable trait for a broadband antenna.

the point where the wire is about 0.65 λ long. Beyond that length, minor lobes increase rapidly until at 1 λ, the pattern consists of two equal lobes at 45° to the wire, and a null broadside to the wire. See **Fig 2**.

We are now entering the realm of the long-wire family of antennas, which includes the V beam and the rhombic. Since we wish to develop gain broadside to the array of dipoles in the curtain, we must avoid the area where gain in that direction is decreased by the presence of minor lobes. The frequency range over which the end-fed wire Zepp may be used as an element in curtain arrays, goes from 0.25 to 0.65 λ, or just over an octave.

To put numbers on it, a 40-foot wire will exhibit reasonable impedance and pattern from below 7 to above 14.5 MHz. This then will become the basis of our curtain array. The first step is to put two of these elements end-to-end and feed the pair with our open line. This forms the classic "Extended Double Zepp." See **Fig 3**.

The 80-foot Extended Double Zepp (EDZ) in itself is an outstanding performer, considering its simplicity. It has a broadside major lobe from 160 through 20 meters, although the losses and SWR on 160 and 80 meters would be high enough to produce noticeable loss of performance compared to a longer dipole. For 40, 30 and 20 meters, there would be significant gain over a dipole and the maximum gain would be at right angles to the wires. On the higher frequency bands, lobe splitting would make the pattern such that broadside gain would be replaced by high gain in other variable directions. If the purpose of the antenna was to maintain a signal in a given direction, the patterns for the higher frequencies would not be suitable.

Note that at 14 MHz, the gain is about 3 dB above that of a single λ/2 dipole. This is achieved by effectively feeding two λ/2 dipoles in phase and separating them by

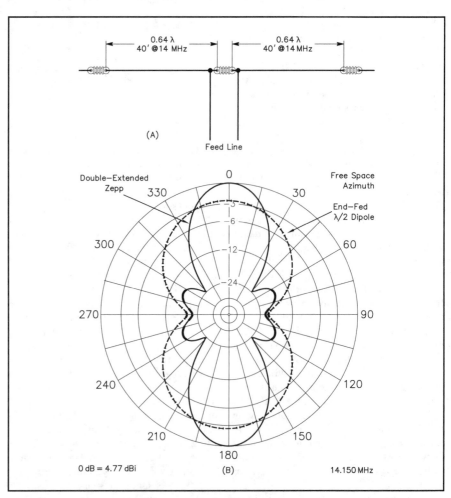

Fig 3—At A, layout of an Extended Double Zepp (EDZ) antenna. At B, the azimuth patterns of the EDZ compared to a λ/2-long end-fed dipole.

some distance. Two λ/2 dipoles mounted end-to-end and fed in phase have only about 2 dBd gain. This is because their "capture areas" overlap considerably, and each one is therefore duplicating the action of the other to some extent.

By moving them farther apart, each one begins to affect a different area of space, and thus the full 3 dB of gain is approached. One can appreciate this effect by considering two dipoles that are mounted side-by-side, a few inches apart. This would not double the gain, since each dipole is affecting an area of space that the other has already completely exploited, either extracting energy from it in receive, or putting energy into it in transmit.

Having gotten the maximum gain possible by horizontal expansion (without great complications), we now look at getting gain by expanding the array vertically. This we do by adding a second EDZ below the first one. (We assume that the first one is at the top of the towers.)

The spacing of these two Zepps was under our control on Palau and we therefore examined various options. See **Fig 4A**. We were tempted to place the two layers a λ/2 length apart. That would allow the feed to be connected at the bottom of the pair, with a twist to maintain the top and bottom sections in phase. However, this approach has two disadvantages. The main one is that having done so, we immediately convert the array into a single band design, and lose our desired broadband performance. The second disadvantage is that λ/2-wave stacking is not optimum from a gain standpoint.

We therefore took the approach of optimizing the stacking distance for maximum gain on the main band of interest, (in our case 14 MHz), and fed it in such a manner as to maintain the broadband frequency performance. This is done by connecting the two Zepps with open line (no twists), and feeding in the center. See Fig 4B. From symmetry considerations you can see that this will force the currents in the four dipoles to be equal, except for small differences due to the effect of the ground. (The lower dipoles are nearer ground and therefore have a somewhat different impedance.)

This brings us to our final design of the basic bi-directional array. The overall appearance of the antenna is a "Lazy H," lying on its side stacked vertically, not on its back stacked horizontally. This design is optimized for 14 MHz, but maintains good pattern and reasonable SWR down to 7 MHz. Its pattern is intact even on 80 and 160 meters, but losses may be pretty high. **Fig 5** shows the azimuth and elevation patterns for the final configuration over real ground at 14.15 MHz.

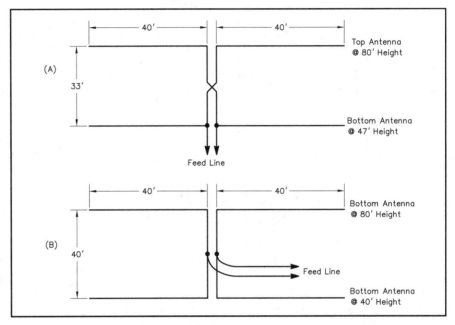

Fig 4—Feeding a stacked EDZ curtain: at A, the feed line between the two antennas is twisted (to reverse phase) and fed at the bottom end of the array. At B, the feed line is not twisted and the feed-line connection is at the center of the transmission line connecting the two antennas in the array.

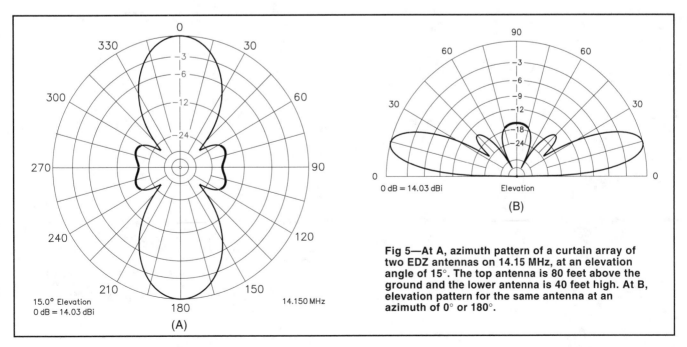

Fig 5—At A, azimuth pattern of a curtain array of two EDZ antennas on 14.15 MHz, at an elevation angle of 15°. The top antenna is 80 feet above the ground and the lower antenna is 40 feet high. At B, elevation pattern for the same antenna at an azimuth of 0° or 180°.

Now we can consider the final step, adding reflectors. There are three ways to do this. If one wants front-to-back ratio on all bands, the best way is to add a screen behind the dipoles. This screen, to be effective up to 14 MHz, would consist of horizontal wires about 6 feet apart and extending a bit beyond the ends of the dipoles. For an 80 foot tower, this would require about 10 wires (the bottom few can and should be deleted to avoid decapitating passers-by!)

The question of how to support the reflector wires must be addressed. One possibility is to use the side of a building that has conductive siding instead of using an array of wires. This approach is probably practical only for a 14 to 29-MHz array. Choosing to use a screen dictates the permanent loss of propagation off the "back" of the array, but is a clean and simple broad-band solution.

A second approach would be to duplicate the "H" behind the first one. This would allow instant reversal of direction and optimization from the shack of the front-to-back ratio at any operating frequency. Such is accomplished by bringing both feed lines into the shack and either feeding the two with an appropriate phase shift. Perhaps a more practical approach is to feed one and insert a variable reactance in the other. This would be adjusted for best front-to-back ratio.

A third approach was chosen for the Palau antenna. This was to hang passive tuned reflectors for 14 MHz behind the fed curtain. See **Fig 6**. This has the advantage of simplicity, but sacrificed the unidirectional pattern on the other two bands. See **Fig 7** for the azimuth and elevation patterns for the Palau antenna. The elevation pattern in

Fig 7B is interesting because the pattern for a single five-element 20-meter beam on a 35-foot boom is overlaid on the pattern for comparison. The stacked curtain with reflectors has more than 2.5 dB gain than the Yagi. The rearward pattern of the Yagi is cleaner, but the forward gain of the curtain is very impressive for a simple antenna!

Had we used trapped reflectors, or a set of reflectors, unidirectional patterns could have been obtained on both bands. A design was modeled for a version with two separate reflector sets, one below the other. The second set of reflectors has not been hung, but can be hung as soon as needed. Our present goal was to gain excellent 14-MHz operation and acceptable 7 and 10-MHz operation. We achieved this with only one set of reflectors. While it is highly optimized for 14 MHz, the final design does a reasonable job on 10 and 7 MHz as well.

Construction

Two existing towers were used to support this antenna. They were providentially aligned almost exactly at a right angle to the desired beam direction. One tower was about 80 feet tall and the other over 200 feet high. The top set of dipoles was therefore put at 80 feet and the other at 40 feet on the shorter tower. The other ends were supported on the taller tower at points chosen to allow the dipoles to hang nearly horizontally. At those points on both towers, a wooden cross-arm 8 feet long was attached. See Fig 6 again.

From the "front" end of the cross arms, the Lazy H was hung, with the center of each dipole fed with 450-Ω ladder line. At the center of this ladder line, another was at-

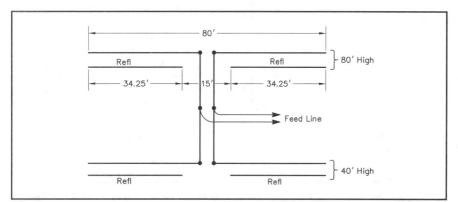

Fig 6—Layout of stacked 20-meter curtain array with passive reflectors located 8 feet behind each antenna. The top antenna is 80 feet high and the lower antenna is 40 feet high.

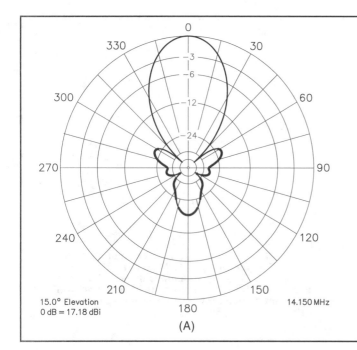

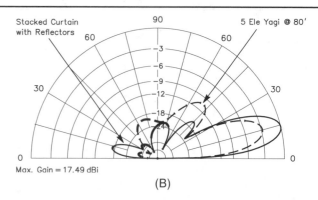

Fig 7—At A, azimuth pattern for curtain array with passive reflectors shown in Fig 6. This is at 14.150 MHz for an elevation angle of 15°. At B, a comparison of elevation patterns for antenna shown in Fig 6 with a five-element 20-meter Yagi at 80 feet height. The curtain array has more than 2.5 dBi gain than the Yagi, which has a 35-foot boom.

tached and led away at about a 45° angle to the ground. This made its way to the shack and the tuner.

The rear end of the arms held the two reflectors, separated by a length of non-conductive guy. Separate insulators were not used at the ends of the reflectors, the support lines themselves being nonconductive. The upper reflectors are supported on one wire and insulating guy combination, and the lower set on another identical one. In this way, unlike the driven side (where the top and bottom sections must be raised and lowered together), the reflectors can be put up one at a time. If you want to use additional reflectors, they should be mounted a foot or so below the 14 MHz ones, by allowing more sag in the support lines. *NEC/Wires* predicts that a 1-foot spacing will allow each reflector to resonate on its own band with little interaction with the other reflector(s).[2]

On the Air

The main problem experienced on the air was that the antenna initially worked much better on receive than on transmit. Packet signals from all over the US were readily copied, but the transmit signal was not dramatically better than that from the comparison antenna, a trapped vertical. While the receive signal was 2 or 3 S units better on the curtain, the transmit signal was only an S unit or so better. After a bit of head scratching, we found that the transmit signal was greatly improved by adding additional line to the balanced feeders.

The additional line moved the impedance to be matched at the transmitter to a part of the Smith Chart where the tuner could be a low-Q configuration. More of the power was going to the antenna and less was being dissipated in the tuner coil. The tuner controls were also less fussy. With this antenna, and indeed, in any high-SWR situation, one should try to use a feed line length that agrees with the characteristics of your tuner.

In the old days, the *ARRL Handbook* showed both series and parallel connections for tuners. Thus, either high or low imped-

ances could be matched, while keeping the loaded Q of the tuner coil in the optimum range. This is a subject much too big to be covered here. See Reference 1 for more discussion.

As with any high-SWR open-line-fed antenna, there is some change in SWR with changing weather conditions. It is helpful to readjust the match during heavy rain, but having done so, the operation is otherwise perfectly normal. The packet station can be left on even during rain storms and works fine except for a reduction in output power with changing SWR.

Variations

Based on the success of the new antenna at KC6WH, a second antenna was built, using the above concepts, for broadcast service at a 50 kW transmitter level. This array was optimized for 10 MHz, and because of the feed method used, cannot be successfully used on other frequencies. Due to the high transmitter power, the array cannot be operated at frequencies where the SWR is more than 3:1 Unlike the KC6WH array, the broadcast antenna does use a λ/2 feeder between the two Zepps, so that it can be fed from the bottom. This simplifies mechanical construction considerably, considering the weight of the #0 wire used in the construction!

By playing with element lengths and impedance of the lines, a 3:1 SWR was obtained at the ground. From there, additional matching brought the SWR to < 1.1:1 for use by the 50-kW transmitter. This antenna is doing very well into Japan and Korea.

Note that on the broadcast design, the elements slope. See **Fig 8** for the layout of this curtain. This was done for two reasons. One was to get the center of radiation of the elements further apart for increased gain, while allowing a λ/2 phasing section to be used. The other advantage is that with lots of "sag" in the top element, the stresses on the end supports was much less. This antenna works in front of a ground screen and thus has little gain off the back. The slanted

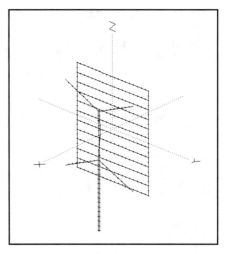

Fig 8—Layout of unidirectional curtain array tailored for 50 kW broadcast station at 9.73 MHz. The feed line is fed directly at the very bottom with the antenna tuner network.

elements could be used with the broadband feed method and any of the three above mentioned methods to get unidirectional pattern. This antenna could be scaled for any ham band of interest where single-band operation is acceptable.

As I mentioned earlier, none of the above ideas are new, but the use of this class of antennas by amateurs seems to be less than would be justified by our results. If you can stand not being able to rotate it, then this may be the answer for your needs. It was for KC6WH.

I wish to thank Mike Baugh, W8AKF, and the gang at KE6WVV for lots of help in reporting performance of the various configurations during testing.

Notes and References

[1]*The ARRL Antenna Book*, 17th Edition (Newington: ARRL, 1994), pp. 25-3 to 25-8.
[2]*NEC/Wires*, available from Brian Beezley, K6STI, 3532 Linda Vista Dr, San Marcos, CA 92069. Tel 619-599-4962, e-mail: k6sti@n2.net.

Amazing Results with a "No-See-Um" Underground HF Antenna

By Ed Wirtz, W7JGM
845 South Eikleberry Court
Sedro-Woolley, WA 98284

W7JGM describes his experiments with several low-profile antennas for local HF contacts. In fact, one is really low profile—it is underground!

I have been experimenting with low-height antennas to produce high-angle, Near-Vertical Incidence Skywave (NVIS) signals for my work on local nets. These types of antennas are useful for local high-frequency ARES and RACES emergency communication and where local zoning ordinances restrict amateur antenna structures. One of my antennas even lies below the surface of the ground!

The useful frequency range of an NVIS antenna is from 2 to about 12 MHz. Higher frequency signals, when radiated at the 70° to 90° angle employed for NVIS, tend to penetrate the ionospheric layers rather than bouncing back to earth. An amateur NVIS antenna is designed strictly for short skip work, not DX. The purpose is to obtain "close-in" coverage in a 500-mile area.

The advantages of a ground-level antenna, or one that is buried just below the surface are:

1. No wind resistance
2. Not as susceptible to lightning strikes
3. Requires little or no maintenance
4. Does not violate local antenna zoning regulations
5. Reduces TVI complaints
6. Greatly reduces power-line noise
7. Low cost to erect—easy installation for one person
8. Radiates a near-vertical skywave signal for statewide coverage
9. Has a very wide 1.5:1 SWR range of 300 to 500 kHz, without retuning the tuner
10. Provides good signals 500 miles away
11. Installs with just a hoe and shovel (or lay it on the ground).

Since I live in a mobile home park, my antenna space is limited. The court owner is kinder than many, since he granted me per-mission to erect a 50-foot TV mast to support a 20-40-80 meter inverted V, plus a couple of 2-meter antennas for VHF voice and packet. I still wanted to run some antenna experiments.

Next to my mobile home is a heavily wooded area where a neighbor's dirt driveway is located. This is also a place where children play. The fir and pine trees there are 100 feet high, spaced about 15 to 20 feet apart. I installed my inconspicuous NVIS antenna in this area.

I made my first experimental NVIS antenna out of a 100-foot length of #10 insulated single-conductor copper house wire, lying on top of the ground. I placed the antenna tuner outside, next to a good ground rod. The random-length antenna ran from there, curving to follow a rather crooked path through the trees. My reason for not running the antenna directly into the radio room was to keep the RF farther away from the HF transceiver (and computers). I ran RG-58 coax from the tuner back to the radio room, a distance of about 45 feet

I used an SWR analyzer to adjust the random-wire antenna tuner for a 1:1 SWR at 3.950 MHz. The MFJ-207 SWR Analyzer and the Autek RF Analyst RF-1 are my favorites for remote tuning without causing interference to anyone else. If you use your transmitter and SWR meter to tune the tuner, you will have to temporarily move them to the tuner location to adjust for a low SWR reading.

Remember, the tuner out at the antenna must have a good RF ground. Use at least an 8-foot ground rod. Do not try to use the coax shield to provide an RF ground from the ground system back at your radio room. The good RF ground must be at the tuner location!

If your tuner does not have a single-wire antenna terminal, you can attach your random wire antenna to a banana plug and insert it in the center of the tuner's SO-239 antenna jack. For future ARES/RACES portable use I wanted minimum weight and bulk and therefore chose the very small MFJ model 16010 tuner. Its range is more limited than larger models but it is sufficient for this application.

With the antenna lying on the ground, I received good signals from up to 500 miles away. However, because of the other families using the area, I decided to bury my antenna just below the surface of the ground. That way, cars could drive over it and the neighbors wouldn't be aware of it. This was intended to be a quick test, just to see if it would work.

A hoe was used to cut a shallow 2-inch ditch for the 100-foot run. I placed the insulated wire in the ditch and loosely covered it with dirt. The end was sealed against moisture using a wirenut and putty. My new antenna was now definitely out-of-sight! I suppose it must have looked like I was burying a "dead" antenna, but it definitely became "live" when I started to use it.

Excellent on-the-air contacts were made with K7ALX, in Gold Beach, Oregon (500 miles away) and K7UPT, in Toledo, Washington (150 miles away). They gave me signal comparisons between the underground antenna and the 50-foot high inverted V. The underground antenna was almost identical to the inverted V—seldom was there a difference of more than 1 or 2 S units on transmit. Many other close-in contacts showed only a 1 or 2 dB difference between the two antennas.

On the inverted V, I heard K7ALX at an S7 level. However, power line noise was almost S6, making reception difficult. On the underground antenna, his signal was S3, but the noise level was barely detectable, making it much easier to listen to him.

I am located in the northwest corner of the state of Washington. I used the underground antenna while I was the net control station for the Washington State Emergency Net, and had no trouble hearing and working stations all around Washington. My underground antenna caused quite a stir, and some skeptics thought I was pulling some sort of joke on them—in spite of my honest reputation! There were many requests to discuss construction details after net time.

I have since used the underground antenna on the Washington Amateur Radio Traffic System, the Northwest Single Sideband and the Columbia Basin nets with good results. CBN net control, KB7HUF, at Eugene, Oregon (350 miles away) gave me my best signal report of 20 dB over S9 while I was running 100 W.

Comparing 1.5:1 SWR Ranges on Both Antennas

When using the inverted V with the tuner adjusted for 1:1 at 3.972 MHz, I found the 1.5:1 frequency range was from 3.949 to 3.999 MHz, or about 50 kHz. With a single tuner setting for the underground antenna, the 1.5:1 range was from 3.680 to 4.175 MHz, almost 500 kHz. The 500 kHz was reduced to about 300 kHz when the first rain came. [This wide a bandwidth indicates that ground losses are substantial for this antenna—exactly what you would expect for an underground antenna!—Ed.]

On 40 meters the tuned frequency was 7.225 MHz and the 1.5:1 range was from 6.715 to 7.595 MHz. On 20 meters the tuned frequency was 14.230 MHz and the 1.5:1 range was from 13.829 to 16.224 MHz. I have not run signal checks on 20 meters, so I don't know how much of my signal is penetrating the F layer and not being reflected back to earth.

If you plan to use this antenna on 160 meters, then I would suggest a larger tuner, such as the MFJ-949 series, since the range of the tiny MFJ-16010 is limited.

Some Problems

With time, some problems developed with the underground antenna. The heavy northwest rains kept packing mud around the antenna and seemed to reduce my signal somewhat. The heavy rains also caused the 1:1 SWR frequency to shift. However, I could always work over a wide range without having to retune because of the wide SWR bandwidth at the tuner. I speculate that the saturated-ground problem could probably be solved to a great extent by using a 100-foot length of PVC tubing to keep the underground antenna dry.

I also noted that the only two weak-signal reports I received were from two stations that were directly off the back end (tuner end) of the antenna. There seems to be a narrow null in the field intensity pattern off that end. If necessary, this could be corrected by changing the antenna direction to shift the null area away from a desired azimuth.

Results with an NVIS Antenna 12 Feet Above Ground

Since the underground NVIS antenna was so successful, I decided to go just above the terrain's danger zone with another 100-foot random wire antenna. It had to be 12 feet high to clear the neighbor's firewood supply truck and also to keep the neighbor's children from touching a "hot" antenna wire.

I used 100 feet of superflex #14 bare antenna wire because it coils up easily, without kinking. I use this wire for my motor home camping and for quick setup at an ARES/RACES site. For tree insulators, I purchased electric-fence insulators (a bag of 25 for $2.75). These incorporate built-in nails—one quick tap with the hammer and the insulator is installed.

This antenna is also relatively close to the ground, electrically speaking, so I had some benefit from power-line noise reduction. Electric-fence insulators are grooved, so the antenna wire just drops in place. (The same insulators are also useful for running RG-58 coax because the groove will accept a small coax and lock it in place at the same time.)

The results with the 12-foot high 100-foot long random wire antenna are almost identical to the 50-foot high inverted V, on both receive and transmit. W7SFT, in Tacoma, Washington (120 miles away) gave me identical signal reports for both antennas and I had the benefit of a reduced noise level on receive.

A 2:1 SWR bandwidth comparison for the V and the 12-foot antennas showed 89 kHz for the V and 150 kHz for the 12-foot high NVIS. Note: insulators must be used on the trees. When I tried laying bare wire on top of nails pounded into the trees, the tree trunks seemed to kill most of the signals.

Another NVIS Possibility Using a Horizontal Mobile Whip

The US Army has been using NVIS-type antenna configurations on their vehicles. They tie down their vehicle whips to prevent hitting bridges and branches and also supposedly to radiate a near vertical incident skywave signal. This technique is said to reduce long skip and to keep their signal in a more confined, local area.

Once you have arrived at your destination, you might experiment with tying back your vertical mobile whip so that it is partly horizontal—this should give you less noise and better local-area coverage.

Antenna Tuners, Baluns and Transmission Lines

Baluns in the Real (and Complex) World

By Frank Witt, AI1H
20 Chatham Road
Andover, MA 01810

AI1H stresses a series of baluns well beyond their specs. How do they perform?

Introduction

Baluns are very useful devices. They can solve the problems we face when we connect our unbalanced transmitters or antenna tuners to balanced antenna systems. Baluns are also useful at the antenna feed point when a coaxial cable feed line is connected to a balanced antenna.

The goal of using a balun in an antenna system is to prevent unintended and undesirable RF currents from flowing. When coaxial cable is used, the undesirable current is the current that flows on the outside surface of the shield. For balanced feed lines, such as ladder lines, the currents in the two parallel conductors should ideally be equal and opposite. Inadequate unbalanced-to-balanced conversion can lead to unequal currents. Radiation from the so-called balanced feed line will then occur.

Commercially available baluns usually provide either a 1:1 or 4:1 impedance transformation, although other ratios are available. Specifications usually apply when the balun is terminated in a resistive termination of a particular value. However, the typical environment of the baluns we use is far from ideal. This study addresses the following questions:

1. What load impedances are encountered by baluns?
2. What if the load resistance departs from design-center load resistance?
3. What are the properties of some typical baluns?
4. Are there significant differences in the performance of current and voltage baluns?[1]
5. Do complex impedance loads stress baluns more than purely resistance loads?
6. How much improvement is possible by putting the balun between the transmitter and the antenna tuner?

Many of these questions may be answered using affordable, low-power SWR and impedance testers with known resistive loads.

Conclusions

I briefly summarize here some conclusions reached after making numerous measurements and calculations. The material supporting these answers is found in the remainder of the article.

1. The load impedances encountered by baluns in typical use range very far from the design-center load resistance.
2. Except for 1:1 current baluns with floating loads, loss will increase as the load impedance departs from the design-center load resistance. The loss can be enough to destroy the balun when operating at its advertised power level limit. For this reason, the balun is often the weak link in an antenna system.
3. 1:1 and 4:1 baluns may be characterized by the parameters of their equivalent circuits, namely, Z_0, $F_{1/8TH}$ and Z_W. These terms are defined in the article. "N:1" baluns tested do not provide an N:1 impedance transformation at the higher frequencies. In addition, loss and balance quality versus frequency and load impedance vary considerably for the baluns tested.
4. Current and voltage baluns perform differently. Voltage baluns have an inherent internal ground connection at both ports and perform best with truly balanced loads. Current baluns are more versatile and robust in amateur service, since our nominal "balanced" loads are often not truly balanced. They do not have an internal ground connection at the balanced end.
5. I believe that complex impedance loads tend to be no more stressful to baluns than resistive loads. The important parameter to look at is "load SWR," which is a measure of how far the load impedance is away from the design-center resistance of the balun. For most applications, baluns may be evaluated with resistive loads; however, these loads should simulate the load SWRs expected in service. Also, it is very important that the load SWRs be achieved with resistances above *and* below the design-center resistance.
6. I observed no improvement in balun loss by placing a particular 1:1 current balun between the transmitter and antenna tuner. The balun loss increases with load resistance. Balance quality was improved at low frequencies but deteriorated at high frequencies. Also, the antenna tuner mechanical design must address the "hot cabinet" problem. If the balun is designed or selected properly, voltage and current levels are

in a narrower range when the balun is next to the transmitter, so the transverse stress on the balun is much less.

Baluns in Antenna Systems

We will examine the entire antenna system to gain insight into the real world environment of the balun. As seen in **Fig 1**, baluns are used in a variety of ways in an antenna system. Most commercial antenna tuners provide an unbalanced output and a balanced output with a built-in balun. Fig 1A shows this case and also the case where an external balun is connected at the unbalanced tuner output. The popular G5RV antenna, which has a prescribed balanced feed line length, sometimes has the balun positioned as in Fig 1B.

A very common arrangement for resonant dipoles with coaxial feed is shown in Fig 1C. If the feed line is symmetrically placed with respect to the balanced antenna or the feed line is not some unfortunate length, a balun may not be required. By unfortunate length, I mean a length which provides a low common-mode impedance at the antenna feed point. For a situation where the coaxial feed line is in the clear between the antenna feed point and system ground at the rig, an unfor-

tunate length would be a near multiple of a half wavelength in free space. Such a length makes the feed line a resonant radiator, which will be excited from the antenna feed point if a balun is not used.

It also produces a current maximum at the cabinet of the unbalanced antenna tuner. This could result in RF in the shack. I have confirmed this model with *MININEC*-based antenna-simulation programs. A good test of whether a balun is worthwhile is to plot the SWR versus frequency around resonance, with and without the balun. If the two plots are essentially the same, the balun is not required. To be safe, I usually use a 1:1 current balun in this kind of application.

As will be seen later, balun performance may be degraded when it is not terminated in its design-center resistance. This condition may be overcome by using the configuration shown in Fig 1D. Here, the antenna tuner reduces the range of impedances seen by the balun. It is possible to use an unbalanced antenna tuner in this application, but the tuner cabinet must be insulated and may have high RF voltage on it. Careful mechanical design can overcome these problems. Although the transverse voltage or current stress on the balun will be less using

this configuration, the balun loss and balance quality may not be improved.

The Real World Environment of the Balun

Baluns are usually specified to work with a certain resistive termination. For 1:1 baluns, the load resistance is usually 50 Ω, and for 4:1 baluns it is usually 200 Ω. What load impedances are encountered by baluns? I show here a few examples where a balun might be used to demonstrate that we typically terminate baluns far from either 50 or 200 Ω. Later the impact of this fact will be presented.

For the balun placement of Figs 1A and 1B, consider a 102-foot center-fed wire, 50 feet above average ground. I chose this length because it is the recommended length for the G5RV antenna.[2,3] These antenna systems are commonly thought of as multiband HF antennas. Let's first connect the 102-foot wire to the balun/tuner with 100 feet of 450-Ω ladder line (Fig 1A). Then, the same antenna wire, now part of a G5RV antenna, is connected to a balun with a 32.3-foot segment of the same 450-Ω line (Fig 1B).

The results are shown in **Table 1A** and **1B**. The antenna impedance was found with the aid of *EZNEC*.[4] I assumed #14 wire, a ground conductivity of 5 mS/m and a dielectric constant of 13. The 450-Ω line is assumed to have a velocity factor of 0.95 and a matched loss of 0.058 dB/100 feet at 1 MHz. The matched loss is assumed to increase in proportion to $\sqrt{frequency}$.

There are lots of data in Table 1, and careful study is warranted. For the nine HF amateur bands, I calculated Z_{TL}, the load impedance seen by the balun, and the loss of the 450-Ω ladder line. The SWR on the line at the point of connection to the balun is also shown. Note that the widely varying complex impedances presented to the baluns are rarely ever close to 50 or 200 Ω. Baluns in these applications are called upon to handle very wide excursions in load impedance.

In fact, Louis Varney, G5RV, has recommended against the use of a transformer/balun at the connection point between the balanced feed line and the coaxial cable. Recognizing that this may lead to feed-line radiation or RF in the shack, he has recommended coiling some of the coaxial cable at the connection point. This may help for some bands, but is not a broadband solution.

In my opinion, the unsolved "balun problem" is a weakness of the G5RV antenna. Also, it's clear from the data that the SWR on a 50-Ω cable between the antenna tuner and the 450-Ω line (the no-broadband-balun case) for the 160, 30, 17 and 10-meter bands would be very high. For those bands the coax cable loss would be excessive.

In this study, I found that a good way of stating the stress on a balun is to evaluate

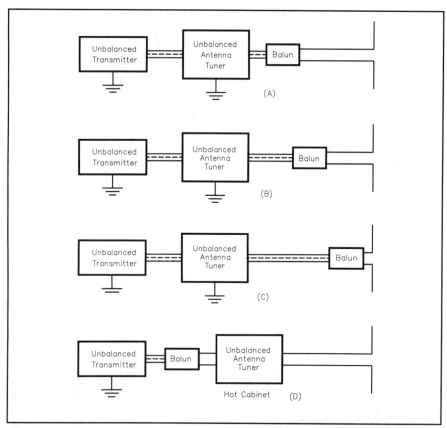

Fig 1—Placement of baluns in antenna systems. At A, the balun either directly follows the antenna tuner or is built into it. The G5RV antenna is an example of the configuration shown in B. For balanced antennas fed with coaxial cable, the connection shown in C is often used. Configuration D provides the most friendly environment for the balun, but not necessarily for a conventional tuner or its user.

the SWR of the load using the balun's nominal load resistance as a reference. I call this measure the "load SWR." That is what is meant in Table 1 by "50-Ω SWR" and "200-Ω SWR." This will become clearer when the measured results on some typical baluns are presented. Let's assume that if the load SWR defined this way is higher than about 8:1, the balun loss will be excessive for applications where the transmitted power level is close to the balun's limit.

We will see that for some baluns and bands, even load SWRs lower than 8:1 can be destructive. In the table, all cases where the load SWR is 8 or less are shown with a bold enlarged number in a shaded box. Fortunately, if the balun is operating well under its maximum safe power level, SWRs higher than 8:1 might be tolerated. It is clear, however, that this 102-foot antenna is not going to perform well on 160 meters with either of these feed systems.

With similar data, others have conjectured that it would be desirable to have a balun with a higher optimum load resistance, such as 450 Ω. Indeed, support for this notion can be seen in Table 1 by looking at the "Line SWR at Balun" columns, since the characteristic impedance of the line is 450 Ω. These numbers are suitable over more bands than for the 50 or 200-Ω cases. They are similar to the numbers one gets if a "450-Ω SWR" column were shown. Unfortunately, if one tries to achieve a balun with a 450-Ω optimum load and also a 9:1 impedance ratio to transform to the 50-Ω impedance level, the balun bandwidth and load range suffers. Other approaches to solving the problem may be more appealing.

Another application to consider is an 80-meter inverted-V dipole fed with coaxial cable, the case shown in Fig 1C. Again using *EZNEC*, I analyzed an antenna whose legs are 64.5 feet long with an apex 60 feet above average ground. The angle between the wires is 90°. At the resonant frequency of 3.72 MHz, the SWR for a 50-Ω feed line is 1:1. At the band edges (3.5 and 4 MHz), the SWR is about 7.5:1. Most commercial 1:1 transmitting baluns can handle this much stress. However, operation on other bands would result in excessive stress at high power levels and undesirable balun loss, as well as excessive coax feed-line loss. Use multiple resonant dipoles connected in parallel with the 80-meter dipole if you want to operate on other bands with a coax-fed antenna.

The antenna impedances shown in Table 1 may be used in similar calculations for other feed system arrangements. I recommend the ARRL program *TL.EXE*, by Dean Straw, N6BV. *TL.EXE* is bundled with late editions of *The ARRL Antenna Book* and *The ARRL Handbook* and is also available on the ARRL BBS. It not only makes detailed transmission line calculations, but also models non-ideal antenna tuners.

Baluns Tested

I evaluated four very different baluns:

1. W2DU 1:1 current balun.[5] This balun is a piece of 50-Ω coax cable threaded through about 50 ferrite beads.
2. W2AU 1:1 voltage balun.[6] This is a trifilar winding over a ferrite rod.
3. Radio Works Model B4-2KX 4:1 current balun.[7] Two 1:1 baluns, each wound on a separate toroid, are connected with their inputs in parallel and their outputs in series.
4. Van Gorden Engineering 4:1 voltage balun.[8] This is a bifilar winding over a ferrite rod.

Balun Properties

What properties of a balun distinguish them from one another? Ideally, we would like to see the balun do its job over the frequency range of interest. I will describe the performance of so-called *wide-band baluns* that are intended to perform well over all the HF bands. There is a class of baluns that are tuned, or are effective only over a narrow frequency band. The characterization techniques I use are also applicable to narrow-band baluns, but I will confine this article to the results with wide-band baluns.

Where possible, all characteristics will be described for the HF bands; that is, 160 through 10 meters. Here are the parameters of interest:

1. *Impedance transformation.* An ideal N:1 balun would provide an input impedance Z_L/N when it is terminated in an impedance Z_L. How well does the balun meet this condition over the HF bands?
2. *Loss.* How much loss does the balun introduce into the antenna system?
3. *Balance Quality.* How well does the balun do its primary job—preventing unbalanced load currents?
4. *Power Limit.* What is the maximum power the balun will handle for various load conditions? Information on power

Table 1A

Characteristics of 102-Foot Center-Fed Horizontal Wire at 50 Feet, Fed with 450-Ω Ladder Line. Length of Ladder Line to Balun at Antenna Tuner = 100 Feet.

Freq MHz	Z_{ANT} Ω	Z_{TL} Ω	Line SWR at Balun	Ladder-Line Loss, dB	50-Ω SWR	200-Ω SWR
1.83	5 − j 1558	5 − j 27	84	10.9	12	39
3.8	42 − j 325	265 − j 1157	13	0.9	107	27
7.1	480 + j 1056	77 − j 129	6	0.5	**6**	**4**
10.1	1884 − j 3043	114 − j 623	12	1.2	71	19
14.1	98 − j 68	107 − j 43	4	0.5	**3**	**2**
18.1	2114 + j 1623	290 − j 738	6	0.9	43	11
21.1	399 − j 1167	72 + j 179	7	1.2	11	**5**
24.9	198 + j 316	1224 − j 491	3	0.5	28	**7**
28.4	3098 + j 565	1022 − j 1210	6	1	49	12

Table 1B

Characteristics of 102-Foot G5RV Center-Fed Horizontal Wire at 50 Feet, Fed with 450-Ω Ladder Line. Length of Ladder Line to Balun = 32.3 Feet.

Freq MHz	Z_{ANT} Ω	Z_{TL} Ω	Line SWR at Balun	Ladder-Line Loss, dB	50-Ω SWR	200-Ω SWR
1.83	5 − j 1558	3 − j 558	172	3.8	2478	693
3.8	42 − j 325	32 + j 89	15	0.5	**7**	**8**
7.1	480 − j 1056	76 − j 174	7	0.1	10	**5**
10.1	1884 − j 3043	60 + j 398	14	0.5	55	17
14.1	98 − j 68	104 − j 103	5	0.2	**4**	**3**
18.1	2114 + j 1623	153 − j 525	7	0.3	39	11
21.1	399 − j 1167	55 + j 93	9	0.4	**5**	**5**
24.9	198 + j 316	136 − j 84	3	0.2	**4**	**2**
28.4	3098 + j 565	1611 + j 1443	7	0.4	58	15

limits is not provided in this article.

For a power-limit guideline, use the manufacturer's specifications and derate them by the loss data provided here. It will be clear from the data presented in this article that a realistic power limit will depend on the value of the balun's load impedance and the operating frequency. Loss results made at low power levels apply at high power levels if the balun is operated in its "linear" region. Actually, this issue is very complex and should also address the issues of core saturation, voltage breakdown and temperature rise.

All of these attributes, except for power limit, may be measured by using low-power SWR testers and known load impedances. I used the Autek Research RF Analyst, Model RF-1,[9] for measuring the magnitude of impedance. For SWR measurements, I used the MFJ HF/VHF SWR Analyzer, Model MFJ-259.[10] I chose this combination because for the units I own, it provided the best accuracy.

The method for making a comprehensive set of measurements on baluns is described. Loss and balance measurements are based on those described in my *QST* article, "How to Evaluate Your Antenna Tuner."[11]

Accuracy of Low-Power SWR Testers

The low-power SWR testers used contain a low-power RF signal source, a frequency counter, and a 50-Ω SWR bridge. The accuracy must be good enough to obtain meaningful data. I used precision $1/4$-W film resistors to determine the accuracy of the units at HF.

A unique feature of the Autek Research RF-1 is that it measures the magnitude of impedance, |Z|. **Fig 2** shows |Z| versus frequency for this instrument. For these measurements, the load resistors are mounted inside PL-259 connectors, since the RF-1 uses an SO-239 connector. To keep the leads short, each resistor is mounted inside the hollow center connector of the PL-259. The data show that the instrument provides very good accuracy for loads from 12.5 to 200 Ω and adequate accuracy for this project for resistive loads ranging from 6.25 to 400 Ω. Other measurements I have made using complex impedance loads reveal that similar accuracy exists for such loads. The instrument truly measures the magnitude of an unknown complex impedance.

For loss measurements, we use data in the SWR range between 1:1 and 2.4:1, so it's important that the instrument be accurate in that range. See **Fig 3** for measured data for the MFJ-259 HF/VHF Analyzer. The performance meets the requirements of this application. The SWR accuracy is very good and is independent of frequency for all HF bands.

Balun Loads

A general balun load is shown in **Fig 4**. It is a three-terminal network, where two of the terminals are connected to the balun and the third (the one connected to Z_3) is grounded. It is important to keep this in mind as we characterize the performance of a balun. If Z_1 and Z_2 are equal, the load is considered "balanced." The value of Z_3 will vary considerably from one antenna system to the next. An ideal current balun will force I_1 and I_2 to be equal, regardless of the value of Z_3. An ideal voltage balun will force V_1 and V_2 to be equal in magnitude and 180° out of phase.

I evaluated the baluns with switchable resistive loads. I used the AI1H Geometric Resistance Box, Model 50U/100B.[12] Loads are provided representing a wide range of resistances. For unbalanced applications, the terminations range from 1.5625 Ω to 1600 Ω. The center-tapped balanced loads ($Z_1 = Z_2$) range from 3.125 Ω to 3200 Ω. Center taps may be grounded or left floating by means of a switch mounted within the load box. Actually, an arbitrary value may be used for Z_3 by adding an external impedance. For my tests, I either float or ground the center taps. The ground path, which ends up being Z_3, is a piece of hookup wire about 8 inches long, with a small claw clip at the end. It is not characterized, but is kept constant during the tests.

The complete set of terminations is seen in **Table 2**. Notice that all load resistance values are in geometric progression. Each resistance is related by a factor of 2 or $1/2$ to its adjacent neighbors. The reason for this

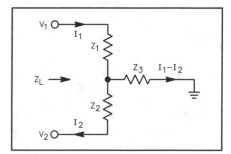

Fig 4—Equivalent circuit of balun load. The voltages V_1 and V_2 are referred to ground.

Table 2

Terminations in AI1H Geometric Resistance Box

Unbalanced Load Resistance, Ω	Balanced Load Resistance, Ω
1.5625	3.125 w/ct
3.125	6.25 w/ct
6.25	12.5 w/ct
12.5	25 w/ct
25	50 w/ct
50	100 w/ct
100	200 w/ct
200	400 w/ct
400	800 w/ct
800	1600 w/ct
1600	3200 w/ct

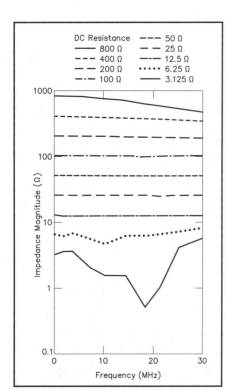

Fig 2—|Z| versus frequency for the Autek Research RF-1 RF Analyst. The impedance is plotted on a logarithmic scale.

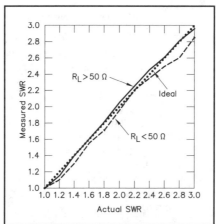

Fig 3—SWR accuracy at HF for the MFJ-259 HF/VHF SWR Analyzer. Ideal behavior is shown as a dotted line. For resistive loads above 50 Ω, use the solid line; below 50 Ω, use the dashed line. These results are independent of frequency over the entire HF range.

arrangement will be seen when the method for measuring loss is described.

Balun Models

To understand the impedance measurements for the baluns, it is helpful to view simplified models of current and voltage baluns. The 1:1 current balun in **Fig 5A** consists of an ideal 1:1 transformer, a winding impedance, Z_w, and a transmission line of characteristic impedance Z_0.[13] The transmission line influences the high frequency properties of the balun, and the winding impedance can affect the characteristics at all frequencies. For a 4:1 current balun (not shown), the model consists of two identical 1:1 baluns with their unbalanced ports in parallel and their balanced ports in series. Z_0 should be regarded as an "equivalent" characteristic impedance. In this paper it is always referred to the unbalanced port of the balun.

Fig 5B shows a model for a 1:1 voltage balun. It consists of a transmission line and an ideal 1:1:1 transformer with a winding impedance. It is connected as a 1:1 autotransformer. Fig 5C shows a model of a 4:1 voltage balun. It is an ideal 1:1 transformer connected as a 4:1 autotransformer. Voltage baluns try to force equal voltages of opposite phase between each balanced-end terminal and ground through the presence of a grounded center tap. The current balun tries to force balanced currents at the load. In the models, Z_w, could have been shown across any of the windings because the transformers in the models are ideal. To approach the performance of an ideal balun, Z_w should be large compared to the load impedance.

At this point, I digress to point out an interesting property of a transmission line which helps us understand the behavior of baluns. Consider a lossless line of characteristic impedance, Z_0, which is $1/8$th wavelength at the test frequency. If the line is shorted at the far end, the impedance at the near end will be $+jZ_0$. If it has no termination at the far end, the impedance at the near end will be $-jZ_0$. Thus the magnitude of the impedance for either condition will be Z_0.

What I have not seen reported elsewhere is that if the line is terminated at the far end with a resistor of any resistance, the magnitude of the impedance at the near end will be Z_0. This means that if we test our balun with resistive loads, we should find a frequency where the magnitude of the impedance measured is independent of the load resistance. I call this useful frequency $F_{1/8TH}$. This provides us with a convenient way to compare some of the high-frequency properties of baluns. Further, the value of Z_0 we find from this procedure will give us information on the center of the load range.

Fig 6A is a plot of |Z| versus frequency

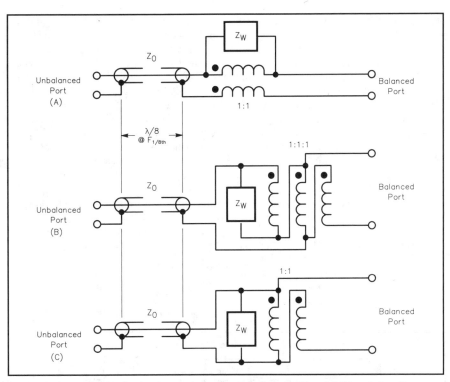

Fig 5—Balun models. At the top, a 1:1 current balun. In the middle, a 1:1 voltage balun. At the bottom, a 4:1 voltage balun. Z_w is the winding impedance. The transformers are ideal transformers. Sources of loss are the resistive part of the winding impedance and loss in the transmission line.

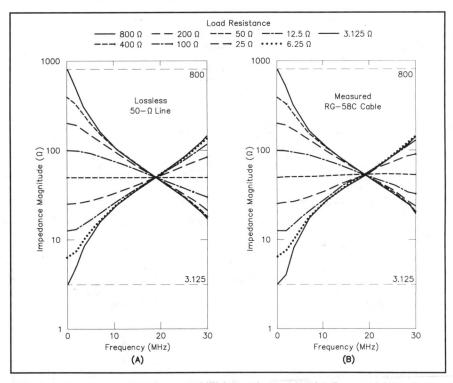

Fig 6—The frequency dependence of |Z| for a 50-Ω cable with $F_{1/8TH}$ = 19 MHz. The impedance is plotted on a logarithmic scale. The graph at A shows the calculated result for a lossless cable. At B, measured results for an RG-58C cable are shown.

for a segment of 50-Ω lossless transmission line terminated at the far end in a series of resistive loads. For this line, $F_{1/8TH}$ is 19 MHz. Note how all the curves pass through the 50-Ω, 19-MHz point. Fig 6B shows the measured results for $44^{1}/_{2}$ inches of RG-58C, 50-Ω coax terminated with the A11H Geometric Resistance Box. The impedance-magnitude measurements were made with the Autek Research Model RF Analyst. Note the great similarity of the two graphs for lossless and real lines.

If we connect an impedance Z_L to the balanced port of an N:1 balun, we should expect to see Z_L/N at the input end. This is only approximately true for the low bands, and is far from true at higher frequencies. A study of Figs 5 and 6 reveals that because of the equivalent transmission line, the impedance-transforming properties of the balun will not match those of an ideal transformer.

The presence of equivalent transmission lines in the models is of no concern so long as the characteristic impedances of the lines are compatible with the applications. In Figs 1A, 1B and 1C, the baluns are connected in tandem with feed lines, so the addition of a short length of line will not be observed. Of more importance is the effect of the winding impedance and other loss mechanisms of the transformer. If the balun is connected between the transmitter and the tuner (Fig 1D), the values of Z_0 and $F_{1/8TH}$ become more important.

In 1983, John Nagle, K4KJ, described several balun tests.[14] He described the usefulness of such concepts as balun characteristic impedance and the equivalent electrical length of the windings. He described tests for finding Z_0 and $F_{1/8TH}$ by measuring the reactance at the unbalanced port when the balanced port is open- or short-circuited. He also showed the value of measuring |Z| versus frequency for various resistive loads as a way to find Z_0. However, he did not recognize that $F_{1/8TH}$ could be derived from the same data.

Impedance-Transformation Test

Now that all these preliminary necessities are taken care of, let's learn something about the baluns. The test setup of **Fig 7** was used to determine the magnitude of the unbalanced input impedance. I kept the leads between the balun and the balanced loads as short as physically possible.

For some frequencies and loads, a balun will have different unbalanced input impedances if the center tap is left floating or is grounded. This difference is one measure of the quality a balun's balance, since for an ideal balun the two conditions would yield the same impedance results. I prefer to use the parameter *imbalance*, which is described later, to assess the balance quality of a balun in a real-world situation.

The measured impedance data for the W2DU 1:1 current balun for resistive loads ranging from 3.125 Ω to 800 Ω is shown on the left side of **Fig 8**. Note that this graph is for the case when the center tap is floating. The $F_{1/8TH}$ frequency is beyond the measuring range of the Autek Research RF-1. This balun has the highest $F_{1/8TH}$ of the four types tested. For a resistive load of 100 Ω, |Z| is independent of frequency. Thus the Z_0 for the W2DU balun equivalent circuit appears to be around 100 Ω. This was a surprise to me, since the W2DU balun is made from cable with a characteristic impedance of 50 Ω!

I was able to explain these results by making the same measurement on a W2DU balun without its external plastic enclosure. See Fig 8B. I fitted the ends of the balun with coaxial connectors. The Z_0 for this unpackaged W2DU balun measured 50 Ω. I concluded from this experiment that the parasitic elements within the W2DU balun package, in the AI1H Geometric Resistance Box and the external leads between the balun and the load box, account for a measured equivalent Z_0 in excess of 50 Ω. The calculated $F_{1/8TH}$ for this balun (unpackaged) is 86 MHz.

In contrast to the other baluns tested here, there is no sign of the effect of winding impedance in the impedance plot. This is to be expected. Since the center tap of the loads is floating, no current flows in the choking impedance. In fact, the winding impedance is not in the transverse circuit at all. When the center tap of the load is grounded, current flows on the outside of the coax shield and the effect of the winding impedance shows up.

From **Fig 9**, it is clear that Z_0 for the W2AU 1:1 voltage balun is between 50 and 100 Ω. Note that $F_{1/8TH}$ is about 23 MHz. From the impedance graph, you can see the effect of winding impedance for loads in excess of 100 Ω for 160 meters and to a lesser extent on 80 meters. The nature of a voltage balun is that the winding impedance has about the same effect whether the load center tap is floating or grounded.

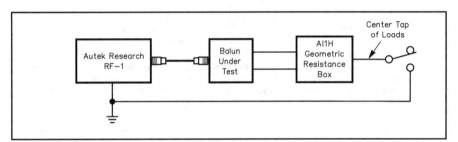

Fig 7—Using the Autek RF-1 and AI1H Geometric Resistance Box to find the impedance transformation properties of a balun.

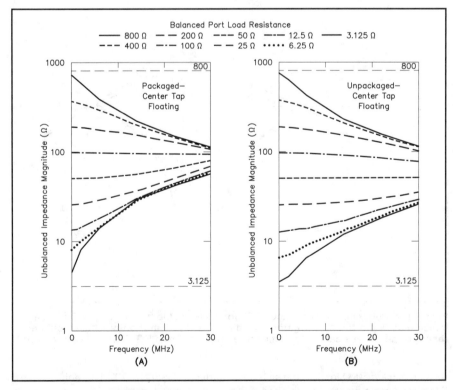

Fig 8—Results for W2DU 1:1 current balun. At A, data for the packaged unit; at B, data for the unpackaged unit. The impedance magnitude is measured with the center tap floating.

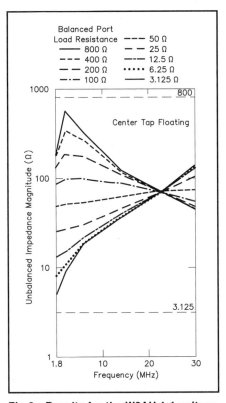

Fig 9—Results for the W2AU 1:1 voltage balun with the load center tap floating.

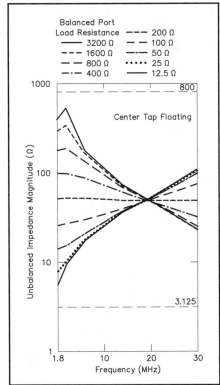

Fig 10—Results for Radio Works Model B4-2KX 4:1 current balun with the load center tap floating.

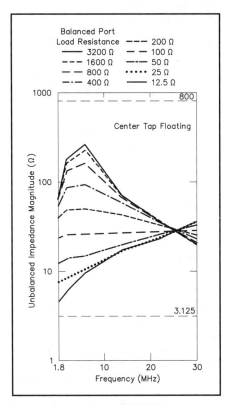

Fig 11—Results for Van Gorden Engineering 4:1 voltage balun with the load center tap floating.

The Radio Works Model B4-2KX 4:1 current balun was tested with resistive loads ranging from 12.5 Ω to 3200 Ω. From **Fig 10**, it is clear that this balun is optimized for operation with a 200-Ω resistive load (Z_0 equals 50 Ω). $F_{1/8TH}$ is about 19 MHz. Note the great similarity between these results and the calculated data plotted in Fig 6 for a 50-Ω cable with an $F_{1/8TH}$ of 19 MHz. The effect of winding impedance on the impedance transforming properties is evident for the high-impedance loads on 160 meters. This is true even though the center tap is floating. In contrast to the 1:1 current balun, the internal connections of the 4:1 current balun force longitudinal currents to flow in one or both of the winding sets for all load conditions.

Z_0 for the Van Gorden Engineering 4:1 voltage balun is between 25 and 50 Ω. See **Fig 11**. $F_{1/8TH}$ is about 26 MHz. Note the winding-impedance effect for high-impedance loads at lower frequencies.

These results validate the usefulness of the equivalent circuits of Fig 5. The models and the values of Z_0, $F_{1/8TH}$ and Z_W offer a concise way of describing most 1:1 and 4:1 baluns. The models shown are not complete, in that exact models would include other network elements. These other components are needed to explain some of the higher frequency characteristics. For example, as we saw in Fig 8, the effect of the leads connecting the balun to the balanced load can affect the external prop-

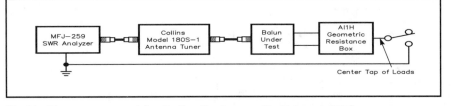

Fig 12—Test arrangement for finding the loss and imbalance of baluns.

erties of the unit. I believe that the approach may be extended to model broadband transformers with arbitrary turns ratios.

Balun Loss Test

The data given in Fig 8 through 11 provide a means for comparing various baluns. Some information may be found regarding the band of operation and the quality of the balun action, but no quantitative information about the loss of the balun for various loads is provided by the impedance transformation test. Previous work has focused on loss measurements when the balun is terminated in its design-center resistance.[15] Fortunately, with the use of a low-loss antenna tuner, such as is shown in **Fig 12**, loss and balance quality measurements are possible. See "How to Evaluate Your Antenna Tuner" for a description of the tests.[16]

Here's an outline of the theory behind the

loss test. I define *loss* to be a measure of the power dissipated in the terminated balun. Assume for the moment that the balun in Fig 12 is terminated in R_L and has no loss.. Assume also that the antenna tuner has no loss. The antenna tuner is adjusted so that the SWR at the input with a 50-Ω reference is 1:1. Then the SWR at the output of the balun will also be 1:1 when the reference resistance for the SWR measurement is equal to R_L, the load resistance.

If the load resistance is halved or doubled, the SWR at the output will be 2:1. The SWR at the input of the antenna tuner (again with the 50-Ω reference resistance) will also be 2:1. This is a direct consequence of the fact that there is no energy dissipated in the perfect antenna tuner and balun. The antenna tuner and balun act strictly as impedance-transforming agents. However, if there is loss in the antenna tuner and/or balun, the

177

SWR at the input will not be 2:1. Follow these steps to obtain the loss:

1. Adjust for SWR = 1:1 with a load resistance of R_L. If you use an SWR meter, which also measures impedance or resistance, adjust to achieve 50 Ω as well.
2. Change the load resistance to $R_L/2$. Record the SWR and call it S_1.
3. Change the load resistance to $2 \times R_L$. Record the SWR and call it S_2.
4. Calculate an estimate of the loss, L_{EST}, in dB, from:

$$L_{EST} = 5 \log \frac{(S_1+1)(S_2+1)}{9\,(S_1-1)(S_2-1)} \quad \text{(Eq 1)}$$

5. Another way to express this loss is as a percentage of the transmitter output power, P_{LOST}. Calculate P_{LOST} from:

$$P_{LOST} = 100 \left(1 - 10^{\frac{-L_{EST}}{10}}\right) \quad \text{(Eq 2)}$$

How accurate is this method? The accuracy is almost totally dependent on the accuracy of the SWR tester. I have confirmed through computer simulation of many electrical circuits with a wide range of loads, that this method of estimating loss is accurate to within 0.3 dB, assuming of course that the SWR tester is perfect. Fortunately, some SWR testers on the market have sufficient accuracy to make this method very practical and useful.

Why worry about the loss? In most cases, the effect on the signal will not be noticed. Even if 50% of the transmitted or received power is lost in the balun, that is only 3 dB, and 3 dB won't affect many HF QSOs. On transmit, however, when the transmitted power is high, the lost power heats the balun and could damage it. The balun is often the weak link in an antenna system.

Keep in mind that the loss measured is the total of the loss of the antenna tuner and the balun. The Collins Model 180S-1 has very low loss, so it is reasonable to assume that most of the high loss reported below is due to the baluns.[17] The range of this tuner does not extend to 160 meters, so loss for only bands 80 through 10 meters is measured. This is unfortunate, because often the loss is higher and the balance quality poorer on 160 meters than on 80 meters.

See **Fig 13** for the measured loss of antenna tuner alone. I tested the tuner with real and complex impedance loads, since in my tests, it is called upon to transform complex impedance loads. The complex impedance loads were chosen to give the same SWR with a 50-Ω reference as the resistive loads. A good way of visualizing the test loads is to use a 50-Ω Smith Chart and recognize that the resistive loads all lie on the chart's resistive impedance line. The complex impedance loads all lie on a perpendicular line that passes through the center of

the Smith Chart. Along this line, the reflection coefficient for a 50-Ω reference is pure imaginary (except at 50 + j 0 Ω).

It is unfortunate that the loss of the antenna tuner cannot be easily separated from the total loss. However, because the antenna tuner used has very low loss, there is enough useful information from which conclusions may be drawn. I should mention that in order to obtain SWR = 1:1 for a small number of the balun/load combinations, I had to augment the tuner with external components.

The loss data for the baluns and Collins antenna tuner are shown in **Fig 14**. *Remember that the loss is the sum of the losses in the balun and in the antenna tuner.* Purely resistive loads with center taps floating were used for this test. The "jaggedness" in the data is due to the fact that there is more than one setting of the antenna tuner that will transform the impedance of the antenna tuner/balun combination to 50 Ω. The loss of the antenna tuner will vary for these settings. I did not take the time to find the lowest loss settings, although I tried to find better settings if the surrounding data points showed lower loss.

Now we see what the impedance magnitude measurements did not reveal. Except for the 1:1 current balun, as we depart from the design-center load resistance, the balun loss clearly increases. The loss can be high enough to destroy the balun if it is operated at high power levels. The acceptable load range is much narrower for the 4:1 baluns than for 1:1

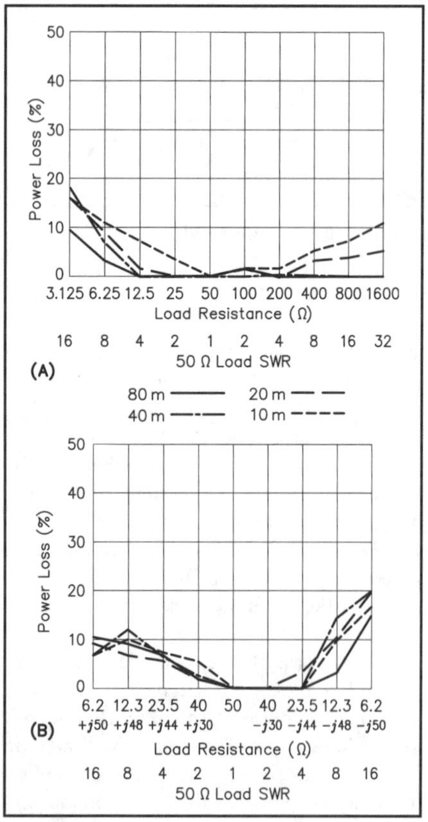

Fig 13—Loss of Collins Model 180S-1 antenna tuner. At A, loss with resistive loads. At B, loss with complex impedance loads.

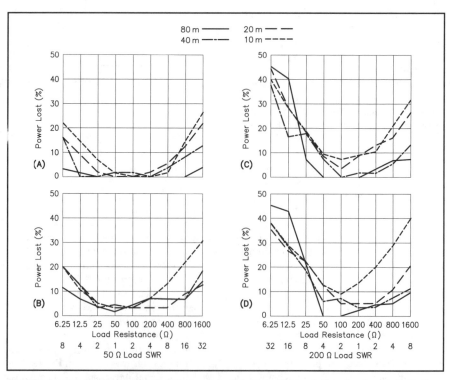

Fig 14—Loss of baluns plus Collins antenna tuner with resistive loads. At A, B, C and D are the results for the W2DU 1:1 current balun, the W2AU 1:1 voltage balun, the Radio Works B4-2KX 4:1 current balun and the Van Gorden Engineering 4:1 voltage balun, respectively.

baluns. Some common loads will result in some 20% of the power heating up the balun and antenna tuner. If the average transmitter output power is a kW, then 200 W is roasting the balun/tuner combination! Keep in mind that the Collins tuner used for these tests is one of the most efficient tuners around.

Notice the correlation between the Z_0 of the balun ($4 \times Z_0$ for the 4:1 baluns) and the optimum load resistance. It points up the importance of Z_0 in balun design, although the range of acceptable Z_0 is wide for most applications.

An interesting question arises for the W2DU 1:1 current balun. Since the center tap of the load is floating for these tests, no longitudinal current flows, and we would expect the balun loss to be very low. It is, since it equals the loss of one foot of coaxial cable. This means that all the loss is in the tuner. I have verified this by replacing the balun with one foot of 50 Ω coaxial cable of the same type used in the W2DU balun. The tuner settings were essentially the same and the loss was the same as that shown in Fig 14A.

Balance-Quality Test

We want to know how well a balun does its job. Ideally, a balun with an unbalanced generator at one end and a balanced load at the other (whose center tap is grounded) should force equal currents (equal in both magnitude and phase) to flow in each leg of the load. The direction of these currents, I_1 and I_2, is shown in Fig 4. The load in that figure is balanced if Z_1 and Z_2 are equal, so let's assume that is true. The current flowing from the center tap of the balanced load to ground is $I_1 - I_2$, which ideally should be zero. Hence an excellent measure of the balance quality is "imbalance" or IMB, which is defined as:

$$IMB = \frac{\text{Current flowing from center tap}}{\text{Average current in the balanced load}}$$
$$\text{(Eq 3)}$$

$$IMB = 2 \times \left| \frac{I_1 - I_2}{I_1 + I_2} \right| \qquad \text{(Eq 4)}$$

IMB has physical significance in an antenna system. For example, if the balun's load is a balanced antenna fed with the balanced feed line shown in Fig 1A, the common-mode radiation from the feed line can be derived from IMB.

For testing baluns and antenna tuners with a balanced output, I devised a method for finding an estimate of the imbalance from SWR measurements at the input of the antenna tuner. Refer to Fig 12 for the test setup. Since the AI1H Geometric Resistance Box is used, the loads are resistances. The impedance between the center tap and ground is zero.

The imbalance test is performed as follows:

1. Adjust the antenna tuner for SWR = 1:1 when the center tap of the load is floating.
2. Ground the center tap and observe the

SWR at the input to the tuner. Record the SWR which we will call S_B.
3. For 1:1 baluns, calculate an estimate of the imbalance, IMB_1, from:

$$IMB_1 = 2 \times (S_B - 1) \qquad \text{(Eq 5)}$$

4. For 4:1 baluns, calculate an estimate of the imbalance, IMB_4, from:

$$IMB_4 = 4 \times (S_B - 1) \qquad \text{(Eq 6)}$$

Where did these estimates come from? S_B is a measure of the relative reflected power caused by the grounding of the center tap of the balanced load. It's plausable that we can derive a measure of balance quality from S_B, but to use it to estimate IMB is a stretch.

However, I was challenged by a question from Roy Lewallen, W7EL, about the quantitative significance of S_B. I was able to show for the 1:1 current balun that if the model at the top of Fig 5 is used (at frequencies where the effect of the transmission line can be ignored) and if the winding impedance is a pure resistance, then the estimate is exact.

What if the winding impedance is complex, which is generally true? Incredibly, the estimate is in error for all loads by less than 11% as long as the magnitude of the winding impedance is greater than the load resistance! This is true for both 1:1 and 4:1 current baluns. I have chosen to use these same imbalance estimates for both current and voltage baluns.

If IMB_1 or $IMB_4 = 0$ (that is, $S_B = 1$), then we know that the balance is perfect, and current baluns will not cause any undesirable feed-

line radiation. If $S_B > 1$, then an imbalance exits in the grounded load. For example, if S_B is 1.2, the imbalance estimate is 0.4 for 1:1 current baluns. The magnitude of the center tap current is 40% of the average load current—not a good situation. For a 4:1 current balun the same value of S_B would lead to an imbalance estimate of 0.8. Just how detrimental balance deficiencies will be depends on the particular antenna system. The imbalance test is "worst case," since Z_3 is never zero.

Shown in **Fig 15** is the imbalance data for the four baluns tested. The voltage baluns show better imbalance than the current baluns. The excellent balance quality for the voltage baluns can be misleading. If a voltage balun is connected to a load that is not perfectly balanced, it will force radiating currents to flow on the transmission line.

Baluns in the Complex (Impedance) World

It is commonly believed that baluns work well with resistive terminations, but that their performance suffers with complex impedance loads. All the balun data presented thus far in this article used purely resistive loads. The difficulty with measurements using complex impedance loads is that they must be balanced, and we must have access to their center taps.

The test arrangement of **Fig 16** was used to measure baluns with complex impedance terminations. Two 50-Ω cables are use in parallel to create a 100-Ω balanced cable. By using

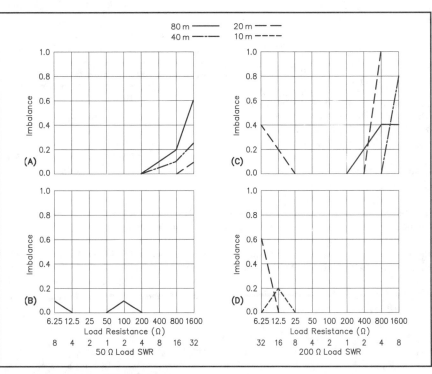

Fig 15—Imbalance for the baluns tested. At A, B, C and D are the results for the W2DU 1:1 current balun, the W2AU 1:1 voltage balun, the Radio Works B4-2KX 4:1 current balun and the Van Gorden Engineering 4:1 voltage balun, respectively.

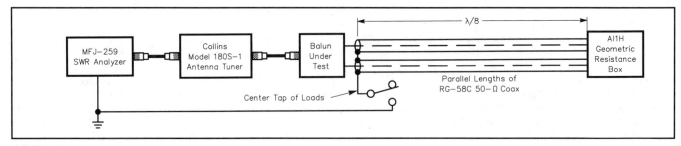

Fig 16—Test arrangement for finding the loss and imbalance of baluns with complex impedance loads. The center tap of the loads in the AI1H Geometric Resistance Box are left floating.

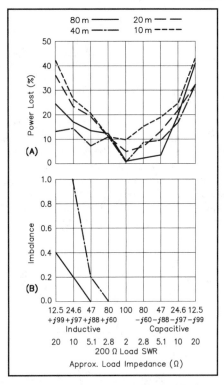

Fig 17—At A, loss for the Radio Works Model B4-2KX 4:1 current balun plus Collins antenna tuner with complex impedance loads. At B, the imbalance is shown. The impedance values shown are approximate.

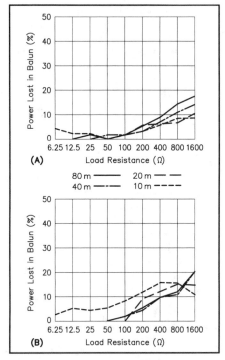

Fig 18—At A, loss of an unpackaged W2DU 1:1 current balun, when the balun is located on the antenna side of the tuner. At B, loss when the balun is next to the transmitter.

cables which are λ/8 at the test frequencies, the AI1H Geometric Resistance Box loads are "rotated" 90° on the constant SWR circles of a 100-Ω Smith Chart. The loss measurements include the loss of the antenna tuner, the balun and the λ/8 balanced transmission line. Since the terminating resistance and the matched loss for the line is known, the loss of the cable may be calculated and subtracted from the total loss.

Fig 17 shows the loss and imbalance results for the Radio Works Model B4-2KX 4:1 current balun with complex impedance loads. Fig 17A depicts the termination impedance and the "200-Ω SWR"—the SWR calculated with a 200-Ω reference. I chose this because the nominal balanced-load resistance for which the balun was designed is

200 Ω. The values come out as they do because a 100-Ω λ/8 balanced transmission line was used. The values shown in the figure adequately stress the balun for this test.

The imbalance when complex impedance loads are used is shown in Fig 17B. Compare the data of Fig 17 with their resistive load counterparts in Fig 14C and 15C. Loss is well correlated with 200-Ω SWR. It does not matter whether the 200-Ω SWR comes about because of real or complex impedance loads. I believe this to be a general result, but the conjecture should be tested on more baluns.

Balun Location

It's clear that as the load SWR increases, the loss of the balun increases. To overcome this problem, some designers have placed a balun between the transmitter and the an-

tenna tuner, as shown in Fig 1D.[18,19] The philosophy behind this approach is to have the antenna tuner do the work of transforming the balanced load impedance to 50 Ω (or 200 Ω, if a 4:1 balun is used) and then to have the balun do the job of making the unbalanced transmitter happy. In this way the balun would be operating with its design-center load, at least in concept. Further, the extreme transverse voltage or current stress encountered when the balun is on the antenna side of the tuner is avoided.

To test this idea, I measured loss and output balance for the arrangement where the W2DU 1:1 current balun is connected between the low-power SWR tester and the Collins antenna tuner. I used the balun without enclosure to minimize any effects of lead parasitics. When the center tap of the loads is floating, the balun behaves exactly like a 1-foot length of transmission line and the loss measured is essentially the same as the loss of the antenna tuner alone.

However, when there is a finite impedance between the center tap and ground, current flows on the outside surface of the balun's coax shield, setting up a magnetic field. Any loss in the ferrite then becomes evident. The worst case is when the center tap is grounded. In these tests, I used an 8-inch length of hookup wire between the center tap and ground.

In contrast to all the other loss measurements I've presented thus far, it is possible to separate out the balun loss from the balun-plus-antenna-tuner loss. The reason is that the antenna tuner settings are either not changed at all (or are changed only slightly) when achieving 50 Ω with and without the center tap grounded. Further, the tuner's load impedance remains nearly the same for the two conditions. This means that the antenna tuner loss does not change for the two test conditions. Simply subtracting the loss measured with and without the center tap grounded yields the balun loss.

Fig 18A plots balun loss when the balun is at the antenna side of the antenna tuner. The loss increases as the load resistance increases. This is evidence that the winding impedance has a significant resistive component. Also note that the loss is greater for the low-frequency bands.

The loss results when the balun is placed between the transmitter and the antenna tuner are shown in Fig 18B. The results for 80 and 40 meters are very similar to the case where the balun is located at the antenna tuner's ouput terminals. Notice the increased loss for 10 meters for all load resistances. At this writing, I do not have an explanation for this effect.

What about imbalance? Imbalance results are shown in **Fig 19**. For 80 and 40 meters, the imbalance has been improved by the movement of the balun to the input side of the antenna tuner. However, on 20 and 10 meters it has gotten worse. I suspect that capacitance of the tuner cabinet to ground (which appears in parallel with one half of the load resistance when the center tap is grounded) may be the cause, but this must be investigated further.

As mentioned earlier, Roy Lewallen, W7EL, has done an interesting analysis of a 1:1 current balun. He published this on the Internet.[20] In it, he showed that the winding impedance is just as critical in determining loss and balance quality when the balun is on the input side of the tuner as it is when it is on the output side. His analysis supports the results I measured with the W2DU 1:1 current balun for 80 and 40 meters. My results revealed that loss and balance properties are similar for the two cases considered. A more detailed model is required to explain the higher-frequency results. Roy's model did not include the effect of a transmission line within the balun.

I checked the 50-Ω load SWR at the input of the tuner when the unpackaged W2DU balun was placed on the transmitter side of the tuner. It was always under 1.5:1 and the magnitude of the input impedance of the tuner ranged from 45 to 55 Ω. Thus the goal of terminating the balun near its design-center load resistance was achieved. This occurred because $F_{1/8TH}$ is high and Z_0 is close to 50 Ω. When other baluns are considered for use between the transmitter and the antenna tuner, it is important to know their $F_{1/8TH}$ and Z_0. Z_0 should be close to 50 Ω. The impact of not being 50 Ω will depend on the value of $F_{1/8TH}$.

It appears that there is no loss or imbalance advantage to using a W2DU 1:1 current balun on the transmitter side of the antenna tuner. In high power applications, of course, the transverse-voltage or current-stress relief obtained makes this placement more desirable. The measurement techniques and principles outlined here can lead to an optimum design of a balun/tuner combination.

Summary

By examining the real world environment of baluns, we have been able to focus on their limitations. This contrasts with much of the other published material on baluns, which generally considers only load impedances at or near the design-center values. In this study I have "followed the data" and have verified some useful models for 1:1 and 4:1 baluns. These models match the measured data for a wide range of loads and frequencies.

I have presented test procedures for measuring balun model parameters, together with loss and balance performance. All of this work was made feasible through the availability of low-power SWR and impedance testers, together with simple measurement techniques for deriving loss and balance quality. The AI1H Geometric Resistance Box provides appropriate balanced loads for all tests.

The presence of equivalent transmission lines and impedance of the windings as parts of the balun model prevent them from behaving like N:1 impedance transformers over the entire HF band. Further, losses increase and balance quality deteriorates significantly for loads typical of those encountered in many real-world balun applications. I recommend that you measure your own antenna tuner/balun combination. You may find the results surprising!

I am indebted to the following folks whose work and suggestions improved the quality of this project: Roy Lewallen, W7EL; Walt Maxwell, W2DU; Christopher Kirk, NV1E; Pete Schuch, WA2UAQ; Dean Straw, N6BV; Fred Griffee, N4FG; Jack Belrose, VE2CV; and Jim Evans, NV1W. Thanks also to Ralph Jannini, KA1FAA who provided the unpackaged W2DU 1:1 current balun for evaluation.

Notes and References

[1] R. Lewallen, "Baluns: What They Do and How They Do It," *ARRL Antenna Compendium, Vol 1* (Newington, ARRL: 1985), pp 157-164.

[2] L. Varney, "The G5RV Multiband Antenna . . . Up-to-Date," *ARRL Antenna Compendium, Vol 1* (Newington, ARRL: 1985), pp 86-90.

[3] J. Belrose and P. Bouliane, "On Center-Fed Multiband HF Dipoles," *ARRL Antenna Compendium, Vol 4* (Newington, ARRL: 1995), pp 103-111.

[4] *EZNEC* antenna simulation software, Roy Lewallen, W7EL, PO Box 6658, Beaverton, OR 97007.

[5] W2DU and W2AU 1:1 Baluns, Unadilla, a division of Antennas Etc, PO Box 4215BV, Andover, MA 01810. Also see W. Maxwell, "Some Aspects of the Balun Problem," *QST*, Mar 1983, pp 38-40, and W. Maxwell, *Reflections* (Newington, ARRL: 1990), Chapter 21 [out of print].

[6] See Note 5.

[7] Radio Works Model B4-2KX Current Balun, The Radio Works, PO Box 6159, Portsmouth, VA 23703, tel 804-484-0140.

[8] Van Gorden Hi-Q 4:1 Balun, Van Gorden Engineering, PO Box 21305, South Euclid, OH 44121, tel 216-481-6590.

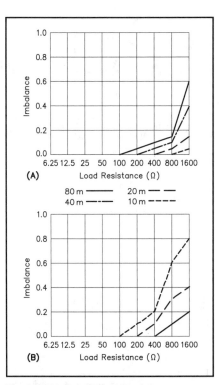

Fig 19—At A, imbalance when unpackaged W2DU 1:1 current balun is on the antenna side of the balun. At B, imbalance when unpackaged W2DU 1:1 current balun is on the transmitter side of the balun.

[9] Autek Research Model RF-1 RF Analyst, Autek Research, PO Box 8772, Madeira Beach, FL 33738, tel 813-886-9515.

[10] MFJ Model MFJ-259 HF/VHF SWR Analyzer, MFJ Enterprises, PO Box 494, Mississippi State, MS 39762, tel 800-647-1800.

[11] F. Witt, "How to Evaluate Your Antenna Tuner - Parts 1 and 2, *QST*, Apr and May 1995, pp 30-34 and pp 33-37, respectively.

[12] AI1H Geometric Resistance Boxes and related items are available directly from the author. Send an SASE for further information.

[13] R. Lewallen, "The 1:1 Current Balun," available over the Internet, ftp.teleport.com/ pub/vendors/w7el. The simplified model of a 1:1 current balun at the top of Fig 5 is the same as that described by Roy, except that he did not include the transmission line segment.

[14] J. Nagle, "Testing Baluns," *Ham Radio*, Aug 1983, pp 30-39.

[15] J. Belrose, "Transforming the Balun," *QST*, Jun 1991, pp 30-33.

[16] See Note 11, *QST*, Apr 1995, pp 32-34.

[17] See Note 11, *QST*, May 1995, pp 34-36.

[18] A. Roehm, "Some Additional Aspects of the Balun Problem," *ARRL Antenna Compendium, Vol 2* (Newington, ARRL: 1989), pp 172-174.

[19] R. Measures, "A Balanced Balanced Antenna Tuner," *QST*, Feb 1990, pp 28-32.

[20] See Note 13.

A High-Efficiency Mobile Antenna Coupler

By Jack Kuecken, KE2QJ
2 Round Trail Drive
Pittsford, NY 14543

KE2QJ explores a subject near and dear to his heart: mobile antenna couplers. He comes up with some eye-opening loss figures for non-optimized designs.

Mobile antenna installations for the HF bands have long been a problem. When working a mobile on a medium to long skywave path we have come to expect feeble signals because of inefficient antenna systems. I want to present some experiments to improve the overall performance of mobile antenna systems.

Recent commercial developments have made some dandy little HF transceivers available. The Icom IC-725 and IC-735, the Kenwood TS-50S and the Ten-Tec Scout are nice HF rigs, suitable for land-mobile use. They are compact, lightweight and run directly on 12 V dc, so they are easily installed on a car.

Unfortunately, our current position in the sunspot cycle, with solar flux numbers running well below 100, has forced operation lower and lower in frequency. While a 6 or 10-meter mobile antenna can be built with little or no matching, a mobile antenna for the 80 or 40-meter bands requires a little effort to obtain reasonable efficiency. The antenna proper can usually be no more than nine or ten feet long because of the need to clear roadside wires and obstacles. For any frequency below 24 MHz, such an antenna is electrically short. At 3.5 MHz, a ten-foot whip is only 0.0356 λ long, roughly 15% of the length required for resonance. Something must be done to make such antennas efficient.

A typical solution, either homebrew or commercial, places a loading coil in the center of the antenna to resonate the antenna for some frequency on one band. On the lower bands, a change of even a few percent in frequency usually requires retuning. Changing bands usually requires changing the loading coil.

Another approach places the impedance matching circuitry at the base of the antenna. Although they have not been common in ham use, remotely tuned or automatically tuned antenna couplers have been used in military and commercial aeronautical, maritime and land mobile HF applications for more than three decades. These devices will match the impedance of a single antenna to 50 Ω at any frequency between 2 and 30 MHz. Frequency and band changing can be accomplished without stopping the vehicle or altering the external part of the antenna.

Fairly recently, some transceivers have been offered with built-in antenna couplers. The Yaesu FT-900AT is one of these. However, most built-in couplers are limited to something like a 3:1 SWR range because of component sizes and voltage and current ratings.

Some remote antenna couplers have appeared on the ham market. This paper deals with efficiency considerations in such couplers. The subject of the center-loaded antenna will be dealt with in a separate paper.

What Determines the Efficiency of an Antenna Coupler?

The efficiency of an antenna coupler working into a small antenna is determined by several items. Please see the **sidebar** "Why Match the Impedance?"

Installed on any vehicle, an electrically small whip antenna will have the characteristics portrayed in **Fig 1**. (See References 1 through 4.) The expression for radiation resistance is from Kraus.[1]

$$R_R = 40 \left(\frac{\pi h}{\lambda} \right)^2 \qquad \text{(Eq 1)}$$

where
h = physical height, m
λ = wavelength, m

Belrose[2] presents a somewhat different expression; however, both produce the same answer. The expression for antenna capacitance is from *The ARRL Antenna Book*.[3] It

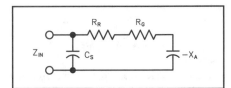

Fig 1—Schematic representation of a short mobile whip, where R_R is radiation resistance of the whip, in series with R_G, the ground-loss resistance, and the whip's capacitive reactance X_A. C_S is the stray shunt capacitance at the whip's mounting base.

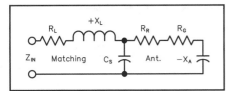

Fig 2— Schematic representation of a short mobile whip, together with a rudimentary matching network composed of series inductive reactance X_L. The series inductor has a loss resistance R_L. The base capacitance C_S provides a stray current path to ground, bypassing the radiating portion R_R of the whip.

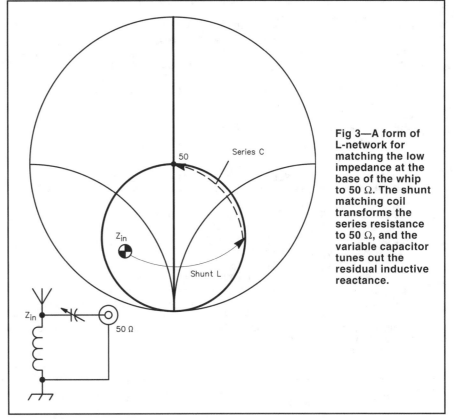

Fig 3—A form of L-network for matching the low impedance at the base of the whip to 50 Ω. The shunt matching coil transforms the series resistance to 50 Ω, and the variable capacitor tunes out the residual inductive reactance.

has been left in English units because of familiarity to hams. A somewhat different formulation in metric units due to G. W. O. Howe may be found in Reference 4.

$$C_A = \frac{17L}{\left[\left(\ln \frac{24L}{D}\right) - 1\right]\left[1 - \left(\frac{fL}{246}\right)^2\right]}$$

(Eq 2)

where

C_A = capacitance, pF
L = rad. length, feet
D = rad. diameter, inches
f = frequency, MHz

In an actual land-mobile vehicular installation there is always a significant amount of ground loss resistance, symbolized by R_G in Fig 1. This loss has been documented by Belrose and Kuecken[5,6] among others. This loss resistance is due to such things as the finite conductivity of the ground, lossy paths through tires, etc. An airplane with an all aluminum skin has less of this loss; however an iron-hull ship can have appreciable ground losses.

A second factor that cannot be ignored is the base capacitance of the antenna. Some stray base capacitance will unavoidably appear in every antenna installation. As shown in **Fig 2** this provides a path for current to flow, which does flow through the matcher but does not flow through the antenna. This current contributes to the losses in the matching network but not to the radiation.

The mechanism by which the impedance transformation is made in the type of matching network used here is shown in **Fig 3** in the Smith Chart presentation. Parallel inductive susceptance is added until the series equivalent circuit of the ensemble has the requisite 50-Ω resistive term. There is a residual inductive reactance term that is canceled with the series capacitor to leave a near-perfect match. The calculation for inductor and capacitor sizes and losses was done with the BASIC language program of

PROG-1.BAS, included on the diskette in the back of this book.

Back in the days when one did these designs with a slide rule and log tables, this was a pretty hairy process. It turns out that most of the resistance in the circuit consists of losses in the inductor. Of course you did not know the inductor size to begin with so you would solve the problem assuming no losses in the inductor. Then you would calculate the losses for the inductor size obtained and redo the problem with the new resistive term. After many iterations you could settle in on a realistic inductor size accounting for the inductor losses. With a computer, such iterations are easily incorporated—see lines 510 to 470 in the program.

What Do the Equations Tell Us?

Without going through all of the gory details of how the program works, let us go on to some of the results and see what conclusions we can form about optimizing the coupler. To begin with, let us look at what happens as a result of the "Q" of the coil. The program was run with Q as the variable parameter and all other parameters held constant. Earlier measurements (Reference 6) using a base-loaded 10-foot whip installed on the back bumper of a mid-sized car showed a ground loss resistance of 15 or 16 Ω between 4 and 6 MHz and climbing slowly after this. A base capacitance of

2.5 pF was assumed as being about as small as practically attainable.

Fig 4 gives the program printouts for two of the readings and **Fig 5** shows the data. The fraction of the radiated power climbs steadily with the Q of the coil. The curvature of the graph is due principally to the presence of the ground loss resistance. Note that if the coil had an infinite Q, the maximum efficiency would only be 0.5/15.5 = 0.032, or 3.2%.

In this series the matching elements went from 30.2 µH and 38.6 pF for a Q of 50, to 18.8 µH and 18.8 pF for a Q of 100. The fraction dissipated in the coil fell from 0.84 to 0.32 over the same range.

In **Fig 6** we see the variation in radiated fraction versus base capacitance. For this series we assumed a constant coil Q of 150 and assumed that the base capacitance had a Q of 500. Here we can see a steady decline across the entire range. As more and more capacitance shunts the antenna, the overall efficiency becomes smaller and smaller.

These results tell us several things:

1. For the maximum efficiency keep the base capacitance as low as possible.
2. Make the inductor Q as high as possible.

A corollary to the first point is that the coupler must be connected to the base of the antenna with the shortest, straightest wire possible. Note that a typical 50-Ω coaxial cable will have about 29.5 pF/foot of capaci-

```
ANTENNA CAPACITANCE IS  27.42471  PICOFARADS
FOR ANTENNA HEIGHT = 3.05  AND RADIUS = .002  METERS
AT FREQUENCY  3.5  MHZ  WITH GROUNDLOSS =  15  OHMS
RADIATION RESISTANCE = .4998645  OHMS AND ANTENNA CAPACITANCE= 27.42471  PF
BASE CAPACITANCE IS  2.5  PF AND COIL Q= 100
ALLOWING FOR BASE CAPACITANCE AND GROUND LOSS
ANTENNA TERMINAL ADMITTANCE IS  5.74723E-06 +J 6.580255E-04  MHOS
 THE COUPLER PARAMETERS ARE
SHUNT INDUCTANCE IS  35.86366  MICROHENRIES
SERIES CAPACITANCE IS  27.75977  PICOFARADS
THE OVERALL EFFICIENCY IS  9.866161E-03  WITH  .6881015  BEING LOST IN THE COIL


 ANTENNA CAPACITANCE IS  27.42471  PICOFARADS
 FOR ANTENNA HEIGHT = 3.05  AND RADIUS = .002  METERS
 AT FREQUENCY  3.5  MHZ  WITH GROUNDLOSS =  15  OHMS
 RADIATION RESISTANCE = .4998645  OHMS AND ANTENNA CAPACITANCE= 27.42471  PF
 BASE CAPACITANCE IS  2.5  PF AND COIL Q= 150
 ALLOWING FOR BASE CAPACITANCE AND GROUND LOSS
 ANTENNA TERMINAL ADMITTANCE IS  5.74723E-06 +J 6.580255E-04  MHOS
  THE COUPLER PARAMETERS ARE
 SHUNT INDUCTANCE IS  38.42649  MICROHENRIES
 SERIES CAPACITANCE IS  23.90517  PICOFARADS
 THE OVERALL EFFICIENCY IS  1.333197E-02  WITH  .5785371  BEING LOST IN THE COIL
```

Fig 4—A portion of printout from PROG-1.BAS, showing parameters calculated for two values of matching-coil unloaded Q. For Q = 100, the overall system efficiency is only 0.98%, with 68.8% of the input power lost in the coil. Clearly a better coil is needed!

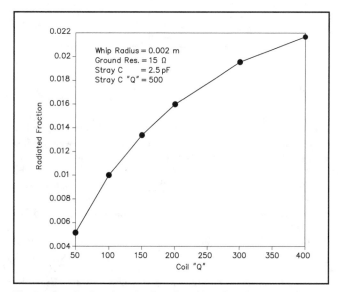

Fig 5—Graph showing the fraction of radiated power versus coil unloaded Q. This is for a 3.5-meter high, 0.002-meter diameter whip on 3.5 MHz, with a ground resistance of 15 Ω and a stray base capacitance of 2.5 pF. The radiation efficiency is only 2.2% for Q = 400.

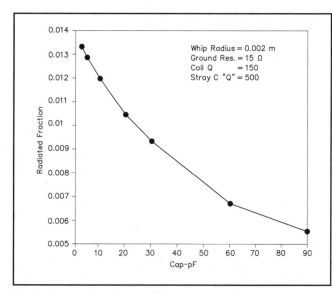

Fig 6—Graph showing the fraction of radiated power versus stray base capacitance for same whip as in Fig 5, with an unloaded coil Q = 150. The base capacitance should be held to the absolute minimum possible.

tance. At low frequencies this capacitance appears in parallel with the load. Only three feet of this cable will cut the radiated fraction by a factor of about 2.5. This tells us something about the idea of having the antenna coupler removed any distance from the antenna itself.

In passing it should be noted that this is not a problem with a center-loaded antenna with a low impedance at the base. For such antennas the built-in coupler in the radio can serve to extend the bandwidth to perhaps twice the width achievable without it. How-

ever, it is useless against a "barefoot" electrically short whip.

The corollaries to the second point will take some more explaining.

Tweaking Up the Hardware

The antenna coupler used in the measurements of Reference 6 is referred to as *Mobile Coupler III*. It is housed in a wooden box measuring 7×12×4.75 inches. The back wall and the end wall adjacent to the antenna itself are L-shaped aluminum plates. A four-sided box with top, bottom, front and one end is

made of 1/2-inch birch plywood, varnished with urethane and lacquered white to match the car. A feedthrough insulator goes through the metal end of the box and a pigtail of braid about two inches long connects the feedthrough and the base of the antenna.

The roller inductor used is homebrew. The process of making a roller coil usually requires access to a lathe and a drill press. The coil form is made from a 4-inch length of white PVC pipe. The pipe has a nominal inside diameter of 1.5 inches. Since PVC pipe is seldom round, it was first turned to a diameter of 1.850 inches OD and then threaded at 10 threads per inch. The coil consists of 32 turns of #14 tinned copper wire.

The coil inductance maximum is 23 µH. Out in the open this coil had a measured Q of 150 to 160 and a stray capacitance of about 2.5 pF. When installed in the coupler, the coil showed a measured Q of about 80 and a stray capacitance of about 9.9 pF.

Optimizing the Mobile Coupler

After making the measurements described in Reference 6, it seemed logical to pursue a course in which the mobile coupler would be optimized as much as possible. The limited Q of the roller coil and the stray capacitance of the coupler looked like logical targets for improvement.

1. The first step was to change the design so that the coil axis was vertical rather than horizontal as in the previous design. There are several desirable features to be obtained from this change. Among other things there is some direct radiation from the coil. Making the coil vertical will align this radiation with the radiation from the whip. Further, in the vertical direction the top of the coil is removed from the ground, minimizing the stray capacitance.

2. The second step was to minimize the use of metal in the coupler housing. Whereas Mobile Coupler III had a back and end made of aluminum plate, and the antenna was fed with a feed-through insulator, the new coupler would utilize an aluminum plate only in the frame holding the roller coil to-

gether and the frame holding the variable capacitor and the drive motors. All sides of the weather housing would be made of dielectric material.

3. The roller coil would be built as large as was deemed practical. This point perhaps deserves more explanation. For single-layer air-core inductors the optimum Q is where the length is slightly greater than the diameter, with a pitch equal to two wire diameters. See Terman Reference 7 and Kuecken Reference 5, Chapter 24. Using one of the formulas for single-layer solenoids, we calculate the parameters of coils designed to give us 23 µH while varying diameter and wire size. The pitches for the wire sizes listed are 10, 8, 6, 4 and 2 turns per inch, reading left to right on the x axis of the Figs, with the wire diameter always being ½ the pitch. The various coils would have 34, 28, 23, 17 and 10 turns respectively.

From **Fig 7** we see that as the coil diameter grows, the length of the wire grows. We also increase the wire diameter with the coil diameter. The DC resistance of the wire is inversely proportional to the cross section area of the wire and therefore to the square of the diameter.

However, due to the skin effect, the RF resistance of the wire is proportional to the circumference of the wire and therefore proportional only to the diameter of the wire. In **Fig 8** we see therefore that the smaller RF resistance of the wire more than offsets the increased length. Other things being equal, the Q of the coil will grow with increasing diameter.

The 9.3-inch diameter coil wound with ¼-inch tubing would provide, at least in theory a Q nearly four times as high as the existing coil. As a practical matter, the diameter of this coil would be far too great for something intended to mount on the rear bumper of a car.

As a more practical size, it was decided to make the diameter 3.5 inches and to wind the coil with #8 solid copper wire (0.128 inch diameter) at four turns per inch. The target inductance of 23 µH called for 23.5 turns. The inductance of this coil as a function of the number of turns active is given in **Fig 9**. The unused turns on the coil are shorted out by the roller. If the unused turns are left unshorted, one is almost certain to find frequencies where the unused portion acts as a Tesla coil transforming the voltage to astronomical levels.

A commonly asked question is "What does the shorted section do to the coil Q?". The Q stays at very respectable levels until major portions of the coil have been shorted out. For purposes of the antenna coupler the droop in Q at small inductances is not particularly significant, since these portions are used only at frequencies where the antenna has become very easy to match and little loss in signal strength results.

How Does One Measure a High Q?

The accurate measurement of a really high Q is not a trivial problem. Typical bridge or impedance measurements are not particularly effective for such measurements. For example suppose that we have a coil with a reactance of 400 Ω and a Q of 400. If we are

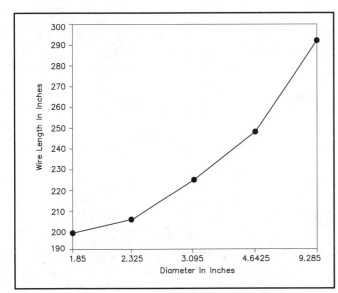

Fig 7— Graph showing length of wire versus diameter in inches for 23 µH coil. The wire diameter is kept constant at one-half the pitch. The winding pitch is related to the number labels on the x-axis: 10, 8, 6, 4 and 2 turns per inch reading from left to right. The various coils have 34, 28, 23, 17 and 10 turns respectively. As the coil diameter increases, the length of wire needed for 23 µH increases.

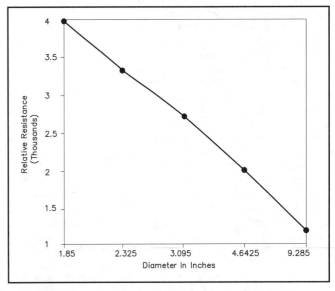

Fig 8—Graph showing the relative loss resistance for a 23 µH coil as the diameter is varied in the same fashion as in Fig 7. Other things being equal, the Q of the coil grows with increasing diameter—practical problems arise, associated with trying to mount huge loading coils on car bumpers!

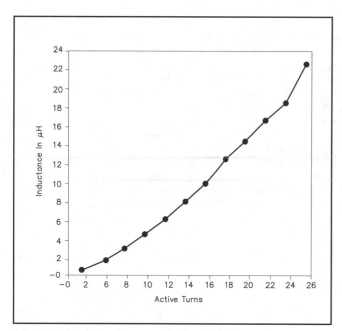

Fig 9—Graph showing inductance versus number of turns for 3.5-inch OD roller-inductor wound with #8 solid-copper wire at 4 turns per inch. The unused turns are shorted out by the roller.

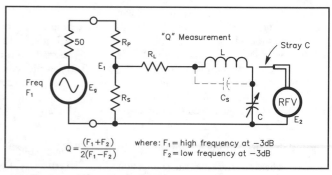

$$Q = \frac{(F_1+F_2)}{2(F_1-F_2)}$$ where: F_1 = high frequency at –3dB
F_2 = low frequency at –3dB

Fig 10—Schematic diagram showing test setup to measure unloaded Q of an inductor.

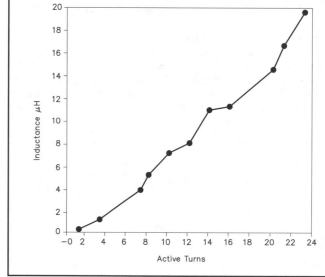

Fig 11—Photo showing homemade roller inductor, mounted next to variable capacitor. Both are driven by motors.

measuring impedance , our instrument must be able to distinguish between $1 + j\,400$ and $0 + j\,400\,\Omega$ to tell the difference between a Q of 400 and one of infinity! If the impedance read $2 + j\,400\,\Omega$, we would assume the Q to be 200—a 100% error.

A more usual way of reading Q is illustrated in **Fig 10**. A signal generator is set to the resonant frequency of the L-C circuit and a current flows through the circuit. Neglecting strays, the Q is the ratio of the voltage across the capacitor divided by the input voltage. This is all well and good except that the values of R_P and R_S must be such that slight changes in the resonant circuit do not disturb the value of E_1. If R_P is $50\,\Omega$ then R_S must be a tiny fraction of an ohm if the reading is not to be in error. In the measurements reported here, a transformer with a one-turn secondary was used for R_S, giving an impedance on the order of $1/64\,\Omega$. A digital signal generator supplied the signal and the E_2 meter reading was made with a spectrum analyzer with a 1 dB/cm sensitivity. The generator frequency was varied between the –3 dB points and the Q was calculated with the formula at the bottom of Fig 10.

How About the Coil Form?

Besides the AC resistance of the wires, the properties of the coil form play some part in the determination of coil Q. The RF properties of the white PVC pipe used to make plumbing do not seem to be widely available! To test this, a PVC capacitor was fashioned by applying some copper foil tape to oppo-

Fig 12—Curve showing inductance in µH versus the number of active turns in the homemade roller inductor wound on a wooden form. This roller inductor had about 20% higher unloaded Q than the roller inductor made of PVC.

site sides of a piece of PVC used to make the coil. This capacitor was then hung across the resonating capacitor and the differences in capacitance and circuit Q were measured. Knowing the area of the capacitor plates and the thickness of the material, the dielectric constant and the Q or loss tangent of the material can be calculated. At 5 MHz these measurements yielded a dielectric constant of 5.38 and a loss tangent of 11.44°. This is a rather mediocre-to-poor dielectric material; however, the extent to which this affected the Q of the coil was not known.

To find out about the losses in the core, I

decided to make another coil of the same diameter and number of turns, using a superior dielectric and also using a construction technique that minimized the effect of the dielectric on the coil. I fashioned a coil form out of well-dried rock maple. This form used two circular end pieces of $1/2$-inch maple with six splines of $3/8 \times 7/8$-inch maple glued into radial slots in the end pieces; the $7/8$-inch dimension extended radially. When the glue had thoroughly set, the form was mounted in a lathe and ground round. A Dremel tool was then mounted on the lathe cross slide, turning a small milling cutter. The lathe was locked up for four turns per inch and the form slowly turned by hand to mill the slots in the splines for the wire. The form was then given three coats of polyurethane varnish, sanding in-between to seal it from moisture. The #8 wire was first wound on a piece of 3.5-inch pipe to give it an initial circular shape and then worked down to a tight fit on the wooden form.

The photo of **Fig 11** shows the finished product after installation of the roller and the end bearings. The coil was mounted in the weather protection box. Drive motors are mounted at the far ends of the capacitor and inductor.

The curve of **Fig 12** shows the inductance of the wood form coil versus unused turns. The relative raggedness compared to the PVC coil is probably due to the tendency of the wire to pull down into tangents between the splines. This probably also accounts for the somewhat smaller inductance.

The Q of this coil is about 20% higher than the Q of the PVC coil. However, the improvement was not so startling that it would rule out use of the PVC coil. In view of the intended use of the coupler, the wa-

ter-resistant advantages of the PVC unit tend to offset the slight electrical advantage of the wooden one.

Capacitor Q

When the wood-form coil was tested it was assembled with a Jennings Vacuum Variable Capacitor. These vacuum capacitors have no sliding electrical joints and possess a remarkably high Q. With the availability of the very-high Q coils it became possible to measure the Q of the air-variable capacitor. The coil was set up and its Q was measured with the vacuum capacitor. The air-variable capacitor was then substituted and the Q remeasured. If one assumes the losses in the Jennings capacitor to be negligible, the Q of the air variable capacitor was calculated to be approximately 500. The unit used is a high-grade capacitor with ball bearings and glass bead insulators. It was built originally for the ARC-5 transmitter. Both the rotor and stator have the plates soldered in. The capacitor does have a wiper to contact the rotor. Despite the lower Q, this variable was used in the working coupler because it was more suited to the motor-driven application than

the vacuum variable. The variable capacitor does remain as a place where some additional loss may be eliminated.

Radiation Tests

The radiation tests were performed in the near field. During the tests, all of the instruments in the car were powered by an inverter running from the car battery, so that no power cables or other wires extended from the car to ground. A Hewlett-Packard 8656B signal generator was used as the signal source with an impedance bridge with indicating meter. At each frequency the bridge was nulled by tuning the antenna coupler so that an impedance very close to 50 Ω was attained. The generator power into the bridge was set to 0 dBm and the input to the coupler was then –6 dBm.

At a point 50 feet from the antenna on the vehicle, the tripod-mounted antenna shown in **Fig 13** was mounted. This is a one-meter antenna with a 50-Ω termination and a balun. It is not very sensitive but it can be calibrated for capture cross-section area with a few simple measurements and some mathematical manipulations. It is normally

Why Match the Impedance?

A common bit of ham folklore holds that "the only reason to match the antenna is to keep the transmitter happy." This really is not so, as we shall show shortly. Let us consider a 10-foot mobile whip with an average diameter of 0.157 inches. At 3.5 MHz this antenna is only 0.0356 λ tall. It looks like a very small resistor (0.5 Ω) in series with a 27.4-pF capacitor. Now let us suppose that we were to simply hang a 50-Ω resistor in parallel with the antenna, namely from the feedpoint to the ground. The transmitter would certainly be satisfied, since the resulting SWR would be 1.03:1. However, the radiated power would be miserably small, nine millionths of a watt for every watt pumped into the 50-Ω resistor.

With such gross mismatches it is possible to simplify the math to give a more intuitive "feel" for what is happening. Suppose that we pump 100 W into the 50-Ω resistor. The RMS voltage across the resistor (neglecting the minuscule effect of the antenna) would be 70.7 V. At 3.5 MHz the capacitive reactance of the 27.4 pF capacitance is $- j$ 1660 Ω, so (neglecting the minuscule effect of the radiation resistance and the phase angle) a current of 70.7/1660 A, or 0.0426 A would flow in the antenna. This current squared times the radiation resistance is the radiated power, or 0.0000907 W, tallying fairly well with the 9 millionths of a watt into the resistor computed previously.

The reason you match the impedance of the antenna is to permit the efficient transfer of power into the antenna. The problem in the preceding case is obviously the $- j$ 1660 Ω of capacitive reactance in the antenna. Suppose that we take a coil that has $+ j$ 1660 Ω reactance at 3.5 MHz and add it in series. Now, if the coil has a Q of 100, there will be 16.6 Ω of loss resistance in it. The input terminals will see a total impedance of:

$Z_A = 0.5 - j$ 1660 + 16.6 + j 1660 $\Omega = 17.1$ Ω

This is only a 2.9:1 SWR and is obviously much easier to feed. Let us see what this has done to the system. First let us suppose that the transmitter will feed 100 W into the system. The antenna current now is:

$I = \sqrt{\dfrac{100}{17.1}} = 2.42$ A

and the radiated power is the radiation resistance times this current squared or:

$P = 2.42^2 \times 0.5 = 2.92$ W

The addition of the loading coil has upped the radiated power from 0.00009 W to 2.92 W, a large difference! Note that something pretty significant has happened to the voltage at the base of the antenna. Again neglecting the small effect of the radiation resistance, the base voltage has become:

$V = 2.42 \times -j 1660 = 4017$ $V_{rms} = 5664$ V_{peak}

The resonant circuit has multiplied the voltage at the antenna 56.8 times over the voltage present at the same power level on 50 Ω.

Another point is worthy of note. The resonant circuit has made the ensemble very frequency sensitive. If the frequency rises by only 0.5%, or 18 kHz at 3.5 MHz, the capacitive reactance falls to $- j$ 1651.45 Ω, the inductive reactance rises to $+ j$ 1668.59 Ω and the net antenna impedance becomes:

$Z_A = 0.5 - j$ 1651.45 + 16.6 + j 1668.59 $\Omega = 17.1 + j 17.1$ Ω

This is the –3 dB point. Lowering the frequency has a similar effect, except that the sign of the reactance is reversed. Because of the very poor power factor, the electrically short antenna carries many reactive volt-amperes. Both the current and voltage are much higher than associated with a resistive load dissipating the same power.

Fig 13—Photo of calibrated receiving antenna mounted on wooden tripod next to Tektronix 2710 spectrum analyzer.

used for field-strength measurements in RFI and EMI measurements. The absolute power of the received signal was measured with a Tektronix 2710 spectrum analyzer. The spectrum analyzer and the signal generator are both calibrated to 0.1 dB and they generally agree within 0.2 dB. With the known calibration of the receiving antenna, the frequency slope in the data due to the cross section of the receiving antenna can be factored out.

Fig 14 shows the result of these measurements in terms of dBW/cm². If the car antenna had no directivity and was equally efficient at all frequencies, the result would be a straight horizontal line. The drop-off at lower frequencies is the result of the reduced efficiency at these frequencies. The wood-form coil is generally best; however both the PVC coil and the wood-form are considerably better than the Mobile III, as expected at the lower frequencies.

In the vicinity of 4.9 MHz there was a pronounced resonance, showing in all three sets of data. The details of this resonance are shown in **Fig 15**. I don't know whether this resonance is caused by the radiation pattern of the vehicle or by some interference effect at the site. However, it was very pronounced and repeatable. On mobile installations intended to cover the full HF range a search for such resonances is probably in order.

Conclusions

As speculated in the conclusion of the study of Reference 6 there was still a great deal of performance being "left on the table." The coupler redesign got back a significant portion of this performance. With proper coupler design, a more efficient but still reproducible coil form and an improved air-variable capacitor, some further improvements are still available.

Notes and References

[1] J. D. Kraus, *Antennas*, 1st edition (New York: McGraw-Hill, 1950), p 137.
[2] J. S. Belrose, VE2CV (ex-VE3BLW), "Short Antennas for Mobile Operation," *QST*, Sep 1953, pp 30-35, 109.
[3] *The ARRL Antenna Book*, 17th edition, p 16-7, Eqs 8 and 9.
[4] J. A. Kuecken, *Exploring Antennas and Transmission Lines by Personal Computer*, (New York: Van Nostrand Reinhold, 1986), Chapters 17-18.
[5] J. A. Kuecken, *Antennas and Transmission Lines*, (Indianapolis: Howard W. Sams & Co., 1969), Chapters 24, 25.
[6] J. A. Kuecken, "Performance Comparison Between the Use of Coil-Loaded Mobile Whips and Antenna Couplers," *The ARRL Antenna Compendium, Vol. 4*, 1994, pp 97-101.
[7] F. E. Terman, *Radio Engineer's Handbook* (New York: McGraw-Hill, 1943), p 74.

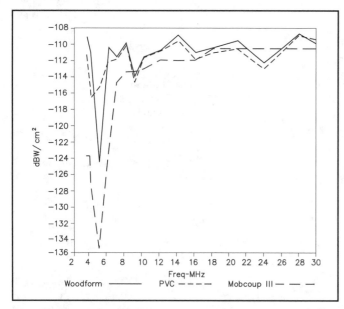

Fig 14—Curves showing radiated power density in dBW/cm² for three different coils: roller inductor mounted on wooden form, roller inductor wound on PVC form and prototype *Mobile III* coupler before optimization of coil Q was attempted.

Fig 15—Curve showing pronounced resonance at 4.9 MHz. The cause for this spurious resonance is not known—it could be in the vehicle itself or due to something at the test site. Such resonances should be investigated for mobile installations meant to operate over a wide frequency range.

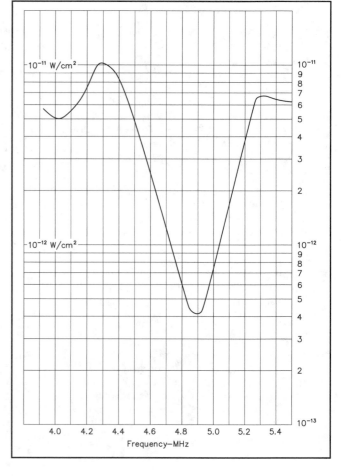

Easy Homebrew Remote Controls

By Jack Kuecken, KE2QJ
2 Round Trail Drive
Pittsford, NY 14534

How would you like to be able to change bands on your mobile antenna without getting out of the car? How about tuning your small loop antenna from the shack? How would you like to remotely tune a single antenna for all bands and have the transmission line operate at low SWR for minimum stress and loss?

The object of this article is to show some simple techniques for remote control, applicable to many of the problems around the ham shack. The systems use parts and pieces readily available to hams and do not require any fancy test equipment.

> *Ever wonder how best to drive that variable inductor or capacitor for your remotely tuned antenna coupler? KE2QJ takes you through the basics of servo design and practice.*

Motors

One of the easiest types of motor to use for remote controls is the permanent-magnet dc motor. A small PM dc motor is easily regulated over a wide range of speed; it can be reversed by simply reversing the polarity of the exciting voltage and can deliver a large amount of torque very smoothly. You will probably want a *gearhead* motor, in which a substantial gear reduction is built into the motor itself. For a motor the size of a D-cell, the rotor will spin at speeds from 5,000 to 12,000 rpm (revolutions per minute). Most of the things you will want to move remotely will work at 60 rpm or less, so you need a 100 to 200:1 speed reduction. Of course, this much gearing down results in a large increase in torque.

Gearhead motors are available from a number of vendors, such as Globe or Pittman, but they are expensive in small quantities. If there is one in the neighborhood, a visit to a surplus house might yield a suitable motor for only a few dollars. Another source that is worthwhile to check out is a VCR repair shop. Most VCRs have several suitable gearhead motors and the motors are seldom the reason for scrapping a VCR. Another practical source is an auto junkyard. Windshield wiper motors are frequently quite suitable.

First Example

Let us suppose that we want to drive a roller coil with a motor. Suppose that the coil is two inches in diameter and has 28 turns. At 60 rpm it will take 14 seconds to turn the coil from one end to the other. Unless it was specifically designed for power drive, this is about as fast as you would want to spin it—especially since you will rarely want to run the coil its entire length. More generally, you will want to move it from one specific setting to another specific setting.

In **Fig 1** we see a PM dc motor (schematically depicted only as commutator and brushes) with a double-pole/double-throw switch for reversing the direction of rotation. The switch would have to be of the center-off variety or the motor would run continuously. We are assuming that we have only a single dc supply, as in an automobile.

Fig 2 shows a practical remote drive. Note that if neither PB1 nor PB2 are depressed the circuit is not energized. Pressing PB1 energizes relay K1 so that terminal B is grounded and terminal A is energized positive by D1. Let us assume this causes the motor to run in a clockwise direction. If PB2 is pressed, K1 is not energized and the limit

switches S1 and S2 are in the position shown in the drawing. In this case, B is energized positive and A is grounded and the motor runs in a counterclockwise direction.

Now we assumed that we were driving a roller coil, and it only has a finite length. Suppose that we hold down PB1 and the coil turns clockwise until it gets to the end. We don't want to run off the end or break some-

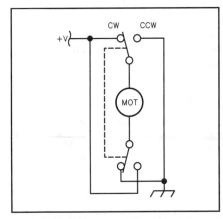

Fig 1—Schematic of permanent-magnet dc motor with switches to control direction of rotation.

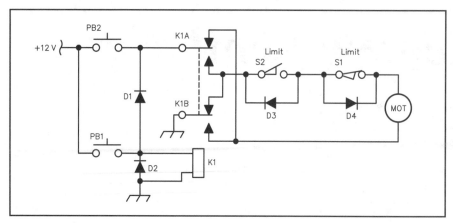

Fig 2—A practical remote drive system, where pushbuttons PB1 and PB2 are used to control the direction of rotation. Limit switches S1 and S2 prevent rotation past the end of the roller coil driven by the motor.

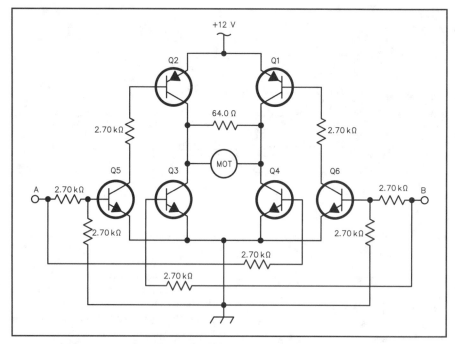

Fig 3—Schematic of a simple solid-state reversing switch system. Terminals A and B are used to control the direction of motor rotation.

thing, so as the roller gets to the end it opens microswitch S2 and the motor stops. Now if we don't do something more, the coil will probably be jammed there, with an open switch in series with the motor! So, D3 has been placed so that if PB1 is released and PB2 is pressed the motor will run in a counterclockwise direction and the coil can back out until S2 closes again. If the counterclockwise rotation is continued to the other end, then S1 opens and a similar situation arises, where D4 permits the unit to run clockwise with S1 open.

Diode D2 is placed across the coil of K1 to absorb the inductive "kick" when the current is interrupted by PB1 opening. The inductive kick can produce a very large voltage spike that can destroy other components. Note also that D1, 3 and 4 must have a current rating large enough to handle the stalled-rotor current of the motor.

The switching system of Fig 2 is perfectly workable and I have used it in a number of applications. However, a solid state switch can be superior in some respects. For one thing, the relay is relatively slow. When PB1 is first energized, the motor is quickly energized to run the motor in a CCW direction until the limit switch breaks the circuit and reestablishes it for CW rotation. This makes the operation a bit "jumpy" and introduces some mechanical stresses into the system.

Fig 3 shows a solid-state reversing switch. If terminals A and B are both at

ground potential, all six transistors are turned off. Assume that we take terminal A to +12 V. This will turn on transistors Q4 and Q5. When Q5 turns on, it draws its current from the PNP transistor Q2 base-emitter junction, thereby saturating Q2. The current path will be through Q2 through the motor and through Q4 to ground. The limit switches and anti-jam diodes have been omitted for clarity.

For motor currents up to 6 A, PNP transistors Q1 and 2 can be power-tab types, equivalent to TIP-42 cases. NPN transistors Q3 and 4 can be power-tab types TIP-41 or 2N3055. Driver transistors Q5 and 6 can be small NPN transistors, such as 2N3904, 2N2222 or similar switching types. For 1 or 2-A motor currents, heat sinking will probably not be required because the transistors are all run either cutoff or saturated, and therefore dissipate very little power.

When the current to the motor is first turned on, there will be an inrush equal to the supply voltage divided by the motor's dc resistance. Once the motor gets rolling it generates a "back EMF" and the current falls sharply. When the motor is suddenly reversed while running, the back EMF and the line voltage will add and a large slug of current is drawn, nearly twice the inrush surge. The transistors should be capable of handling this inrush surge at twice the rated voltage.

The 64-Ω resistor is added to absorb the inductive kick when the switch is suddenly opened. The presence of this conductive path will tend to limit commutating spikes as well. To help with radio noise, a 0.1-μF capacitor, with a voltage rating at least three times the supply voltage, is placed in parallel with the motor at the motor terminals.

Servo Logic

We have seen how we can make a motor turn the coil in either direction and stop before it runs off the end. Now, what about making it stop at some particular place? This is the purpose of the servo logic. For this let us examine the circuit in **Fig 4**.

Suppose that we get one of those fancy 10-turn potentiometers with a counter dial reading 0 to 1000 and mount that on the control panel. Next, we attach another potentiometer to the roller coil shaft. We'll assume for a moment that the coil and potentiometer have the same number of turns. Now, if we attach the outside ends of the two potentiometers together and excite both from some voltage source and rotate the coil until the tap voltage on the coil potentiometer is equal to the tap voltage on the panel potentiometer, we know that the reading on the panel potentiometer dial is equal to the coil position.

For example, if the panel dial reads 500 we know that the roller is halfway along the

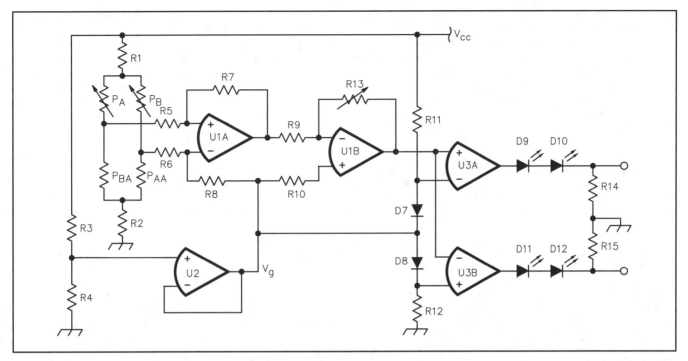

Fig 4—Schematic of a servo-logic system to control rotation of a roller coil. PA is a 10-turn potentiometer on the front panel of the control box, with a "follow-up" potentiometer PB coupled to the remote roller inductor.

coil. If it reads 750, it is three-quarters of the way along the coil. Let's see how we make this work.

To understand the operation, let's break up the circuit. Op amps like to have a bipolar supply; that is, with both positive and negative voltages. In many situations one has only a one-sided supply—like a car, which has only a positive voltage. We could build an inverter to generate a voltage negative with respect to ground; however, it's easier if we don't have to resort to this. The circuit of Fig 4 uses a *virtual ground*. U2 acts as a voltage follower. If R3 = R4, the output voltage of U2 will be just one-half of the V_{CC}. Since V_{CC} is applied to the top of the R11, D7, D8, R12 chain and the junction of D7 and D8 is held to V_G by U2, the bottom of R11 will be one diode drop above V_G and the top of R12 will be one diode drop below V_G.

We can let PA be the potentiometer on the panel and PB be the potentiometer on the coil. U1A is configured as a difference amplifier, referenced to V_G. If there is no difference between the tap voltages of PA and PB, the output of U1A will be equal to V_G. If the tap voltage of PB is higher than that of PA, the output of U1A will be higher than V_G by the same amount. (Note that R5 = R6 = R7 = R8.) In other words, amplifier U1A simply reproduces the difference in the tap voltages.

U1B is another matter. This is configured as an inverting amplifier, also referenced to V_G. The output of U1B relative to V_G will be equal to $-R13/R9 \times (V1A - V_G)$. The reason

for making R13 variable will be discussed shortly. This amplifier simply allows us to make the difference voltage as large as practical.

Servo mechanisms have a variety of operating mechanisms. The servo we are dealing with here is described as a "bang-bang" servo. Note that if the motor voltage were made linearly proportional to the error voltage, the motor would go slower and slower as it approached the target and in fact would never get there. High-class servos like those used on chart recorders and process controllers usually determine the response with *Proportional, Integral and Differential* or PID programs. In contrast, your garage door opener simply turns the motor on full speed and proceeds to open or close the door as fast as it can. It doesn't stop until it gets there and hits the switch. Like our "bang-bang" servo, it is highly nonlinear. The size of the signal applied to the motor is always maximum.

The final portion of the servo logic is the comparator and switch drive. Note that these two op amps have no feedback and therefore exhibit a very high gain, probably in excess of a million. For U3A the inverting terminal is one diode drop above V_G. If V_{UNBAL} is any lower than this, this amplifier output is clamped to the lower rail, probably about 1 V above ground. If V_{UNBAL} rises even a millivolt above the anode voltage on D7, the output of U3A will jump to the positive rail.

For U3B the arrangement is reversed, since the non-inverting terminal is clamped

at one diode drop below V_G. In order for the output of U3B to go high, V_{UNBAL} must go below the cathode voltage of D8. When one of the two op amps goes high, it excites either the A or B switch input, thereby driving the motor in one direction or the other.

Over the range where V_G is between the anode voltage of D7 and the cathode voltage of D8, the outputs of both U3A and U3B are as near the bottom rail as they can get. With some op amps and comparators this may be very close to the bottom rail, but for ordinary op amps like the 741, this voltage is likely to be about a volt or a volt and a quarter. This span of low voltage is termed the *dead band* of the servo.

To prevent the small voltage present when the comparators are in the deadband from partially exciting the switch, each comparator has two LEDs in series with a resistor. These LEDs have a forward drop of about 1.5 V to light them and virtually no current flows until the output of the comparator rises above about +3 V. If the LEDs in each leg are brought to the front panel, they serve as indicators of when the servo is moving, and in which direction. When both LEDs are out the servo has reached its target.

In a non-linear servo, the dead band is very important. In our example, as the coil approaches the target value set on the front panel pot, the motor is running at full bore. Depending upon the amount of *coast* in the system, the coil and motor inertia may carry the unit across the deadband and into an overshoot region, where the servo reverses and tries to slam the system backward. If the deadband is too small, the entire system may wind up by *hunting* or dithering back and forth across the target point.

As a rule of thumb, if the system has lost half of its velocity crossing the deadband it will be marginally stable. In this system the actual deadband is fixed by the two diode drops. However, the value of V_{UNBAL} for a given amount of position error is controlled by R13, so that the deadband is actually adjustable. When the system is up and running, one can throw a small error into the system and increase the value of R13 until the system starts to hunt. Then you may decrease the gain until the system is stable. Hunting can be detected by the fact that both LEDs are lit.

The amount of deadband can be measured by determining how far the control must be moved in order to make the system move. In a practical roller coil application I find that a count of three or four out of 1000 is a typical dead band. Note that the "zippier" the system is, the wider the deadband must be.

Speed Control

There are many places where a remote control requires varying the speed of the drive from very fast to very slow. For example, when tuning an air-variable capacitor, you would like the motor to be able to swing the capacitor through its range fairly rapidly. Then when you see the desired dip (or peak), you would like to be able to slow it down to a very slow speed to precisely tune it. We will talk about speed controls for PM dc motors here.

The speed of a PM dc motor is essentially linearly proportional to the voltage appearing across the armature—in other words, double the voltage to double the speed. The characteristic of these motors is that they tend to have some dc resistance when the motor is stationary. As the motor turns, it develops a Counter Electromotive Force (CEMF) proportional to the motor speed. If you can spin the motor shaft by hand, you can measure the CEMF with a voltmeter. The CEMF bucks the supply voltage and limits the current. The torque output of the motor is proportional to the current.

A typical small motor is the Siemens 990412052405. This is a small plastic-cased gearhead motor about the size of a flashlight D-cell. It is used to run the windshield wipers in a Volkswagen. It has a no-load rotor speed of 14,500 rpm, which is reduced to a shaft output speed of 60 rpm by a planetary gear train. The motor has a *stalled rotor* resistance of 3.9 Ω. With a 6-V supply, typical characteristics of this motor are shown in **Table 1**.

The motor can develop a very substantial torque in the locked or stalled rotor condition. **Fig 5** shows three relatively common ways of controlling the speed of a PM dc motor from a constant voltage supply. Fig 5A is far and away the worst, since it robs the motor of almost all of its starting torque. Let us suppose that we want the motor to run at 30 rpm with a 5.93 inch-ounce torque load. At 30 rpm the CEMF would be 5.65/2 = 2.825 V. At 5.93 inch-ounce, the current would be 0.15 A and the IR drop of the motor itself would be 0.56 V. Then the drop in the speed control resistor would be (6 – 2.825 –

0.56) = 2.615 V. The resistance value would be 2.615/0.15 = 17.43 Ω.

Now let's consider what happens when the motor tries to start from a stalled condition. With 17.43 Ω in series with the motor's 3.9 Ω, the starting current would be limited to 6/21.3 = 0.28 A and the starting torque would be (0.28/0.15) × 5.93 = 11.13 inch-ounces, less than twice the running torque! If we attempt to control the motor at an even slower speed, the starting torque would be even smaller!

Fig 5B is better in this regard. The op-amp draws enough base current from U1 to make the voltage at point B equal to the potentiometer voltage at point A, regardless of how much current the motor is drawing. Suppose again that we want the motor to run at 30 rpm with the 5.93 inch-ounces torque load. In this case the voltage at B would be the sum of the CEMF or 2.85 plus the IR drop of 0.56 V, or 3.41V. If the constant voltage regulator can hold the voltage at this level, then the motor would draw a locked rotor current of 3.41/3.9 = 0.874 A. The starting torque would be (0.874/0.15) × 5.93 = 34.6 inch-ounces, or just under six times the running torque.

The difference in performance between these types is very dramatic. The series resistor type of Fig 5A has a nearly constant current. If set for a slow speed, any significant change in torque load will stall the motor since it cannot draw more current to give more torque. By comparison, the control of Fig 5B is much stiffer. If the torque load increases slightly, the motor simply slows slightly and draws a bit more current. However the drive of Fig 5B does lose starting torque as it is set to slower and slower speeds. The decrease is not quite as fast as linear, however it does tend to limit the speed range to about 5:1—at very slow settings the motor will not have enough torque to break away from *stiction* and start. It is characteristic of such systems that they can be made to slow down from fast speeds, but from the slow speed settings they cannot start.

The system shown in Fig 5C is much supe-

Table 1

Characteristics of Siemens 990412052405 Motor, 6 V Supply

	Current	IR Drop	CEMF	Speed
No Load	0.095 A	0.35 V	5.65 V	60 rpm
5.93 in oz.	0.15 A	0.56 V	5.45 V	57.8 rpm
60.88 in oz	1.54 A	6.0 V	0 V	0 rpm

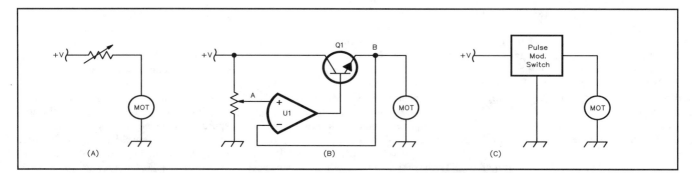

Fig 5—Three methods for controlling motor speed, in order of desirability. At A, a simple series resistor limits speed, but also severely limits starting torque. At B is a better solution, but the starting torque is still limited when pot A is set for slow-speed operation. At C is shown the best solution, a pulse-modulated system that gives good speed control and starting torque.

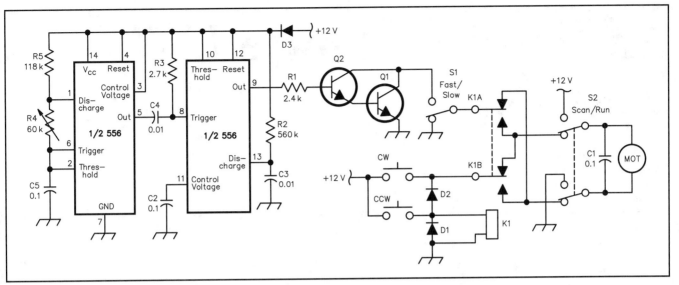

Fig 6—Schematic of a practical circuit used in several antenna couplers built by KE2QJ. This circuit incorporates pulse-modulation for accurate speed control and good starting torque, with relay control of the direction of rotation.

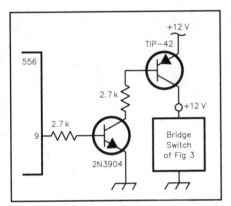

Fig 7—Schematic of a solid-state switch for controlling direction of rotation, as was shown in Fig 3.

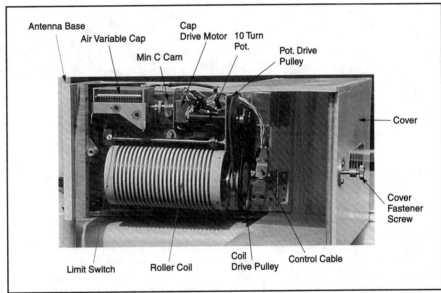

Fig 8—Photo of one of KE2QJ's remote antenna couplers mounted on the back of his car. The box is made of plywood.

rior to the constant voltage arrangement in that it can be made to start at speeds only 5 to 10% of the running speed with a given torque load. In this arrangement a pulse modulation is applied to the driving voltage; that is, the voltage is turned on and off to control the speed. The pulse modulation may be constant pulse frequency with variable pulse width or a constant pulse width with variable frequency. Within limits, this means that the motor has nearly constant (maximum) starting torque at essentially any control speed.

The circuit of **Fig 6** is a practical circuit used on several antenna couplers I have built. Pins 1 through 7 of one side of a 556 timer form a variable-frequency oscillator, with frequency controlled by R4. The pulse train supplied by pin 5 drives a constant pulse width generator (pins 8 through 14). When the frequency is low, the duty cycle is low and at maximum frequency the output

at pin 9 is high most of the time. The low speed functions best when the pulse rate is one or two pulses per second. The motor ratchets along very slowly.

The Darlington pair Q1 and Q2 provide a ground path for the relay switch similar to Fig 2. With the fast/slow switch in the fast position, the motor runs at full line speed. The scan/run switch allows the motor to run continuously in one direction when neither the CW nor the CCW switch is pushed. This is an optional feature of the antenna coupler tuning system and may not be desired in some applications.

If you want to use the solid state switch

of Fig 3, it would be better to substitute a PNP transistor at the output of pin 9 to turn the voltage on/off to the bridge, as shown in **Fig 7**. Switching from the high side preserves the ground reference for the electronic switching bridge.

Summary

I have demonstrated some simple techniques for implementing a closed-loop servo and a speed control with simple, available devices. **Fig 8** is a photo of one of my remote-tuned antenna couplers. How about trying a remote-tuned antenna coupler or an auto-tuning final for your rig?

An Improved Single-Coil Z-Match

By Charles A. Lofgren, W6JJZ
1934 Rosemount Avenue
Claremont, CA 91711
clofgren@benson.mckenna.edu

W6JJZ reviews the latest and greatest developments on the popular "Z-Match" antenna tuner.

The pioneering article on the balanced antenna tuner known as the "Z-Match" appeared in 1955.[1] At the time I was a high school student who'd been licensed for less than two years and was still buying *QST* at the newsstand. Once I spotted the design, I built my own unit (on a plywood "chassis"). In the years since, I've tried several variations on the basic Z-Match design,[2] and most recently have worked on improving the "single-coil" version. This article presents a single-coil design covering 80 through 10 meters and offers some suggestions for further experiments.[3]

Background and Theory

As described in 1955 by Allan King, W1CJL, the original Z-Match was built around the multiband tank circuit that came into use circa 1950 to reduce bandswitching chores in the tube rigs of the period. This multiband tank uses two separate resonant circuits to tune simultaneously through two frequency ranges—for example, 3.5 to 10.5 MHz and 10 to 30 MHz—and thus cover the full HF spectrum. Because of the two separate circuits, King's Z-Match required separate low-band and high-band output links.

By the time King published his design, R. W. Johnson, W6MUR, had described another version of the multiband tank circuit.[4] This used a tapped single inductor. Although not incorporated into the Z-Match at the time (at least not in any published design that's come to my attention), Johnson's circuit provides the basis for the single-coil Z-Matches appearing in recent years. The "single-coil" label is a slight misnomer, however, because the Z-Match also includes one or more output links.

Whether in its classic form as described by King or in the recent single-coil versions,[5] the heart of the Z-Match is essentially an L-network. This can be seen by referring to **Fig 1**, which is the single-coil design described in this article. The input capacitor, C1, functions as the series arm of the L-network. The multiband tank circuit formed by C2 and L1 serves as the parallel or shunt arm of the network. In operation, the tank circuit is detuned to the high-frequency side of resonance, thereby presenting an inductive reactance between the output side of C1 and ground. In a "normal" L-network having its shunt arm on the output side, the load would appear in parallel across the shunt element. Here, output is taken through an output link. (For a fuller description of the operation of a basic

Z-Match circuit, see my earlier *Antenna Compendium* article, referenced in Note 1.)

Improvements

The single-coil Z-Match is just about the simplest balanced-output antenna tuner around. (It also allows unbalanced output.) The design eliminates complicated tapping or switching arrangements and bandswitching is accomplished by adjusting the main tuning capacitor in the multiband tank circuit. Although my initial tests showed several problems with the single-coil design, these are easily cured with minor modifications.

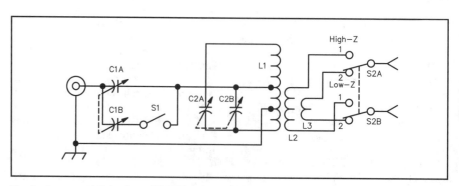

Fig 1—Improved "single-coil" Z-Match antenna tuner schematic for 3.5 to 30 MHz coverage. Two output links are used for high and low-impedance output loads.

One problem with Z-Matches in general is limited matching range. In Fig 1, the option of switching in additional capacitance at C1 extends the range, particularly on the lower bands. Similarly, the two output links considerably broaden the impedance-matching range. These modifications add two simple switches, but in practice they pose little inconvenience.

Another problem in Z-Match design is that efficiency tends to fall off unless the output link or links are tightly coupled to the tank coil. In the circuit in Fig 1, the necessary tight coupling is achieved by interwinding the turns of the output links between the turns of L1. The toroid core helps too. The availability of a separate high-impedance output link, with more turns, serves the same purpose.

A third problem is that the output balance may deteriorate under some load conditions, particularly high impedance loads. This is more likely with the tight coupling necessary for best efficiency. Output balance can be improved in the single-coil Z-Match by changing the ground point on the tank coil. This too has been done in the circuit in Fig 1. The result may be regarded as a "semi-balanced" circuit. The links are symmetrical around the ground point on L1, but some imbalance remains in the tank circuit itself. Aside from improving output balance and altering the settings of C1 and C2 at which a match occurs, the change in the ground point on the coil does not affect the operation of the circuit.

To check balance, both Dean Straw, N6BV, and I have run tests on the design using the Geometric Resistance Box described by Frank Witt, AI1H,[6] and we have found the results generally satisfactory. Other tests with a current probe show that in most instances with "real-life" antennas and balanced feed systems, the circuit in Fig 1 results in feed line currents that are balanced within 1 dB (current on one side of the line versus current on the other side). To the extent problems with imbalance arise, they appear where one would suspect—on the higher frequencies with high-impedance loads, where small stray capacitances present asymmetrical leakage paths for common mode current. This is true, too, with other tuner designs.

In any event, the problems resulting from current imbalance on feed lines can be exaggerated. While imbalance may produce some feed-line radiation, this is still radiation and except with highly directive antennas is not a serious concern. In most real-world situations with non-resonant dipoles and the like, there has to be substantial imbalance—several dB or more—before the station at the other end begins to tell the difference. What may be more serious is that imbalance may be accompanied by other unwanted effects. It may be a symptom indicating, for example, that part of the antenna system is functioning as an end-fed wire worked against ground, with RF current also flowing into a lossy ground system and being converted to heat. Or "RF in the shack" may appear. So imbalance is worth minimizing, but it is not the bugbear sometimes suggested.

For unbalanced feed, simply ground the output terminal that connects through S2 to the lower side of the output links. Use the ungrounded output terminal to feed either a single wire or the center conductor of a coax line (grounding the shield).

Published single-coil designs tap the line from C1 into the tank coil L1 at various and differing points. The fact is that any tap point on L1 for the connection to C1 is a compromise, given the wide range of likely operating conditions and frequencies. Tests show that using the center tap on L1, where one section of C2 is also tapped into L1, is about as good a compromise as you will find for both efficiency and impedance range.

Construction and Operation

The single-coil Z-Match can be built for either QRP or QRO operation, although availability of variable capacitors with sufficient plate spacing is a limiting factor for power level. (The avid tuner builder keeps a sharp eye out at swap meets and flea markets for units that may someday be useful!) As a guide, subminiature variables with minuscule plate spacing will handle 20-30 W before beginning to arc. Standard-size "broadcast" variables should take 100 W or more, depending on the load. In this regard, my neighbor and fellow QRPer, Cam Hartford, N6GA, briefly had a "big" rig hiding in his closet. We were able to run 120 W of RF through one of my single-coil units into a complex 800-Ω load before the tuner's broadcast variables began to arc. At this power level, the main inductor, wound on a T-200-6 core, showed no sign of heating.

The components in Fig 1 are:

C1 and C2: 330 pF per section or greater.
L1: 24 turns enameled wire on a T-130-6 or T-200-6 core, tapped at 6 and 12 turns from the bottom; or 22 turns enameled wire on a T-157-6 core, tapped at 5 and 11 turns from the bottom. (See below regarding wire size and modification of the number of turns.)
L2: 10 turns enameled wire, interwound between the turns of L1, with 5 turns on each side of the ground tap on L1. (This is the high-impedance link.)
L3: 4 turns enameled wire, also interwound between the turns of L1, with 2 turns on each side of the ground tap on L1. (This is the low-impedance link.)

The dual-section capacitor at C1 can be replaced with a single-section capacitor and a switched padder capacitor (silver mica, 300 pF or more, depending on the value of C1 itself). If a three-section variable is used at C1, wire S1 so that it can switch all three sections into the circuit. Or a DPDT switch with a center-off position may be used at S1 to switch between one section of C1, all sections, or all sections plus a padder. Whatever it is, the switching arrangement should allow one section of C1 to be used alone, for situations requiring low minimum capacitance.

Both C1 and C2 float above ground. At QRO levels, this requires insulated shaft couplings, at least at C2. At QRP levels, insulated knobs suffice.

One additional precaution is advisable with C2. If the Z-Match is built on a metal chassis or in a metal cabinet, keep a quarter-inch or preferably more spacing between the frame of C2 and the chassis or cabinet. This helps maintain output balance by reducing stray capacitance from the bottom end of the tank circuit to ground.

Match the wire size for L1 to the core actually used—AWG #18 for a T-200 core, #22 or #24 for the smaller cores. Select a wire size for the links that allows interwinding between the turns of L1. Depending on the minimum capacitance at C2, it may be necessary to decrease the number of turns on L1 by two (one off each end) in order to get the desired frequency coverage. This can be done after the completed unit has been tested for coverage. Just keep the ground tap mid-way between the center-tap and the bottom of the coil, and center the links around the ground tap.

For QRP levels, any of the cores indicated above are more than adequate (although wire spacing gets a bit tight with the T-130 core, and Q probably suffers). For power levels of 100 W or more, use the T-200 core. For more power-handling capacity still, a T-225-6 would be advisable, although I have not used one myself in the circuit.

Small toggle switches at S1 and S2 are suitable for low or moderate power. Heftier switches should be used for high power. (A possible problem at QRO levels with inadequate contact spacing at S2 is a Tesla-like effect from the contacts connected to the high-impedance link, when tuning a low-impedance load using the low-impedance link.)

While adjusting the Z-Match, keep these two points in mind:

1) Whenever possible, use the high-impedance link, even when you can get a match with either link. This loads the tank circuit more heavily, and may produce significantly better efficiency (up to a dB or so, depending on the exact load that the tuner sees).

2) In cases where you can tune 30 meters

and sometimes 20 meters at both the low-capacitance end of C2 (the high end of the low-frequency range) and the high-capacitance end (the low end of the high-frequency range), use the low-capacitance setting. This gives a lower C/L ratio and again better efficiency.

No special instructions are necessary otherwise for adjustment, except to note that the tuning of C2 can be quite sharp. (For this reason, a vernier drive on C2 is a convenience, although by no means necessary.) Initial peaking on receiver noise often simplifies adjustment.

Further Variations

Two possible modifications are worth mentioning. One involves the main inductor. Toroid cores allow miniaturization, and also maintain their Q when packed into cramped quarters with nearby metal surfaces. However, you may want to experiment with an air-wound inductor. If properly spaced away from surrounding objects, this should have higher Q and perhaps higher efficiency (depending on the coupling of the links).

For an air-wound inductor at L1 in the single-coil Z-Match, the point to keep in mind is that frequency coverage depends not only on the coil's inductance, but also its length/diameter ratio. This is because of the role of mutual coupling between the two halves of the coil. Johnson's 1954 article includes data and formulas for calculating coil dimensions for single-coil multiband tank circuits.[7] Alternatively, dimensions can easily be found through trial and error using pieces of coil stock, a two-section variable capacitor, and a dip meter. (Simply breadboard the multiband tank consisting of C2 and L1 in Fig 1.) Either way, aim at coverage running approximately from 3.2 to 16 MHz on the low-frequency range of the circuit and 10 to 50 MHz on the high-frequency range. (Exceeding the desired range of 3.5 to 30 MHz, with substantial overlap between the two subranges, is important, given the unpredictable nature of load impedances.)

The air-wound links (L2 and L3) can either be interwound between the spaced turns of L1, or placed concentrically around L1 with one smaller and one larger in diameter. For purposes of reducing stray capacitive coupling, the latter is probably the better arrangement. Keep in mind, however, the countervailing consideration: tight coupling is important for best efficiency.

A second variation is actually another tuner. The basic Z-Match circuit need not be multiband. **Fig 2** shows a simple single-band version. This particular circuit is one that I've built to accompany a NorCal 40, a single-band 40-meter QRP rig. It allows field use of a variety of antennas based on little more than a roll of light wire. (These include a center-fed Zepp with impromptu 6-inch spacers for the feed line and a full-wave loop terminated at the tuner itself.)

In Fig 2, the capacitors are single-section subminiature 365-pF variables. For 40-meter coverage, the tank coil, L1, is 14 turns (center-tapped) on an FT-114-63 or FT-114-67 core. Link L2 consists of 12 turns interwound with L1, and L3 is four turns, interwound with L1 and L2.

The circuit in Fig 2 would also be useful

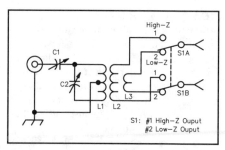

Fig 2—Single-band improved Z-Match antenna tuner schematic.

with a balanced-feed antenna dedicated to use on a single band. C1 and C2/L1 should be tailored to the band and power level. Here, too, the builder might experiment with an air-wound coil in place of a toroid. (And here, unlike the case of the multiband Z-Match, there is no need for a particular length/diameter ratio.)

Acknowledgments

For a thick and constantly growing file of Z-Match articles from Great Britain and "down under," I am indebted to Herbert "Pete" Hoover III, W6ZH, and Fred Bonavita, W5QJM. Cam Hartford, N6GA, helped with various tests, and John Dundas, AB6DG, offered useful suggestions.

Notes and References

[1]A. King, W1CJL, "The `Z-Match' Antenna Coupler," QST, May 1955, pp 11-13, 116-118. For a review of King's design, see C. Lofgren, "The Z-Match Coupler—Revisited and Revised," The ARRL Antenna Compendium, Vol 3, pp 191-195. During the 1950s, both Harvey Wells and World Radio Laboratories commercially produced the Z-Match, and King, who was an engineer for Harvey Wells, described the Harvey Wells version in his article.

[2]See C. Lofgren, "The Z-Match Coupler—Revisited and Revised," in Note 1 and C. Lofgren, "Beyond the Z-Match: The IBZ Coupler," Communications Quarterly, Winter 1995, pp 27-32.

[3]Some of this work appeared by C. Lofgren, "The Z-Match: An Update," QRP Quarterly, Jul 1995, pp 10-11.

[4]R. W. Johnson, "Multiband Tuning Circuits," QST, Jul 1954, pp 25-28, 122.

[5]For several versions based on articles from Australia and New Zealand, see William Orr's column in CQ, Aug 1993, pp 50-53.

[6]F. Witt, "How to Evaluate Your Antenna Tuner, QST, Apr 1995, pp 30-34.

[7]See Note 4 above. Photocopies are available from the ARRL Technical Dept secretary for a fee. Call 860-594-0200 or fax request to 860-594-0259.

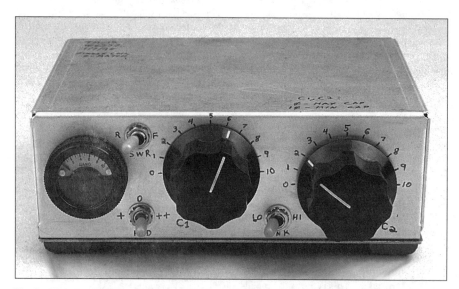

Fig 3—Photo of QRP tuner constructed by W6JJZ. It measures 6 × 2½ × 4 inches.

A Better Kind of Automatic Antenna Tuner

By Eric P. Nichols, KL7AJ
PO Box 56235
North Pole, AK 99705

Automatic antenna tuners have been around in various forms for many years. However, the design of a fully automatic tuner, capable of operating over a wide frequency range, is not a trivial matter.

To get an idea of what is involved, think about what you do intuitively when you are tuning an antenna tuner for minimum SWR at the transmitter, using only your wits and an SWR indicator. We'll use the case of a three-control tuner, like a pi-network, and a random-wire antenna. Your mental process probably goes something like this:

1. Let's see; last night when I shut down, I was on 40 meters. Now I want to operate on 20, so I'll set the transmitter to 14.292. The last time I was on 20, I remember that the inductor was set at switch position "C," the input capacitor was around 10 o'clock and the output capacitor was at 3 o'clock. Let's set things there and see what happens.
2. Give it a bit of RF. Yuck! The reflected power meter is pinned. Well, let's crank the capacitors and see what happens. Nothing.
3. Well, maybe "D" on the inductor will work. Good guess. Now we're down to a mere 6:1 SWR.
4. Let's try the capacitors again. I'll just twirl the input capacitor around. Very little change.
5. So, let's try the output cap. Ahh, that's better. I get something that resembles an actual dip in the SWR. I'll leave it there.
6. Now, let's try that input cap again. Yes, the SWR is coming down some more.
7. "Rock" the input and output capacitors five or six times to find a null. 1.15:1 SWR—that's good enough for government work.
8. Now, let's pour on the coals and see what happens. Yep, it looks pretty good. Now it's QSO time!

Now, wouldn't you just love to try to write a computer code to perform what you

> *Have you ever wanted to automate your antenna tuner? Here are some details from KL7AJ about how to accomplish this not-so-easy task.*

just did? Sounds like great fun, if it can be done at all. We have crossed over the line into actual artificial intelligence, not just good BASIC programming. Now the obvious question is, "Do any automatic antenna tuners *really* go through all those mental gyrations?"

The answer is "Yes," but you aren't likely to encounter such a number cruncher in the average ham shack. (Advanced scientific programs such as National Instrument's *LABVIEW* actually make such coding pretty simple. At my place of employment we use this technique for tuning and matching a very-high-power plasma induction coil, where the impedance is constantly, and very quickly, changing over very wide ranges.)

But for the plain vanilla ham shack, there must be a better way. Indeed there is and we'll get into that better way shortly.

First, let's define the problem and discover why it is so difficult to make an antenna tuner "think" like you do. We intuitively (most of us, anyway) want to do whatever we can to get the SWR to a minimum. Theory tells us that we will have maximum power transfer when the SWR is 1:1, indicating a 50-Ω, non-reactive load.

Now the real problem comes in when we have a mismatch; that is, an SWR other than

1:1. With a high SWR, all we know is that there is a mismatch—we haven't a clue as to which way the impedance is mismatched. An SWR bridge only tells you it's "broke." It doesn't tell you what to do to fix it.

So, knowing nothing besides SWR, all we (or our computer) can do is make an educated guess as to which way it's off, and see what happens when we change something. (As you might imagine, computers are notoriously poor at making educated guesses.)

The real problem in this scenario is our almost total reliance on the SWR bridge as a useful instrument. What we really need is an indication of some parameter in an antenna and feed-line system that will always tell us which is the correct way to go to fix any mismatch. If we can find such a parameter, we don't have to make any guesses. What's needed?

1. The absolute magnitude of the ratio of voltage to current
2. The phase angle of voltage to current

Oddly, both of these parameters are just as easy, or even easier to measure than SWR, and the techniques are much older. The nice thing is that if we can measure the phase angle and the E/I ratio, we can always know which knobs to twiddle to make everything happy.

Simplest Method

There are probably as many different circuits for matching Impedance A to Impedance B as there are hams interested in performing such feats. Because the options are so plentiful, it is easy to forget the well-proven fact that it is possible to match *any* impedance to another impedance with just two components—an inductor and a capacitor in an L-configuration. This fact has not been lost on the military, which has used the ubiquitous L-network for decades to match everything from random wires to tank whips.

Granted, sometimes using just an L-network results in some very unrealistic values for the L or the C, but nevertheless, it is possible to get a perfect match with just two components. Let's look into the means by which we can achieve this simple match, to discover the simplest and most direct method of adjusting them.

A Resonant Voltage Divider

I think it's best to think of an L-network as a resonant voltage divider. We basically have two reactances in series across a voltage source. The voltage at the midpoint of the reactances is proportional to the vector difference between the reactances. (It is sometimes baffling to take a voltage measurement across each of the reactances with an oscilloscope, for instance, and discover that the two voltages add up to something more than the applied voltage. This looks like it defies all kinds of basic voltage laws, but doesn't.)

To avoid confusion, let us always keep our circuit resonant. There are several good reasons for doing so. First, it can generally be shown that SWR rises more quickly with increasingly *reactive* loads than it does with mismatched resistive loads. The marvelous book *Reflections* by Walter Maxwell, W2DU, explains this point very nicely. Second, the mathematical analysis is much simpler with purely resistive circuits; that is, circuits with zero phase shift.

With these ground rules, I invite you to stare really hard at the L-network shown in

Fig 1. If we look into the left side of the network, we encounter a single terminal. It doesn't matter what is attached to the far side of the network. It could be a resistor, a capacitor, another inductor, or any number of series or parallel combinations of all three.

The input of the network at any given frequency can have only one value of reactance and resistance. If we make that inductor infinitely variable over all possible values (including negative values) we will find one place at which the input of the network looks like a pure resistor. Also, a little experience should tell us that the Q of the inductor is expected to be higher than the Q of the capacitor. The capacitor is, in effect, shunted or partially shorted out by the load, whether that load is purely resistive or not. If the inductor's Q is higher, you would expect that adjusting it would have the greatest overall effect on bringing the tuner to resonance.

Please avoid the ill-conceived notion that a system can be resonant at one location and out of resonance somewhere else. There was another brilliant article in *QST* back in 1956 by Byron Goodman, W1DX. It was entitled "My Feedline Tunes My Antenna."[1] It should be required reading by every radio amateur. In essence, one thing the article

proves is that a circuit, no matter how complex, has one and only one value of reactance at a given frequency. You may chisel that statement in granite! In other words, a network is either resonant or it's not.

The obvious first step to making an effective antenna tuner, is to get things resonated as effectively as possible. The secret to this is to adjust the *highest-Q* component first. I might add a little corollary. Perhaps I could call it the "Nichols corollary." *In an antenna tuner, the highest-Q components should be used for tuning and the lowest-Q components should be used for matching.*

This doesn't say that perfect matching cannot occur if the corollary is violated. However, this can result in a very inefficient match. I have seen some antenna tuners that under certain conditions can match very nicely to their own internal losses, with or without an antenna attached! Obviously, it's best to avoid this sad state of affairs.

The easiest way to measure resonance, the condition where the voltage and current are in phase, is with a phase detector like that shown in Fig 2. There are no doubt more elegant phase detectors than this, but I like something I can play around with a bit. The circuit topology here is actually stolen from an old, automatic-direction-finder (ADF)

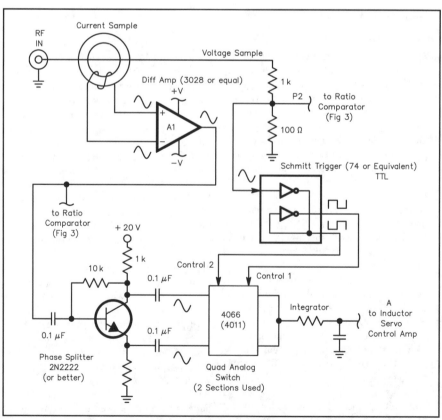

Fig 2—Schematic of a phase detector, utilizing a current sample from a toroidal transformer and a voltage sample from a resistive voltage divider. When the output is zero, the phase shift is zero. This means that the circuit reactance is zero—that is, the circuit is resonated.

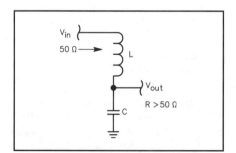

Fig 1—An L-Network, drawn to emphasize that it is "a resonant voltage-divider."

schematic, updated from the original vacuum-tube form.

We must take a voltage sample and a current sample to determine the impedance. This is done at the input of the tuner. Toroidal transformers make really nice current samplers, with one caveat. You should actually measure the phase shift through the toroid with a dual-trace oscilloscope over the entire operating frequency range so you don't have any surprises. It's acceptable to have some phase shift through the transformer, so long as it is pretty flat across the band. If not, you may want to change cores, or use fewer turns in the secondary, although this may require a bit more gain in the buffer stage. If you really are running enough power, you might want to bypass the buffer and phase splitter entirely, running the secondary of the transformer directly into the 4066 switch.

For the sake of argument, let's call the inductor our tuning component and the shunt capacitor our "matching" component. (Of course, there is always some interaction between the two functions, but each control will always have a predominant function.) We can therefore instruct the roller-inductor servo to seek a zero phase-shift point, as measured with our clever little circuit.

As a benefit of using this kind of feedback, we actually achieve a simpler mechanism than a point-and-shoot type of tuner—we don't need positioning potentiometers and all the calibration that entails.

Fig 3 shows the amplitude comparator circuit. It is designed to ignore phase differences between the voltage and current samples. It only looks at the ratio between the amplitudes. The information derived from this comparator is applied to the matching-capacitor servo. It causes the capacitor to seek the condition where the two sampled amplitudes are equal. or where they are at some pre-determined ratio.

The fact that there is interaction between the two functions doesn't matter in this circuit. In action, the two servos seem to "talk" back and forth, as though they had minds of their own. They quickly seek the resonant, matched condition.

What about more complex networks, for example, a pi-network? With a pi- network, there can be several combinations that give a perfect match. With an L-network, there is one and only one. Why bother with the extra complexity?

The only thing you may need is to be able to turn the L-network around. This is necessary for resistances less than the 50 Ω at the tuner's input. So, you should incorporate a "flopover" function for the L-network.

The extra computing "brains" required to automatically tune a pi-network, where the tune and match functions may be very poorly differentiated are not worth the extra trouble. There is probably room for controversy on this point. Of course, more complex tuning networks can offer more harmonic suppression than L-networks. However, I don't believe that antenna tuners should be leaned on to assure spectrally pure signals. I think my signal should adequately clean long before it gets to my antenna system, of which I consider my tuner an integral part!

Parting Comments

I've intentionally made the circuit design somewhat generic. You will no doubt want to tailor some of the components to your particular power level and frequency range.

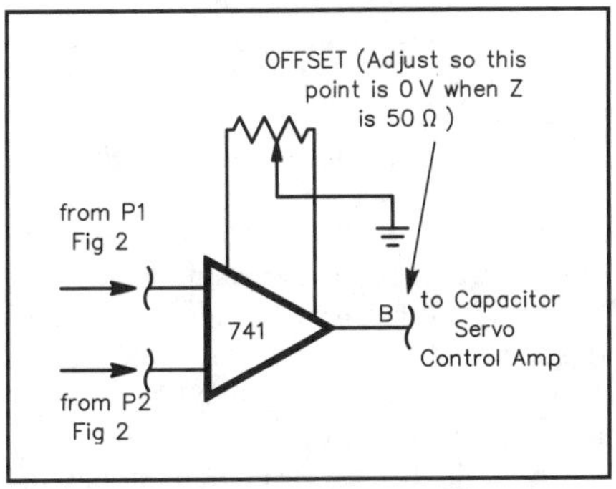

However, the principles of operation are what I want to demonstrate.

Fig 4 shows my modified interface to the tuner design by John Svoboda, W6MIT.[2] It is difficult to improve much on W6MIT's servo system. I added my phase and amplitude sampler to his system by replacing the "follow pot" wiper connections for both servos and connecting them to points A and B of my sampler circuitry.

Incidentally, when I first used my particular version of this tuner built around W6MIT's excellent servo system with a random-wire antenna, I noticed that the inductor servo tended to wander during my QSO. At first, I thought I might have been getting RF into the servo electronics. After a bit of investigation, I found that a breeze had come up—my wire was swaying in the air. The tuner was actually compensating very nicely for the small change in reactance of my wandering dipole! This is probably more of a testimony to the effectiveness of W6MIT's dithering circuit than it was to my phase sampler, but it certainly was interesting.

Notes and References

[1]Byron Goodman, W1DX, "My Feedline Tunes My Antenna," Mar 1956 *QST*, p 49.
[2]John Svoboda, W6MIT, "A Servo-Controlled Antenna Tuner," *The ARRL Antenna Compendium, Vol 2* (Newington: ARRL, 1989), pp 175-181.

Fig 3—Simple amplitude comparator used to detect absolute value of impedance. When the output is zero the impedance is equal to the desired value, in this case, 50 Ω non-reactive, because the circuit has been resonated.

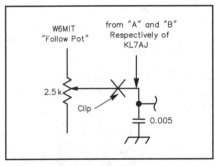

Fig 4—Modifications to W6MIT's servo system to replace follow-up potentiometer system with KL7AJ's phase and amplitude detectors.

Unique Ladder-Line to Wire-Antenna Feedpoint Termination

By Michael Pellock, NA6J
M Bar P Ranch
4955 School House Road
Catheys Valley, CA 95306

For years I've built antennas using ladder-line and for years I've been frustrated to find my ladder line lying broken and tangled under the antenna after a major wind storm. Some of the mechanical solutions I've tried over the years have been either time consuming to fabricate, overly costly, or both.

I finally came up with a method that is very simple and very inexpensive. It uses a piece of Romex insulation to support the ladder line at the feedpoint and 12 inches below, yielding a strong mechanical connection. I suspect that you can find the parts in your junk box. (If not, you might be able to persuade an electrician to give them to you for free!)

The feedpoint termination starts out with a 17-inch long piece of 10-2G Romex electrical house wire. This contains two #10 insulated wires, with an uninsulated ground wire. You will actually use only the outside insulated sheath. Do not try to use smaller size Romex, such as 14-2G or 12-2G—the insulation used on these Romex wires is not strong enough to stand up under adverse weather conditions. Pull out all three wires from the Romex by clamping each wire in a vise and pulling it out of the insulated sheath.

Next, cut out one plastic section from the end of the "window" ladder line, leaving about three inches of insulated wire for attachment to the antenna. Using a leather punch or an awl, punch a hole to clear a #6 screw in the center of the first "window" section below the top three inches of insulated wire. Punch another hole in the sixth window section, approximately 10³/₄ inches below the first hole.

Remove the center ribs from the plastic insulator using a coping saw. Smooth the cutaway section using a flat or square ¹/₂-inch file. See **Fig 1**, a photo of the modified insulator. Now, insert the two ladder-line wires through the holes in the ends of the modified insulator and pull them up tight. These wires will be attached later to the antenna wires.

See **Fig 2**, which shows the completed assembly. Wrap one end of the 17-inch

> ## "Darn ladder line broke again!" Sound familiar? NA6J describes his simple solution to that frustrating problem.

piece of Romex insulation around the center of the insulator. Punch out a hole to clear a #6 machine screw using the leather punch, lining it up using the hole previously punched in the ladder-line insulation.

Fasten together this portion of the assembly with a 6-32 × ¹/₂-inch pan-head machine screw. Use flat fiber washers against the Romex on both sides. (If a round-head screw is used rather than a pan-head screw, install an additional ¹/₄-inch OD flat metal washer between the head of the round-head screw and the fiber washer.)

Weave the Romex insulation through the ladder line as illustrated and punch a #6-size

hole in the Romex insulation, lining it up using the second hole punched in the ladder line insulation. Fasten this part together using another 6-32 machine screw and associated hardware. I dabbed both screw ends (threaded part) with Elmer's weather-tight wood glue to prevent the nuts from loosening.

Finally, attach the antenna wires to the plastic insulator and the ladder-line wires, making sure you solder the connections properly. This completes the assembly, which should be able to stand up under any kind of weather because the Romex insulation is very tough.

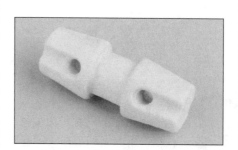

Fig 1—Photo of modified plastic insulator. The ribs in the center were removed with a coping saw and a flat file to create a groove approximately ⁵/₈ inch wide and ¹/₈ inch deep.

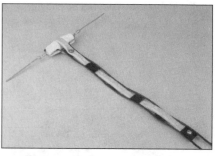

Fig 2—Photo of completed assembly, showing Romex insulation threaded through windows in ladder line and secured by a loop around the modified plastic insulator. The #6 machine screws toward the top and bottom of the Romex hold it to the ladder line for a strong mechanical connection.

Notes

Notes

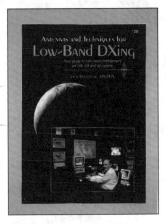
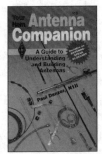
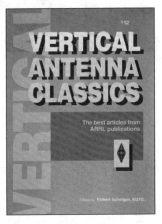

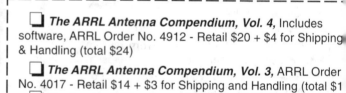

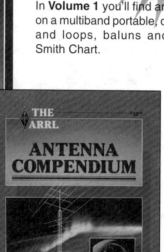
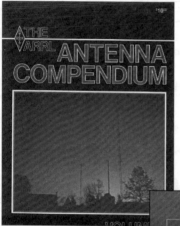

FEEDBACK

Please use this form to give us your comments on this book and what you'd like to see in future editions, or e-mail us at **pubsfdbk@arrl.org** (publications feedback).

Where did you purchase this book?
❏ From ARRL directly   ❏ From an ARRL dealer

Is there a dealer who carries ARRL publications within:
❏ 5 miles   ❏ 15 miles   ❏ 30 miles of your location?   ❏ Not sure.

License class:
❏ Novice   ❏ Technician   ❏ Technician Plus   ❏ General   ❏ Advanced   ❏ Amateur Extra

Name _____   ARRL member? ❏ Yes ❏ No

_____   Call Sign _____

Daytime Phone (   ) _____   Age _____

Address _____

City, State/Province, ZIP/Postal Code _____

If licensed, how long? _____

Other hobbies _____

_____

Occupation _____

From _____

_____

_____

EDITOR, ANTENNA COMPENDIUM VOL 5
AMERICAN RADIO RELAY LEAGUE
225 MAIN STREET
NEWINGTON CT 06111-1494

------------------------- please fold and tape -------------------------